GMPLS Technologies

Broadband Backbone Networks and Systems

OPTICAL ENGINEERING

Founding Editor
Brian J. Thompson
University of Rochester
Rochester, New York

1. Electron and Ion Microscopy and Microanalysis: Principles and Applications, *Lawrence E. Murr*
2. Acousto-Optic Signal Processing: Theory and Implementation, edited by *Norman J. Berg and John N. Lee*
3. Electro-Optic and Acousto-Optic Scanning and Deflection, *Milton Gottlieb, Clive L. M. Ireland, and John Martin Ley*
4. Single-Mode Fiber Optics: Principles and Applications, *Luc B. Jeunhomme*
5. Pulse Code Formats for Fiber Optical Data Communication: Basic Principles and Applications, *David J. Morris*
6. Optical Materials: An Introduction to Selection and Application, *Solomon Musikant*
7. Infrared Methods for Gaseous Measurements: Theory and Practice, edited by *Joda Wormhoudt*
8. Laser Beam Scanning: Opto-Mechanical Devices, Systems, and Data Storage Optics, edited by *Gerald F. Marshall*
9. Opto-Mechanical Systems Design, *Paul R. Yoder, Jr.*
10. Optical Fiber Splices and Connectors: Theory and Methods, *Calvin M. Miller with Stephen C. Mettler and Ian A. White*
11. Laser Spectroscopy and Its Applications, edited by *Leon J. Radziemski, Richard W. Solarz, and Jeffrey A. Paisner*
12. Infrared Optoelectronics: Devices and Applications, *William Nunley and J. Scott Bechtel*
13. Integrated Optical Circuits and Components: Design and Applications, edited by *Lynn D. Hutcheson*
14. Handbook of Molecular Lasers, edited by *Peter K. Cheo*
15. Handbook of Optical Fibers and Cables, *Hiroshi Murata*
16. Acousto-Optics, *Adrian Korpel*
17. Procedures in Applied Optics, *John Strong*
18. Handbook of Solid-State Lasers, edited by *Peter K. Cheo*
19. Optical Computing: Digital and Symbolic, edited by *Raymond Arrathoon*

20. Laser Applications in Physical Chemistry, *edited by D. K. Evans*
21. Laser-Induced Plasmas and Applications, *edited by Leon J. Radziemski and David A. Cremers*
22. Infrared Technology Fundamentals, *Irving J. Spiro and Monroe Schlessinger*
23. Single-Mode Fiber Optics: Principles and Applications, Second Edition, Revised and Expanded, *Luc B. Jeunhomme*
24. Image Analysis Applications, *edited by Rangachar Kasturi and Mohan M. Trivedi*
25. Photoconductivity: Art, Science, and Technology, *N. V. Joshi*
26. Principles of Optical Circuit Engineering, *Mark A. Mentzer*
27. Lens Design, *Milton Laikin*
28. Optical Components, Systems, and Measurement Techniques, *Rajpal S. Sirohi and M. P. Kothiyal*
29. Electron and Ion Microscopy and Microanalysis: Principles and Applications, Second Edition, Revised and Expanded, *Lawrence E. Murr*
30. Handbook of Infrared Optical Materials, *edited by Paul Klocek*
31. Optical Scanning, *edited by Gerald F. Marshall*
32. Polymers for Lightwave and Integrated Optics: Technology and Applications, *edited by Lawrence A. Hornak*
33. Electro-Optical Displays, *edited by Mohammad A. Karim*
34. Mathematical Morphology in Image Processing, *edited by Edward R. Dougherty*
35. Opto-Mechanical Systems Design: Second Edition, Revised and Expanded, *Paul R. Yoder, Jr.*
36. Polarized Light: Fundamentals and Applications, *Edward Collett*
37. Rare Earth Doped Fiber Lasers and Amplifiers, *edited by Michel J. F. Digonnet*
38. Speckle Metrology, *edited by Rajpal S. Sirohi*
39. Organic Photoreceptors for Imaging Systems, *Paul M. Borsenberger and David S. Weiss*
40. Photonic Switching and Interconnects, *edited by Abdellatif Marrakchi*
41. Design and Fabrication of Acousto-Optic Devices, *edited by Akis P. Goutzoulis and Dennis R. Pape*
42. Digital Image Processing Methods, *edited by Edward R. Dougherty*

43. Visual Science and Engineering: Models and Applications, *edited by D. H. Kelly*
44. Handbook of Lens Design, *Daniel Malacara and Zacarias Malacara*
45. Photonic Devices and Systems, *edited by Robert G. Hunsberger*
46. Infrared Technology Fundamentals: Second Edition, Revised and Expanded, *edited by Monroe Schlessinger*
47. Spatial Light Modulator Technology: Materials, Devices, and Applications, *edited by Uzi Efron*
48. Lens Design: Second Edition, Revised and Expanded, *Milton Laikin*
49. Thin Films for Optical Systems, *edited by Francoise R. Flory*
50. Tunable Laser Applications, *edited by F. J. Duarte*
51. Acousto-Optic Signal Processing: Theory and Implementation, Second Edition, *edited by Norman J. Berg and John M. Pellegrino*
52. Handbook of Nonlinear Optics, *Richard L. Sutherland*
53. Handbook of Optical Fibers and Cables: Second Edition, *Hiroshi Murata*
54. Optical Storage and Retrieval: Memory, Neural Networks, and Fractals, *edited by Francis T. S. Yu and Suganda Jutamulia*
55. Devices for Optoelectronics, *Wallace B. Leigh*
56. Practical Design and Production of Optical Thin Films, *Ronald R. Willey*
57. Acousto-Optics: Second Edition, *Adrian Korpel*
58. Diffraction Gratings and Applications, *Erwin G. Loewen and Evgeny Popov*
59. Organic Photoreceptors for Xerography, *Paul M. Borsenberger and David S. Weiss*
60. Characterization Techniques and Tabulations for Organic Nonlinear Optical Materials, *edited by Mark G. Kuzyk and Carl W. Dirk*
61. Interferogram Analysis for Optical Testing, *Daniel Malacara, Manuel Servin, and Zacarias Malacara*
62. Computational Modeling of Vision: The Role of Combination, *William R. Uttal, Ramakrishna Kakarala, Spiram Dayanand, Thomas Shepherd, Jagadeesh Kalki, Charles F. Lunskis, Jr., and Ning Liu*
63. Microoptics Technology: Fabrication and Applications of Lens Arrays and Devices, *Nicholas Borrelli*

64. Visual Information Representation, Communication, and Image Processing, *edited by Chang Wen Chen and Ya-Qin Zhang*
65. Optical Methods of Measurement, *Rajpal S. Sirohi and F. S. Chau*
66. Integrated Optical Circuits and Components: Design and Applications, *edited by Edmond J. Murphy*
67. Adaptive Optics Engineering Handbook, *edited by Robert K. Tyson*
68. Entropy and Information Optics, *Francis T. S. Yu*
69. Computational Methods for Electromagnetic and Optical Systems, *John M. Jarem and Partha P. Banerjee*
70. Laser Beam Shaping, *Fred M. Dickey and Scott C. Holswade*
71. Rare-Earth-Doped Fiber Lasers and Amplifiers: Second Edition, Revised and Expanded, *edited by Michel J. F. Digonnet*
72. Lens Design: Third Edition, Revised and Expanded, *Milton Laikin*
73. Handbook of Optical Engineering, *edited by Daniel Malacara and Brian J. Thompson*
74. Handbook of Imaging Materials: Second Edition, Revised and Expanded, *edited by Arthur S. Diamond and David S. Weiss*
75. Handbook of Image Quality: Characterization and Prediction, *Brian W. Keelan*
76. Fiber Optic Sensors, *edited by Francis T. S. Yu and Shizhuo Yin*
77. Optical Switching/Networking and Computing for Multimedia Systems, *edited by Mohsen Guizani and Abdella Battou*
78. Image Recognition and Classification: Algorithms, Systems, and Applications, *edited by Bahram Javidi*
79. Practical Design and Production of Optical Thin Films: Second Edition, Revised and Expanded, *Ronald R. Willey*
80. Ultrafast Lasers: Technology and Applications, *edited by Martin E. Fermann, Almantas Galvanauskas, and Gregg Sucha*
81. Light Propagation in Periodic Media: Differential Theory and Design, *Michel Nevière and Evgeny Popov*
82. Handbook of Nonlinear Optics, Second Edition, Revised and Expanded, *Richard L. Sutherland*
83. Polarized Light: Second Edition, Revised and Expanded, *Dennis Goldstein*

84. Optical Remote Sensing: Science and Technology, *Walter Egan*
85. Handbook of Optical Design: Second Edition, *Daniel Malacara and Zacarias Malacara*
86. Nonlinear Optics: Theory, Numerical Modeling, and Applications, *Partha P. Banerjee*
87. Semiconductor and Metal Nanocrystals: Synthesis and Electronic and Optical Properties, *edited by Victor I. Klimov*
88. High-Performance Backbone Network Technology, *edited by Naoaki Yamanaka*
89. Semiconductor Laser Fundamentals, *Toshiaki Suhara*
90. Handbook of Optical and Laser Scanning, *edited by Gerald F. Marshall*
91. Organic Light-Emitting Diodes: Principles, Characteristics, and Processes, *Jan Kalinowski*
92. Micro-Optomechatronics, *Hiroshi Hosaka, Yoshitada Katagiri, Terunao Hirota, and Kiyoshi Itao*
93. Microoptics Technology: Second Edition, *Nicholas F. Borrelli*
94. Organic Electroluminescence, *edited by Zakya Kafafi*
95. Engineering Thin Films and Nanostructures with Ion Beams, *Emile Knystautas*
96. Interferogram Analysis for Optical Testing, Second Edition, *Daniel Malacara, Manuel Sercin, and Zacarias Malacara*
97. Laser Remote Sensing, *edited by Takashi Fujii and Tetsuo Fukuchi*
98. Passive Micro-Optical Alignment Methods, *edited by Robert A. Boudreau and Sharon M. Doudreau*
99. Organic Photovoltaics: Mechanism, Materials, and Devices, *edited by Sam-Shajing Sun and Niyazi Serdar Saracftci*
100. Handbook of Optical Interconnects, *edited by Shigeru Kawai*
101. GMPLS Technologies: Broadband Backbone Networks and Systems, *Naoaki Yamanaka, Kohei Shiomoto, and Eiji Oki*

GMPLS Technologies

Broadband Backbone Networks and Systems

Naoaki Yamanaka
Keio University
Yokohama, Japan

Kohei Shiomoto
NTT
Tokyo, Japan

Eiji Oki
NTT
Tokyo, Japan

Taylor & Francis
Taylor & Francis Group
Boca Raton   London   New York

A CRC title, part of the Taylor & Francis imprint, a member of the Taylor & Francis Group, the academic division of T&F Informa plc.

Published in 2006 by
CRC Press
Taylor & Francis Group
6000 Broken Sound Parkway NW, Suite 300
Boca Raton, FL 33487-2742

© 2006 by Taylor & Francis Group, LLC
CRC Press is an imprint of Taylor & Francis Group

No claim to original U.S. Government works
Printed in the United States of America on acid-free paper
10 9 8 7 6 5 4 3 2 1

International Standard Book Number-10: 0-8247-2781-9 (Hardcover)
International Standard Book Number-13: 978-0-8247-2781-9 (Hardcover)
Library of Congress Card Number 2005041375

This book contains information obtained from authentic and highly regarded sources. Reprinted material is quoted with permission, and sources are indicated. A wide variety of references are listed. Reasonable efforts have been made to publish reliable data and information, but the author and the publisher cannot assume responsibility for the validity of all materials or for the consequences of their use.

No part of this book may be reprinted, reproduced, transmitted, or utilized in any form by any electronic, mechanical, or other means, now known or hereafter invented, including photocopying, microfilming, and recording, or in any information storage or retrieval system, without written permission from the publishers.

For permission to photocopy or use material electronically from this work, please access www.copyright.com (http://www.copyright.com/) or contact the Copyright Clearance Center, Inc. (CCC) 222 Rosewood Drive, Danvers, MA 01923, 978-750-8400. CCC is a not-for-profit organization that provides licenses and registration for a variety of users. For organizations that have been granted a photocopy license by the CCC, a separate system of payment has been arranged.

Trademark Notice: Product or corporate names may be trademarks or registered trademarks, and are used only for identification and explanation without intent to infringe.

Library of Congress Cataloging-in-Publication Data

Yamanaka, Naoaki.
   GMPLS technologies : broadband backbone networks and systems / Naoaki Yamanaka, Kohei Shiomoto, Eiji Oki.
      p. cm. – (optical engineering ; 86)
   ISBN 0-8247-2781-9 (alk. paper)
   1. Internet. 2. Computer networks--management. 3. Telecommunication--traffic management. 4. Packet switching (Data transmission) 5. MLPS standard. I. Shiomoto, Kohei. II. Oki, Eiji, 1969- III. Title. IV. Optical Engineering (Marcel Dekker, Inc.) ; v. 86.

TK5105.875.I57Y35 2005
621.382'16--dc22
                                  2005041375

Taylor & Francis Group
is the Academic Division of T&F Informa plc.

Visit the Taylor & Francis Web site at
http://www.taylorandfrancis.com

and the CRC Press Web site at
http://www.crcpress.com

Contents

Chapter 1
Broadband and Multimedia 1
1.1   Multimedia Network 1
1.2   Connection and Communication Mechanism 5
References 10

Chapter 2
Basic Mechanisms of Connection-Oriented Network 11
2.1   Basics of Connection-Oriented Communication 11
2.2   Basics of Connectionless Communication 15
2.3   Communication by TCP-IP 22
References 30

Chapter 3
Connection-Oriented Communications and ATM 33
3.1   Transmission Method in ATM 33
       3.1.1   GFC (Generic Flow Control) Only in UNI 36
       3.1.2   VCI/VPI (Virtual Path Identifier/Virtual
               Channel Identifier) 36
       3.1.3   PT (Payload Type) 37
       3.1.4   CLP (Cell Loss Priority) 37
       3.1.5   HEC (Header Error Control) 38
3.2   ATM Adaptation Layer 38
       3.2.1   AAL Type 1 39
       3.2.2   AAL Type 2 41
               3.2.2.1   Packetizing the CPS 42
               3.2.2.2   Multiplexing the CPS Packets 42
               3.2.2.3   Timer for Cell-Construction Delay 43
       3.2.3   AAL Types 3/4 43
               3.2.3.1   Frame Assembly and MID Multiplexing Functions 43
               3.2.3.2   Error Detection 43
       3.2.4   AAL Type 5 44
               3.2.4.1   Frame Assembly 45
               3.2.4.2   Error Detection 46
3.3   Permanent Connection and Switched Connection 46
3.4   Traffic Engineering in ATM 47
       3.4.1   Connection Admission Control (CAC) 48
       3.4.2   Usage Parameter Control (UPC) 49

	3.4.3	Priority Control	51
	3.4.4	Traffic-Shaping Control	53
	3.4.5	Packet Throughput and Discarding Process	53
	3.4.6	Congestion Control in ATM	55
3.5	Bearer Class and Services		57
	3.5.1	Constant Bit Rate (CBR)	57
	3.5.2	Variable Bit Rate (VBR)	57
	3.5.3	Unspecified Bit Rare (UBR)	59
	3.5.4	Available Bit Rate (ABR)	59
	3.5.5	ATM Burst Transfer (ABT)	60
3.6	OAM Function in ATM		61
References			66

Chapter 4
Internet Protocol (IP) 69
4.1 IP Forwarding 69
    4.1.1 IP Header 69
    4.1.2 IP Address 70
    4.1.3 Forwarding Table (Routing Table) 71
4.2 IP Routing 74
    4.2.1 Hierarchy of Routing Protocols 74
    4.2.2 Categorization of Routing Protocol by Algorithm 75
        4.2.2.1 Distance-Vector-Type Protocol 76
        4.2.2.2 Path-Vector-Type Protocol 82
        4.2.2.3 Link-State-Type Protocol 83
4.3 Example of Routing Protocol 87
    4.3.1 OSPF 87
        4.3.1.1 Principle 87
        4.3.1.2 Link State 88
        4.3.1.3 Scalability and Hierarchization 89
        4.3.1.4 Aging of LSA (Link-State Advertisement) 91
        4.3.1.5 Content of LSA 91
    4.3.2 BGP-4 95
        4.3.2.1 Principle 95
        4.3.2.2 BGP Message 96
        4.3.2.3 Path Attributes 100
        4.3.2.4 Rules of Route Selection 103
        4.3.2.5 IBGP and EBGP 104
        4.3.2.6 Scalability 105
References 107

Chapter 5
MPLS Basics 109
5.1 Principle (Datagram and Virtual Circuit) 109
    5.1.1 Bottleneck in Searching IP Table 109
    5.1.2 Speeding Up by Label Switching 111
5.2 LSP Setup Timing 114
    5.2.1 Traffic Driven 114
    5.2.2 Topology Driven 115
5.3 Protocol (Transfer Mechanism of Information) 115
    5.3.1 MPLS Label 115
    5.3.2 Label Table 116
    5.3.3 Label Stack 118
    5.3.4 PHP 119
    5.3.5 Label Merge 121
5.4 Protocol (Signaling System) 122
    5.4.1 Label-Assignment Method
          (Downstream Type, Upstream Type) 122
    5.4.2 Label-Distribution Method (On-Demand Type
          and Spontaneous Type) 123
    5.4.3 Label-Assignment/Distribution Control Method
          (Ordered Type/Independent Type) 123
    5.4.4 Label-Holding Method (Conservative/Liberal) 124
    5.4.5 Loop-Protection Method (Path Vector/Hop Count) 124
    5.4.6 Hop-by-Hop-Type LSP and Explicit-Route-Type LSP 125
References 125

Chapter 6
Application of MPLS 127
6.1 Traffic Engineering 127
    6.1.1 Problems with IGP 127
    6.1.2 Separation of Forwarding from Routing by MPLS 127
    6.1.3 Source Routing 130
    6.1.4 Traffic Trunk 132
    6.1.5 Restricted Route Controlling 134
    6.1.6 Setting Up the ER-LSP with RSVP-TE 135
6.2 Routing to External Route within AS 146
    6.2.1 Route Exchange by Border Gateway Protocol (BGP) 146
    6.2.2 Routing to External Route within AS 148
    6.2.3 Solution by MPLS 149
6.3 Virtual Private Networks (VPN) 150
    6.3.1 Overlay Model and Peer Model 151
    6.3.2 Virtual Routing and Forwarding (VRF) 152

    6.3.3   MP-BGP 153
    6.3.4   Notification of Outer Label and Inner Label in VPN 156
References 157

Chapter 7
Structure of IP Router 159
7.1   Structure of Router 159
       7.1.1   Low-End-Class Router 160
       7.1.2   Middle-Class Router 161
       7.1.3   High-End-Class Router 163
7.2   Switch Architecture 165
       7.2.1   Classification of Switch Architecture 165
       7.2.2   Input-Buffer-Type Switch 167
             7.2.2.1   FIFO Input-Buffer-Type Switch 167
             7.2.2.2   VOQ Input-Buffer Type 168
             7.2.2.3   Maximum Size Matching 169
             7.2.2.4   Maximum Weighting Size Matching 171
             7.2.2.5   Parallel Interactive Matching (PIM) 171
             7.2.2.6   *i*SLIP 172
             7.2.2.7   Application of *i*SLIP to Three-Stage
                       Cross Network Switch 175
7.3   Packet Scheduling 177
       7.3.1   FIFO (First-In First-Out) Queuing 177
       7.3.2   Complete Priority Scheduling 178
       7.3.3   Generalized Processor Sharing 178
       7.3.4   Packetized Generalized Processor Sharing 179
       7.3.5   Weighted Round-Robin (WRR) Scheduling 180
       7.3.6   Weighted Deficit Round-Robin (WDRR) Scheduling 181
7.4   Forwarding Engine 182
       7.4.1   Route Lookup 182
       7.4.2   Design of Route Lookup 183
       7.4.3   Trie Structure 184
       7.4.4   Patricia Tree 185
       7.4.5   Binary Search Method 187
       7.4.6   Route Lookup with CAM 188
References 190

Chapter 8
GMPLS (Generalized Multiprotocol Label Switching) 191
8.1   From MPLS to MPλS/GMPLS 191
8.2   General Description of GMPLS 193
8.3   Separation of Data Plane from Control Plane 199
8.4   Routing Protocol 199
       8.4.1   OSPF Extension 199
       8.4.2   TE Link Advertisement 201

8.5	Signaling Protocol		204
	8.5.1	RSVP-TE Extension of RSVP-TE and GMPLS	204
	8.5.2	General Label Request	206
	8.5.3	Bidirectional Path Signaling	209
	8.5.4	Label Setting	210
	8.5.5	Architectural Signaling	212
8.6	Link Management Protocol		213
	8.6.1	Necessity of LMP	213
	8.6.2	Types of Data Link	214
	8.6.3	Functions of LMP	214
		8.6.3.1 Control-Channel Management	215
		8.6.3.2 Link-Property Correlation	216
		8.6.3.3 Connectivity Verification	217
		8.6.3.4 Failure Management	219
8.7	Peer Model and Overlay Model		219
	8.7.1	Peer Model	220
	8.7.2	Overlay Model	221
References			222

Chapter 9
Traffic Engineering in GMPLS Networks 223

9.1	Distributed Virtual-Network Topology Control		224
	9.1.1	Virtual-Network Topology Design	224
	9.1.2	Distributed Network Control Approach	225
		9.1.2.1 Virtual-Network Topology	225
		9.1.2.2 Design Goal	226
		9.1.2.3 Overview of Distributed Reconfiguration Method	227
		9.1.2.4 Distributed Control Mechanism	228
		9.1.2.5 Heuristic Algorithm for VNT Calculation	229
	9.1.3	Protocol Design	230
		9.1.3.1 GMPLS Architecture	230
		9.1.3.2 Forwarding Adjacency in Multilayer Path Network	232
		9.1.3.3 Switching Capability	232
		9.1.3.4 Protocol Extensions	233
	9.1.4	Performance Evaluation	234
		9.1.4.1 Effect of Dynamic VNT Change	234
		9.1.4.2 Utilization	236
		9.1.4.3 Dynamic Traffic Change	237
9.2	Scalable Multilayer GMPLS Networks		241
	9.2.1	Scalability Limit of GMPLS Network	241
	9.2.2	Hierarchical Cloud-Router Network (HCRN)	245
		9.2.2.1 HCRN Architecture	245

		9.2.2.2	CR Internal-Cost Scheme and Network Topology 246
		9.2.2.3	Multilayer Shortest-Path-First Scheme 248
	9.2.3	Performance Evaluation 249
		9.2.3.1	Scalability 249
		9.2.3.2	Effect of Multilayer Network Hierarchization 252

9.3 Wavelength-Routed Networks 253
    9.3.1 Routing and Wavelength Assignment (RWA) Problem 253
    9.3.2 Distributedly Controlled Dynamic Wavelength-Conversion (DDWC) Network 255
        9.3.2.1 DDWC Network with Simple RWA Policy 255
        9.3.2.2 Optical Route Selection 256
        9.3.2.3 Extended Signaling Protocol of RSVP-TE 257
    9.3.3 Performance of DDWC 259

9.4 Survivable GMPLS Networks 264
    9.4.1 A Disjoint-Path-Selection Scheme with Shared-Risk Link Groups 264
    9.4.2 Weighted-SRLG Path Selection Algorithm 266
    9.4.3 Performance Evaluation 268
        9.4.3.1 Fixed α 269
        9.4.3.2 Adaptive α 273
        9.4.3.3 Link-Capacity Constraints 274

9.5 Scalable Shared-Risk-Group Management 276
    9.5.1 SRG Concept 277
    9.5.2 SRG-Constraint-Based Routing (SCBR) 279
        9.5.2.1 Admission Control at Link Using Backup-SRG Concept 280
        9.5.2.2 SCBR 282
    9.5.3 Distributed Routing Calculation 284
    9.5.4 Performance Evaluation 285
        9.5.4.1 Shared Restoration versus Protection 285
        9.5.4.2 Effect of Bumping of Existing Backup LSPs 286
        9.5.4.3 Link Protection versus Node Protection 287
        9.5.4.4 Hierarchy 288

9.6 Demonstration of Photonic MPLS Router 290
    9.6.1 Integration of IP and Optical Networks 290
    9.6.2 Photonic MPLS Router (HIKARI Router) 290
        9.6.2.1 Concept of HIKARI Router 290
        9.6.2.2 HIKARI Router Characteristics 292
        9.6.2.3 Optical-Layer Management Characteristics 293
        9.6.2.4 Implementation of MPλS Signaling Protocol 294
    9.6.3 Photonic Network Protection Configuration 294
    9.6.4 Demonstration of HIKARI Router 295

References 298

Chapter 10
Standardization 303
10.1 ITU-T (International Telecommunication Union-T) 303
10.2 IETF (Internet Engineering Task Force) 305
10.3 OIF (Optical Internetworking Forum) 308
10.4 ATM Forum 316
      10.4.1 Committees 316
      10.4.2 Future Activity 318
10.5 MPLS Forum 318
10.6 WIDE Project 319
      10.6.1 Internet Area 320
      10.6.2 Transport Area 320
      10.6.3 Security Area 320
      10.6.4 Operations/Management Area 321
      10.6.5 Applications Area 321
10.7 Photonic Internet Laboratory 321
      10.7.1 PIL Organization 322
      10.7.2 MPLS-GMPLS, Multilayer, Multiroute
             Interworking Tests 323
References 325
Appendix 10.1 ITU Topics 326
Appendix 10.2 IETF Working Groups 327
A2.1 Applications Area 327
A2.2 General Area 328
A2.3 Internet Area 328
A2.4 Operations and Management Area 329
A2.5 Routing Area 330
A2.6 Security Area 331
A2.7 Sub-IP Area 331
A2.8 Transport Area 332

Index 335

Preface

This book describes MPLS (Multi-Protocol Label Switching) and GMPLS (generalized MPLS) concepts and technologies. Both MPLS and GMPLS will be key technologies in the evolution of the next generation of reliable Internet Protocol (IP) backbone networks. This book covers broadband services, IP, IP router technologies, MPLS, and GMPLS, and it addresses issues regarding the standardizations of these technologies. It was written in the hope that it would serve as a useful reference book for industry researchers/engineers and as a textbook for graduate students and senior undergraduates.

This book is targeted mainly toward the networking/communications arena. Given the explosive growth of Internet traffic, there is a compelling need to build high-speed, reliable, and secure backbone networks. MPLS was developed 5 years ago to provide connection-oriented services in IP networks, and now it is being introduced in backbone networks. GMPLS is a protocol that is applied to the TDM (time-division multiplexing) layer, the wavelength path layer, and the fiber layer by generalizing the label concept of MPLS that has been applied to the packet layer for transferring IP packets. MPLS and GMPLS are key technologies for next-generation networks. However, many engineers and students have to search the literature for technical papers on these topics and read them in an ad-hoc manner. To the best knowledge of the authors, this is the first book on the market that explains both MPLS and GMPLS concepts and technologies in a systematic manner.

This book addresses the basics of the network architectures, protocols, and traffic-engineering techniques that are used to operate MPLS and GMPLS networks. The book is based on the material that the first author, Dr. Yamanaka, has been teaching in industry workshops and universities for the past 3 years, and feedback from these classes has been invaluable in the development of this work. Dr. Yamanaka is using a draft of this book as the text for a graduate course he is teaching, "Information and Communication Networks." The course is very popular in both academia and industry because the next-generation network is merging toward MPLS and GMPLS technologies, and people want to understand these important communication network protocols. They want to know the principles of operation and their associated technologies so that they can either design better IP routers or be in a better position to choose a better network to meet their customers' needs.

Given the scarcity of books about high-speed networking technologies, we believe that this book will serve as a useful addition to the literature. As leading researchers at NTT, the authors have first-hand experience in developing real network systems. Thus, the book describes not only fundamental and theoretical technologies, but also state-of-the-art systems and protocols.

TARGET AUDIENCE

This book can be used as a reference for industry workers whose jobs are related to IP/MPLS/GMPLS networks. Engineers who design network equipment or who work for network service providers can benefit from this book because it explains the key concepts of MPLS and GMPLS and describes the techniques for building an MPLS/GMPLS-based network. This book is also a good text for electrical-engineering/computer-science graduate and senior undergraduate students who want to understand MPLS and GMPLS networks so that they can better position themselves when they graduate and look for jobs in the information technology (IT) field.

ORGANIZATION

This book begins by introducing the reader to broadband services to provide a foundation for understanding the requirements in broadband networks in Chapter 1. Once the reader has a complete picture of broadband services and networks, connectionless and connection-oriented services are described in terms of traffic characteristics and user requirements. The chapter concludes with a discussion of Asynchronous Transfer Mode (ATM), IP, and MPLS.

Chapters 2 and 3 describe the basics of connection-oriented networks. In Chapter 2, the concepts of control plane and user plane are explained, followed by a discussion of connection-oriented and connectionless communications. Two connection-oriented communication protocols — Transport Control Protocol (TCP) and ATM — are introduced. Chapter 2 concludes with a discussion of TCP/IP protocol and a description of the flow-control mechanism. In Chapter 3, the ATM protocol is explained. Studying the ATM protocol provides a foundation for discussion of MPLS and GMPLS, which are extensions of the ATM concept. One of the most important functions of ATM is traffic engineering, and ATM functions such as connection admission control (CAC), usage parameter control (UPC), and priority control are discussed. Traffic engineering in MPLS networks is based on these ATM functions.

Chapter 4 describes the basics of IP. IP is the third-layer protocol that consists of an IP forwarding protocol, which is a technology to transmit the IP packets according to the routing table, and an IP routing technology to create the routing table. The IP forwarding technology is first described, and then IP routing technologies are presented.

Chapters 5 and 6 describe MPLS. Chapter 5 presents the basics of MPLS and explains why MPLS is a better alternative than IP packet-based forwarding. A concept of label-switched paths (LSPs) is explained, and then a signaling technique to set up an LSP is introduced. Chapter 6 describes MPLS applications. MPLS executes transmission of IP packets based on its label. By using the label, various applications that have been difficult to realize in conventional IP datagram transmission become possible, such as traffic engineering, improvement of routing efficiency of external routes within an autonomous system, and virtual private networks.

Chapter 7 describes IP router structures. The major functions of IP routers are categorized into two function groups: a data-path function group and a control function group. We describe the basic technologies of the data-path function, such as switch architecture, packet scheduling, and forwarding engine.

Chapter 8 describes GMPLS, which is a protocol that is applied to the TDM (Time Domain Multiplexing) layer, the wavelength-path layer, and the fiber layer by generalizing the label concept of MPLS that has been applied to the packet layer for transferring IP packets. GMPLS makes it possible to execute a distributed control, which is a feature of MPLS, thereby simplifying the operation. It is also possible to totally engineer the traffic based on the traffic information or topology information of each layer and to improve the utilization efficiency of networks. We describe the prehistory of development from MPLS to MPλS/GMPLS, the outline of GMPLS and its protocol, and the application example of GMPLS.

Chapter 9 describes several important studies on traffic engineering in Generalized Multi-Protocol Label Switching (GMPLS) networks. These studies address the topics of multilayer traffic engineering, survivable networks, and wavelength-routed optical networks, and the chapter concludes with a demonstration of a GMPLS-based router.

Chapter 10 discusses standardization issues regarding IP, MPLS, and GMPLS.

Acknowledgments

This book would not have been possible without the help of many people. We would like to thank them for their efforts in improving the quality of the book. We have done our best to accurately describe MPLS and GMPLS concepts and technologies. We alone are responsible for any remaining errors. If any errors are found, please send an e-mail to yamanaka.naoaki@ieee.org so that we can correct them in future editions.

The entire book is based on research work that was done at NTT Labs. Our sincere thanks to Dr. Masao Kawachi (NTT), Dr. Haruhisa Ichikawa (NTT), and Dr. Ken-ichi Sato (Nagoya University) for their support of our research and their valuable advice. We gratefully acknowledge Dr. Satoru Okamoto (NTT) and Daisaku Shimazaki (NTT), who contributed material to some chapters. We are very grateful to Professor Tomonori Aoyama (University of Tokyo), Professor Iwao Sasase (Keio University), and Professor Masao Nakagawa (Keio University) for their suggestions on writing a textbook for graduate students.

Naoaki would like to thank his wife, Asako, and his children, Yuki, Takuto, and Noe, for their love and support. Kohei wants to thank his wife Yuko and daughter Ayano, who gave him encouragement as he wrote this book. Eiji wishes to thank his wife, Naoko, his daughter, Kanako, and his son, Shunji, for their love.

Authors

Naoaki Yamanaka, Ph.D., is a Professor, Dept. of Information and Computer Science, Keio University, Tokyo, Japan and is a representative of Photonic Internet Labs, which is a GMPLS interworking consortium. He has been active in the development of carrier backbone networks and systems based on MPLS and ATM, including terabit/sec electrical/optical backbone switching, for NTT. He is now researching future optical IP networks and optical GMPLS router systems. He has published more than 112 peer-reviewed journal and transaction articles, written 82 international conference papers, and been awarded 174 patents, including 17 international patents. Dr. Yamanaka received Best of Conference Awards from the 40th, 44th, and 48th IEEE Electronic Components and Technology Conference in 1990, 1994, and 1998; the TELECOM System Technology Prize from the Telecommunications Advancement Foundation in 1994; the IEEE CPMT Transactions Part B: Best Transactions Paper Award in 1996; and the IEICE Transaction Paper Award in 1999. He is technical editor of *IEEE Communication Magazine*, broadband network area editor of *IEEE Communication Surveys*, editor of *IEICE Transactions*, as well as TAC chair of the Asia Pacific Board of the IEEE Communications Society. Dr. Yamanaka is an IEEE fellow and a member of the IEICE.

Eiji Oki, Ph.D. Eiji Oki received B.E. and M.E. degrees in Instrumentation Engineering and a Ph.D. degree in Electrical Engineering from Keio University, Yokohama, Japan, in 1991, 1993, and 1999, respectively. In 1993, he joined NTT, Tokyo Japan. From 2000 to 2001, he was a Visiting Scholar at Polytechnic University, Brooklyn, New York, where he was involved in designing tera-bit switch/router systems. He is now engaged in researching and developing high-speed optical IP backbone networks as a Senior Research Engineer with NTT Network Service Systems Laboratories. Dr. Oki was the recipient of the 1998 Switching System Research Award and the 1999 Excellent Paper Award presented by IEICE, and the 2001 Asia-Pacific Outstanding Young Researcher Award presented by IEEE Communications Society. He co-authored a book, "Broadband Packet Switching Technologies," published by John Wiley, New York, in 2001. He is a member of the IEICE and the IEEE.

Kohei Shiomoto, Ph.D., is a Senior Research Engineer, Supervisor, at NTT Network Service Systems Laboratories, Japan. He joined the Nippon Telegraph and Telephone Corporation (NTT), Tokyo, Japan in April 1989, where he was engaged in research and development of ATM traffic control and ATM switching system architecture design. From August 1996 to September 1997, he was engaged in research on high-speed networking as a Visiting Scholar at Washington University in St. Louis, MO, USA. From September 1997 to June 2001, he was directing architecture design for high-speed IP/MPLS label switch router research project at NTT Network Service Systems Laboratories, Tokyo, Japan. Since July 2001 he has been engaged in the research fields of photonic IP router design, routing algorithm, and GMPLS routing and signaling standardization at NTT Network Innovation Laboratories. He received the B.E., M.E., and Ph.D. from Osaka University, Japan in 1987, 1989, and 1998. He is a member of IEICE, IEEE, and ACM.

1 Broadband and Multimedia

1.1 MULTIMEDIA NETWORK

The terms "broadband" and "multimedia" have recently been refocused yet again. These network technologies are going to be implemented in ways that are vastly different than what was envisioned only 20 years ago.

At the end of 2002, the number of Internet users in Japan was nearly 56 million and increasing steadily at the rate of 20% per year. The broadbanding in subscriber lines that was expected to break through with the deployment of FTTH (fiber to the home) using optical-fiber technology has already been achieved using DSL (digital subscriber line) technology earlier and more cost effectively than FTTH. DSL is currently utilized by 15% of users in Japan. That is, the broadband age is now going to open on the basis of the Internet because of the growing number of subscribers as well as the improved quality of the broadband information flow using DSL.

On the other hand, "multimedia" refers not just to the conventional network for telephone services, but to a whole range of expanded services, as shown in Figure 1.1. Internet services have expanded from e-mail to an information-access paradigm of client-server type by providing so-called multimedia platforms through the World Wide Web (WWW). Today, the Internet traffic map includes P2P services or file-exchanging services in which an individual can be both an originator and a receiver of information. Telephone service has advanced to Fax and, more recently, to mobile telephone, and further consolidation with data services using terminals such as an I-mode or L-mode has been rapidly progressing. Business networks have been diversified and highly developed from a simple form of dedicated point-to-point communication to complicated and sophisticated services supporting the various demands for bandwidth, location, quality, etc.

Now let us look at the actual networks to provide these services. The nature of the traffic in multimedia is different in every service, as seen in the following examples.

- *Type of connection*: There is a 1:1 type of connection such as telephone service, and a 1:n type of connection (multicasting connection) such as TV broadcasting, CATV, or delivery of advertisements. There is also an n:m type of connection such as a teleconference service that is executed by connecting multiple remote sites.
- *Volume of traffic (bandwidth and amount)*: The volume of traffic is 64 kbps in telephone services and exceeds 30 Mbps in HDTV services.

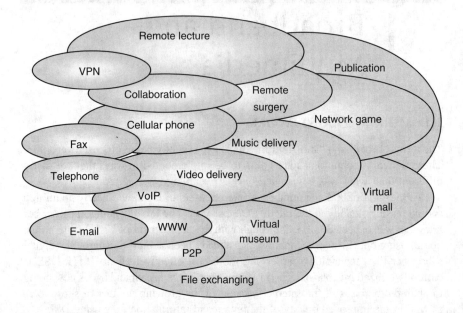

FIGURE 1.1 Multimedia services.

High-volume data files or video contents can exceed 1 Gbyte. In contrast, the amount of data used for remote checking of a residential gas meter is just one packet per month.
- *Hold time*: The average hold time is just 3 min in the case of telephone service, but it is longer than 2 h in the case of communication service through CATV.
- *Symmetric property*: Telephony is basically a symmetric communication having the same traffic volume in both directions. But when the telephone line is used for Internet services, downstream traffic volume increases by about 1000 times compared with upstream traffic.
- *Burstlike nature*: The traffic rate in telephone service is a constant 64 kbps. But when the telephone line is used to access Web sites, although a large amount of data is downloaded for a time just after accessing a Web page, almost no data is downloaded while the page is viewed. That is, burstlike data communication is executed in the case of Internet access.
- *Connection type*: Telephony is usually a 1:1 connection, but sometimes, as in a party line, it involves a 1:n or n:m connection. On the other hand, e-mail and WWW access is 1:1, and CATV service is a 1:n multicast service.

Figure 1.2 shows a typical network structure of multimedia services. It is expected that, although services are provided through multiple media, the infrastructure's backbone will be developed as an integrated and universal network.

Broadband and Multimedia

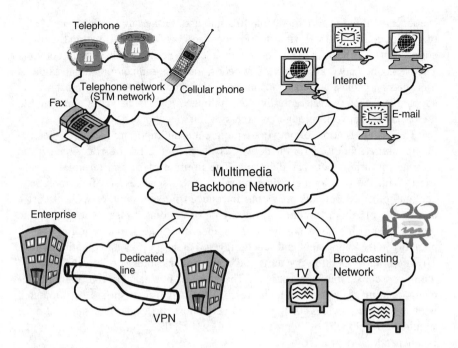

FIGURE 1.2 Multimedia network.

A multimedia communication network fundamentally consists of end-user terminals or telephone equipment; access lines such as ADSL, coaxial cables of CATV, or optical fiber cables; backbone network; network equipment such as servers; and the operating systems that control the data transmission. Figure 1.3 shows a conceptual diagram illustrating the basic principle of a multimedia communication network. In the typical case of accessing the WWW, an end terminal

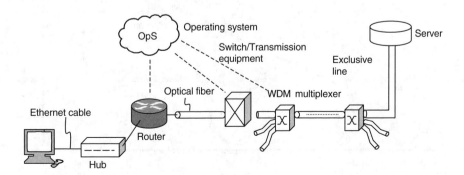

FIGURE 1.3 Concept of communication network.

is connected through Ethernet cable to a hub and, farther, to a small router (edge router). (Details of how IP [Internet Protocol] packets are routed within the router are described in Section 1.2.) The router is connected to the backbone network. The backbone network consists of switches, transmission equipment, and optical transmission cables, and the transmission technology that has been most widely adopted is WDM (wavelength division multiplexing). The server is connected to the backbone network through an exclusive line.

The system lifetime of network equipment is an important factor in selecting the optimum technology for the network system. Figure 1.4 illustrates the system lifetime of various network elements. The lifetime of an end terminal is very short. With the development or evolution of new services and functions, new versions of end terminals are being introduced into the market with almost a one-year cycle time. In contrast, the laying of conduit (tubes to accommodate optical fiber cable) is a costly civil engineering project that requires careful planning in its deployment and scheduling and in developing an operating policy. The lifetime of operating systems and optical fiber cables is relatively long and requires a considerable investment. Therefore, it is clear that the network elements relating to services, which are constantly evolving, should be populated with lower-cost devices having a relatively short life cycle, and the backbone should be developed with a protocol having a long life cycle or with a form that is independent of protocol.

Given this background on the nature of broadband networks, an alternative to a physical backbone based on WDM technology — a universal backbone based on MPLS (Multiprotocol Label Switching) technology — has recently been developed and actively introduced into an actual network. Figure 1.5 shows a

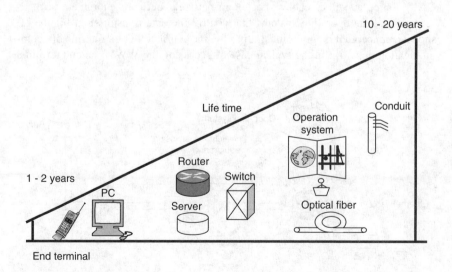

FIGURE 1.4 System lifetime of network elements.

Broadband and Multimedia

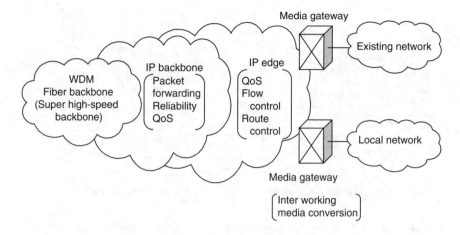

FIGURE 1.5 Hierarchical structure of multimedia network and integration by MPLS technology.

functional image of this network. Various services, access points, and local networks are accommodated within this interworking unit, which is called a "media gateway." Figure 1.5 also lists the conditions that are required for each component. Because traffic is converged to the backbone based on WDM, the MPLS-based backbone network must be high speed, highly reliable, and cost effective. That is, the backbone should be built on the basis of IP on the universal physical network of WDM. Therefore, like the physical network, this IP backbone must also have a high QoS (quality of service), including a high-speed packet-forwarding function, high reliability, short delay, low loss, etc. The IP edge also requires "routing" functions to execute QoS and flow controls that can accommodate different levels of service and provide optimum path control.

1.2 CONNECTION AND COMMUNICATION MECHANISM

The basis of communication is to establish a communication path between users (or user terminals) so that they can exchange information. There are two methods of establishing a communication path, and each has its good points and bad. It is important to understand both of these methods. In this section, we describe both methods from various viewpoints.

Figure 1.6 shows the concept of connection-oriented and connectionless communications. Connection-oriented communication establishes a connection prior to receiving or sending information and reserves the required communication resources such as bandwidth, etc. Therefore, this method is superior in that it is possible to reserve the required quality of communication. On the other hand, the connectionless communication is a method to send information by converting it

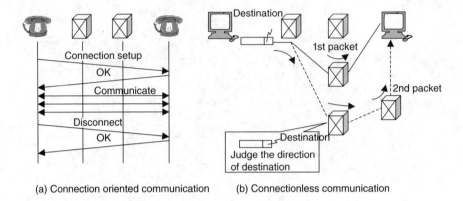

FIGURE 1.6 Connection-oriented communication and connectionless communication.

into multiple packets, each of which has a destination address, and by making each node execute a routing function according to the destination address. Connection-oriented communication is represented by telephony and ATM (Asynchronous Transfer Mode) technologies, and connectionless communication is represented by IP packet-forwarding technology adopted in such IP networks as the Internet. Because connectionless communication does not reserve network resources, it is also possible to use the network resources effectively by dynamically allocating them when communication is not executed. Therefore, it is better suited for burst transmission or data transfer. At the same time, it has a disadvantage in that the quality of communication is not assured because the network resource is not reserved.

We can obtain a better understanding of the basic communication mechanism by taking an ATM system as an example to illustrate the connection-oriented communication. In ATM communication, a label (VCI/VPI [virtual-channel identifier/virtual-path ide ntifier] value of the ATM header) used for routing in each link is attached to a connection when setting up the connection. During communication, routing is executed based on this label by relaying the label (label swap) to transfer information. Figure 1.7 shows the basic principle and features of connection-oriented ATM communication. A packet is transferred to an output port or to the next ATM switch according to the label attached to the ATM packet, or cell. At the same time, it is possible to use different VCI/VPI values in the next link as the VCI/VPI value of the label. As shown in the case of Figure 1.7, the label is swapped from 7 to 5, for example.

In the case of ATM, packet length is fixed to 53 bytes, and label swapping and routing by label are executed by hardware. The route during communication does not change, and it is possible to address multimedia by transmitting many cells when the volume of information is large and by transmitting a few cells when the volume of information is small. Although excellent traffic-control

Broadband and Multimedia

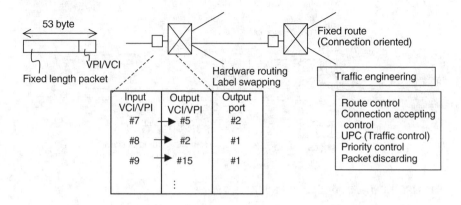

FIGURE 1.7 Basic principle and features of ATM (connection oriented).

(traffic engineering) technology has been established in ATM, in this section we describe only the keywords and their meanings. The details of traffic-control technology in ATM, which is also effectively utilized in MPLS (Multiprotocol Label Switch)-based systems, are then discussed.

Route control: selects explicitly the route to communicate a certain connection and controls congestion or traffic load balance.

Connection-accepting control: judges whether the request for setting up a new connection should be accepted or not and controls the quality of communication by limiting the traffic flow so that traffic exceeding the specified amount may not flow into the network.

UPC (traffic) control: UPC (usage parameter control) determines the traffic volume of communication in advance between user and network and, using this value as a threshold, controls the quality of communication so that traffic exceeding the threshold value that has been contracted may not flow into the network.

Priority control: sets an order of priority to the packets being communicated and controls the quality of communication by discarding packets according to the order of priority when the network is congested.

Packet discarding: discards the packets when the network is congested. Which packets should be discarded is determined by considering the effect of the discard and the frame of its upper level.

Now that we understand the basics of ATM, we can describe the MPLS that transfers the IP packets with extremely high efficiency [1–6]. Transmission with MPLS is executed basically in a connection-oriented way after the IP packets have been "cut through" (see explanation below). Figure 1.8 illustrates the fundamental transmission mechanism of MPLS. Because IP forwarding is

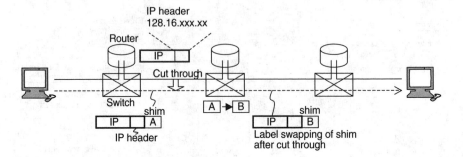

FIGURE 1.8 Basic principle of MPLS.

executed in a connectionless way, the IP packets are usually transferred hop-by-hop through routers according to their destination addresses. For example, when the volume of IP traffic between some users is very high, IP packets with the same address are forwarded repeatedly. Thus, it is possible to transfer the succeeding IP packets by label swapping, i.e., by giving a label to the pair of source address (SA) and the destination address (DA) for IP packets having the same destination address using a technique called "cut through." In the example of Figure 1.8, IP packets are transferred after a label consisting of 20 bits called a "shim header" was attached (Push) to them at the input-side router (ingress router) and, although it is not illustrated in this figure, this label is removed (Pop) at the output-side router (egress router). Transfer is executed simply by label swapping (A → B), as described previously. The process of label swapping is easily accomplished compared with searching the output ports (next-stage routers) by IP address. In IP forwarding, there is no guarantee that the succeeding packets after a certain packet will be transferred along the same route. In contrast, after cut through in MPLS, the packets are transferred by just one route, and, as described in the basic operation mechanism of ATM, it is possible to control the route to be selected (i.e., to determine the route explicitly). Because the MPLS path is connection oriented, there is sufficient bandwidth available to guarantee the quality of communication and therefore to offer new services or applications.

Figure 1.9 shows a comparison between connection-oriented communication (ATM) and connectionless communication (IP). Using this figure, we describe the characteristic features of MPLS. Figure 1.9(a) and Figure 1.9(b) show the features of ATM as the connection-oriented communication and of IP as the connectionless communication, respectively. Because MPLS can communicate by combining the two types of IP transmission — ATM and IP — it is possible to combine their individual advantages. Consequently, we can understand that MPLS is an excellent communication method having the characteristic features of both the connection-less type and the connection-oriented type of communications, as illustrated in Figure 1.10.

Broadband and Multimedia

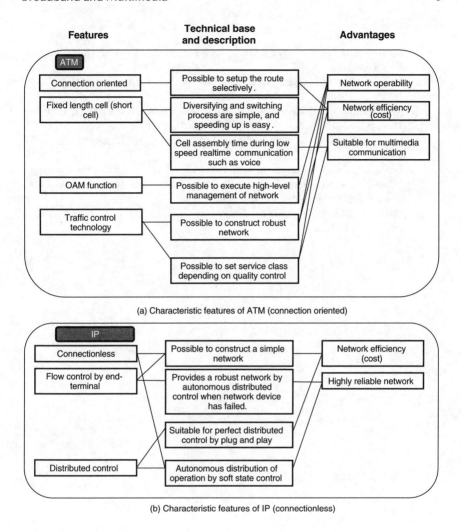

FIGURE 1.9 Characteristic features of ATM (connection oriented) and IP (connectionless).

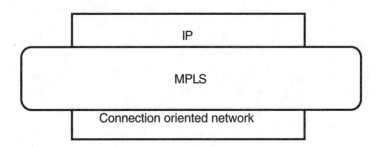

FIGURE 1.10 Features of MPLS.

REFERENCES

1. Kercheval, B., *TCP/IP over ATM: a No-Nonsense Interworking Guide,* Prentice Hall, New York, Dec. 7.
2. Esaki, H., Ohashi, N., Nakagawa, I., and Nagami, K., *MPLS Textbook,* IDG, Tokyo, Japan (in Japanese), July 2002.
3. Awduche, D.O., MPLS and traffic engineering in IP networks, *IEEE Commun. Mag.,* 37, 42–47, 1999.
4. Swallow, G., MPLS advantages for traffic engineering, *IEEE Commun. Mag.,* 37, 54–57, 1999.
5. Pepelnjak, I. and Guichard, J., *MPLS and VPN Architecture, a Practical Guide to Understanding, Designing and Deploying MPLS and MPLS-enabled VPNs,* Cisco Press, Indianapolis, IN. Jan. 2000.
6. Davie, B. and Rekhter, Y., *MPLS, Technology and Applications,* Morgan Kaufmann Publishers, Oxford, UK, May 2000.

2 Basic Mechanisms of Connection-Oriented Network

2.1 BASICS OF CONNECTION-ORIENTED COMMUNICATION

This section describes the basics of connection-oriented communication [1, 2]. As shown in Figure 2.1, the network is divided logically into two types: the C plane (control plane) and the U plane (user plane). The C plane controls the network, and the U plane transmits/receives signals in a free exchange of information.

The communication mechanism is as follows. First, a connection is established from the calling-side (transmission side) terminal through network nodes to the receiving-side terminal on the C plane, and then the calling-side terminal receives an acknowledgement that communication was successful. In this way, information is transmitted using the route through the network that was set up as a pipe for communication. Media other than the physical route obtained in the C plane can also be used in this information transfer. When communication is terminated, a direction to release the connection is sent to the receive side of all the resources along the route. Of course, it is also possible to disconnect the connection from the receive side.

All of these operations are done on a protocol stack, as shown in Figure 2.2. This figure illustrates the operation of ATM-SVC (ATM-switched virtual circuit) service. The transmission-side terminal requests destination, bandwidth, quality level, etc. against the network according to the signaling protocol for Q.2931 UNI (user network interface) call control. In the network, communication is executed by using a B-ISUP protocol on MTP3, which is a UNI controlling protocol, and resources for an ATM node and a label (VCI/VPI — virtual-channel identifier/virtual-path identifier) are reserved in the case of ATM. To the receive-side terminal, a request for establishing the connection is sent again by Q.2931 UNI signaling. On the other hand, the U plane used for communication executes communication by using ATM cells, and disconnection is executed on the C plane using Q.2931 and B-ISUP, just like the previously described setup process.

Now we will look at the signaling protocol in more detail [3]. A long packet for signaling is adapted on ATM by using AAL (ATM adaptation layer). UNI/NNI signaling of Q.2931, B-ISUP, and MTP-3b is transferred by using a sublayer for adaptation, as shown in Figure 2.3. The technical details are as follows. The UNI signal is defined as either Q.2931 of ITU-T of the International Standardization

12   GMPLS Technologies: Broadband Backbone Networks and Systems

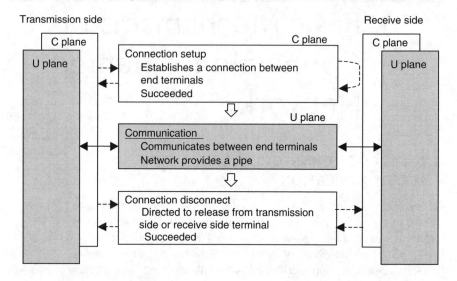

FIGURE 2.1 Basic principles of connection-oriented communication.

Organization or as the UNI specification of the ATM forum. The ATM forum's specification is basically compliant with the ITU-T standard and is mostly used in private networks. The NNI signal is processed based on B-ISUP (ITU-T recommendation Q.2761 and Q.2764). B-ISUP was expanded to broadband use from ISUP (ISDN user part), which is a signal-repeating protocol used in telephony,

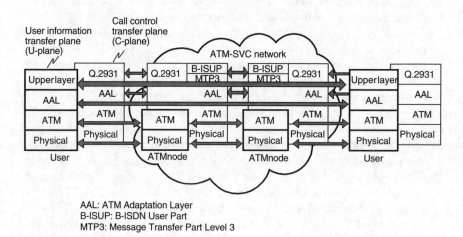

FIGURE 2.2 Protocol stack of connection-oriented communication (ATM-SVC service).

Basic Mechanisms of Connection-Oriented Network

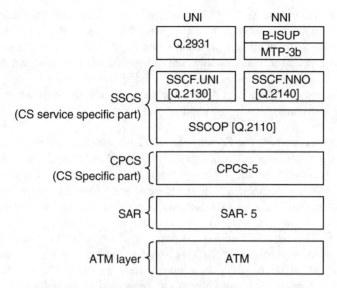

CPCS: Common Part Convergence Sub-layer
SAR: Cell dividing/assembling sub-layer
SSCF: Service Specific Coordination Function
SSCOP: Service Specific Connection Oriented Protocol

FIGURE 2.3 Protocol stack of ATM signal.

and MTP-3b (ITU-T recommendation Q.2210), which controls the signal link. That is, MTP-3b is a protocol for establishing the signal link, and signal control is executed by B-ISUP, as previously described.

As signaling information in the C plane, VPI = 0 and VCI = 5 are prepared by default for signaling each user in UNI. These values of VPI = 0 and VCI = 5 are special numbers and should not be used in the main signal or anywhere else. The ATM cells having these special values are recognized as signaling and are dealt with by the C-plane control process. That is, at the time of setting up the connection, signaling information (message) is transferred by using this VCI. As shown in Figure 2.3, AAL Type 5 (SAR-5, CPCS-5) is used as AAL for signaling, and SSCOP (Q.2110)+SSCF.UNI (I.2130) is used for its service-dependent part (SSCS).

SSCOP (Service-Specific Connection-Oriented Protocol) is a protocol that performs a function corresponding to LAPD (Link Access Procedure on D channel), which is a layer-2 protocol used in conventional N-ISDN, and that executes resending control of frames improperly transferred, flow control, etc. Unlike the conventional go-back-N method, which resends all the information after the frame in which error or loss of information has occurred, SSCOP has a selective resending function that resends only the frames that have not yet arrived in the receive side.

Moreover, flow control in SSCOP is executed in such a way that the receive side notifies the transmission side about the receivable number of frames as a credit number, and the transmission side then transmits the number of frames specified by this credit number. Under a flow control by fixed window, such as LAPD, the transmission side transmits the fixed number of frames that have been acknowledged for transfer. That is, LAPD cannot control the transmission of frames reflecting the status of the receive-side buffer. This shortcoming is eliminated in SSCOP.

SSCS (Service-Specific Convergence Sublayer) is a protocol that has SSCOP, which is a new protocol that is not seen in the upper level of AAL, and SSCF.UNI provides the basic function that is equivalent to conventional LAPD for Q.2931. Figure 2.4 shows the sequence of signaling as seen from the user terminal. As described previously, the user sends a request to set up a connection to the network and executes communication after the connection is established. Let us look at the technical side of SSCF in more detail.

The protocol sequence of SSCF is basically the same as the one used for a telephone call. If we consider the sequence corresponding to a telephone call, SETUP is a request for a call connection corresponding to dialing the counterpart's telephone number, and the SETUP message arriving at the counterpart

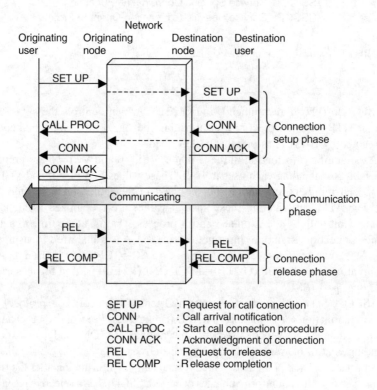

FIGURE 2.4 Signaling sequence of connection-oriented UNI (Q.2931).

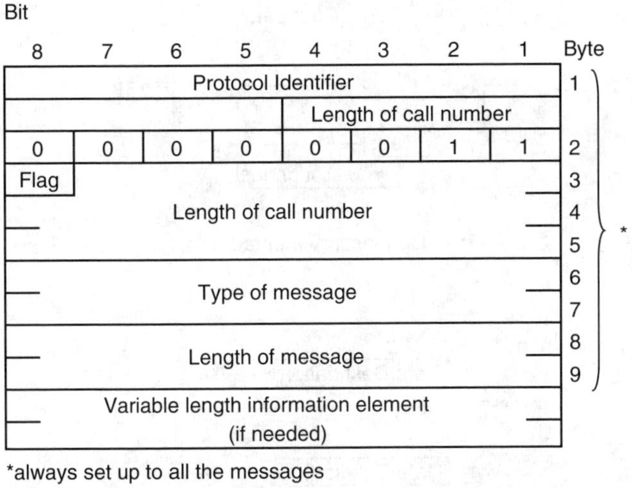

Basic Mechanisms of Connection-Oriented Network

Bit								Byte
8	7	6	5	4	3	2	1	
Protocol Identifier								1
				Length of call number				
0	0	0	0	0	0	1	1	2
Flag								3
Length of call number								4
								5
								6
Type of message								
								7
Length of message								8
								9
Variable length information element (if needed)								

*always set up to all the messages

FIGURE 2.5 Basic message format of Q.2931.

(destination user) corresponds to the ring at the destination terminal (or telephone). When the destination terminal acknowledges (off-hooks the handset) and sends back the CONN (CONNect) message, the network recognizes that the destination terminal is ready to communicate (start receiving information). When this acknowledgment is received by the originating terminal, two-way communication becomes possible.

The originating side enters the calling state (receives the calling signal from the receiver side) after the network receives the CALL PROC (CALL PROCessing) message and the CONN message. If there were no CALL PROC messages, it would not be possible to judge whether the SETUP signal from the originating side had been transmitted to the destination side or whether SETUP was being initiated. Thus, this signaling sequence is a necessary and sufficient function of UNI.

Each message of SETUP and CALL PROC has a format, as shown in Figure 2.5. In the SETUP message, the necessary information of bandwidth, service-quality parameter, bearer class, destination address to be transferred, etc. is written into the variable-length information element.

2.2 BASICS OF CONNECTIONLESS COMMUNICATION

Now that we understand that connection control is the most important thing in connection-oriented communication, we can move on to the basics of connectionless communication [4–7]. In connectionless communication, information is transferred by attaching a destination address to the data (Figure 2.6). This is a very flexible method, much like the home delivery of a newspaper, where the

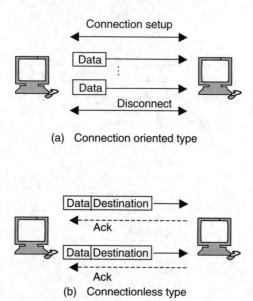

FIGURE 2.6 Connectionless communication.

destination address must be unique. However, this is also a wasteful approach in that each data packet, even if it contains only a few bits, must have an attached address that is checked at each router. The inefficiency of this approach is magnified when a message is segmented into many packets, all bearing the same destination address, which must be checked repeatedly by the routers.

From the user's point of view, the Internet appears to be a seamless global network (Figure 2.7). A user can send a message to the network and have it delivered to a desired counterpart at any time by attaching a destination address to the message. The user gives an e-mail address such as yyy123@xx.zzz.co.jp, for example, instead of the real IP (Internet Protocol) address, and the destination

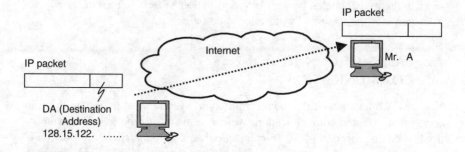

FIGURE 2.7 Internet communication.

Basic Mechanisms of Connection-Oriented Network

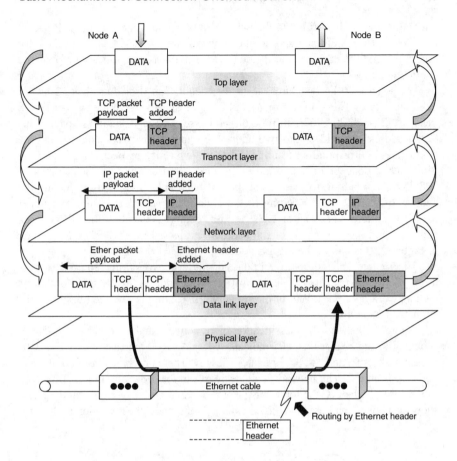

FIGURE 2.8 Example of data transfer in Ethernet.

IP address is automatically attached to each packet of the message. Let us look at how a message is transmitted over the Internet.

Figure 2.8 shows a model of data transmission (e-mail) from Node A to Node B. Data on the top layer has a TCP (Transmission Control Protocol) header attached when it arrives at the transport layer. The transport layer checks whether the data was received normally or not, executes resending of failed data, and controls the transmission rate. Attaching the IP address of destination is a task of the network layer. An IP header is attached to the TCP packet as shown in the figure. However, because the IP address is a logical address, a MAC (Media Access Control) address (an Ethernet address in the example of Figure 2.8) that shows a real, physical address is attached in the data-link layer to notify this physical address to the Ethernet cable as a physical transmission line. While in Node B of the receive side, the Etherpacket that was addressed to itself is terminated, and its

18 GMPLS Technologies: Broadband Backbone Networks and Systems

internal IP packet is notified to the network. On the network layer, the IP packet and the TCP packet included in the payload are extracted and checked for normality, and then the content of the payload (data) is read out to complete the communication.

Let us look at the IP address in greater detail. The IP address has 32 bits or 128 bits in case of IP v.4 (version 4) and IP v.6, respectively. In the case of IP v.4, there are $2^{32} = 4,294,967,296$ possible addresses in the IP address space. This 32-bit address is divided into two address groups: the network address and the host address corresponding to upper bits and lower bits, respectively. Figure 2.9 illustrates a concept of this address system. All users having the address of 128.25.***.** belong to the area at the left-hand side, and all users having the address of 128.12.***.** belong to the area at the right-hand side. The upper part of the address (128.25 or 128.12 in this example) is called the "IP network address," and the lower part of the address (expressed as ***.**, or 112.1, etc. in Figure 2.9) is called the "IP host address."

The routers located at the entrance to these networks are required to pass through only the packets that have an IP network address that matches one of its own network addresses. Within the subnetwork, routing is executed based on the address of each host.

Figure 2.10 shows the classification of IP network and host addresses. Corresponding to the header bits, "0" indicates Class A, "10" indicates Class B, "110" indicates Class C, and "1110" indicates Class D. There are three classes in the end node: A, B, and C. The network scale decreases in the order of Class A, B, and C. In the case of Class A, the top 8 bits (1 of 8 bits is used for recognizing the class) show the network address, and the remaining 24 bits show the host address. In other

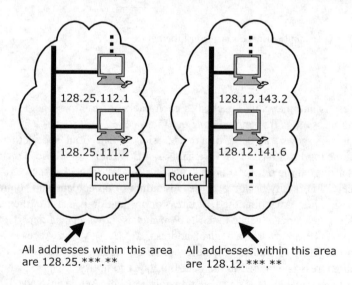

FIGURE 2.9 Network address and IP host address.

Basic Mechanisms of Connection-Oriented Network

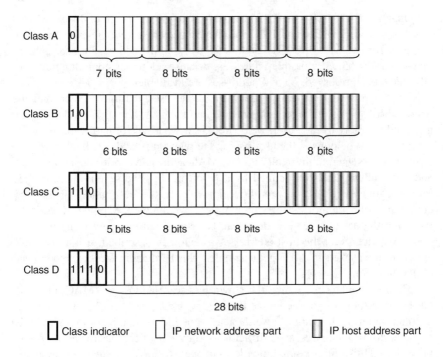

FIGURE 2.10 IP network address and host address.

words, this network can accommodate as many as 2^{24} terminals. The effective part of the network address is 14 bits and 21 bits, in the case of Class B and Class C, respectively. In the case of Class D, the remaining 24 bits are used as the network address, and this address is used for IP multicasting.

Taking the Class B packet as an example, we will describe the IP address in more detail (see Figure 2.11). In this case, because the top 2 bits are "10," excluding these 2 bits from the total 16 bits, it is possible to take as many as $2^{14} = 16,384$ addresses as the IP network address, while the host address can take as many as $2^{16} = 65,536$ addresses. This means that the network address can take

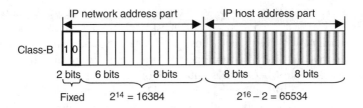

FIGURE 2.11 Format of Class B IP address.

the range from 128.000 to 191.255, corresponding to the range of 14-bit address space. Because 128.0.0.0 and 191.255.0.0 have been reserved, the exact available range is from 128.1.0.0 to 191.254.0.0. As for the host address that can take as many as $2^{16} = 65,536$ possible addresses, because addresses containing all 0 or all 1 have been reserved, 65,534 addresses are available.

To understand how packets reach the target user, we must understand the operation of the lower layers shown in Figure 2.8. Although IP addresses are allocated uniquely to each user, we do not know what type of hardware (personal computer, etc.) belongs to the user having the relevant IP address. Each hardware terminal is assigned a physically unique MAC address when it is manufactured, and this address has been defined by a 48-bit field as the second layer of OSI reference model. That is, even if we know only the destination IP address, communication cannot be executed unless we know the physical address. Although the destination IP address can be recognized uniquely using both the network address and the host address, as previously described, it is impossible to deliver the packets to the target user unless this hardware address is recognized at the same time.

Most computers recognize their own hardware addresses (MAC address), but because they do not recognize the hardware address (MAC address) of the destination terminal, and because the hardware addresses are not hierarchically arranged as IP addresses, it is difficult to know all of the hardware addresses prior to establishing a communication channel. In any case, it is not realistic to remember all the counterparts' hardware addresses. The Address-Resolution Protocol (ARP) solves this problem by obtaining the hardware address from the IP address.

The ARP is a protocol to obtain the hardware address of the next node (the destination hardware or the router to relay the packet) that should receive the relevant packet using the destination IP address as a clue. Because the IP address and the hardware address always have a relationship of 1:1, the relationship between the source IP address and the destination IP address does not change no matter what physical networks intervene between them. Thus, it becomes possible for nodes to communicate with each other through a coordination of the hardware address and the IP address.

Let us look at the mechanism of ARP. Figure 2.12 shows an example of a packet transferring from Node A (IP address = 128.11.000.0) to Node B (IP address = 128.25.111.1). First, an ARP packet is broadcast as an ARP request packet to ask for the MAC address of the destination IP address of 128.25.111.1. These broadcast ARP request packets then arrive at all of the relevant terminals. A terminal whose IP address matches the broadcast IP address returns an ARP response packet to the source terminal. To this ARP response packet, the MAC address information for the IP address is included. (In this case, the IP address is 128.25.111.1, and the MAC address is 0900.1111.0001.) The Node A that receives this ARP response packet now knows that the destination IP address of 128.25.111.1 has a MAC address of 0900.1111.0001, and it becomes possible to transmit the data.

Basic Mechanisms of Connection-Oriented Network 21

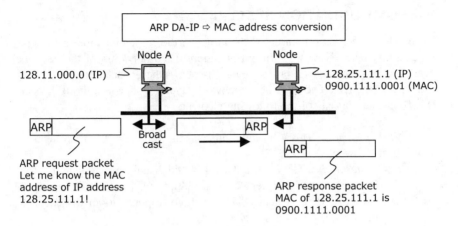

FIGURE 2.12 Address Resolution Protocol (ARP).

Figure 2.13 shows the contents of an ARP packet. An ARP packet is broadcast after the originating Ethernet address, the source IP address, and the destination IP address have been added to it. For this broadcast ARP request packet, the Ethernet address (0900.1111.0001 in this example) corresponding to the destination IP address is inserted and returned as an ARP response packet. In this way, using the APR, it is possible to obtain the MAC address of the destination IP address. This MAC address is as important as the destination address when we transmit information.

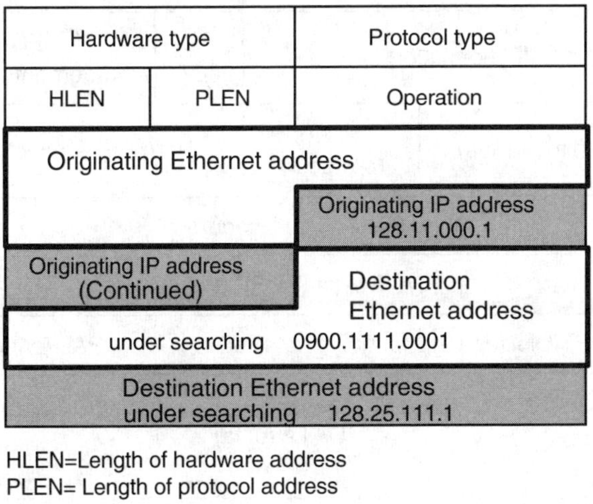

Hardware type		Protocol type
HLEN	PLEN	Operation
Originating Ethernet address		
		Originating IP address 128.11.000.1
Originating IP address (Continued)		Destination Ethernet address
under searching		0900.1111.0001
Destination Ethernet address under searching 128.25.111.1		

HLEN=Length of hardware address
PLEN= Length of protocol address

FIGURE 2.13 Contents of ARP packet.

2.3 COMMUNICATION BY TCP-IP

The previous discussion focused on the basics of the communication method on the data-link layer. Now we turn our attention to the operation of the network layer as well as the transport layer. Figure 2.14 shows the IP packet format (IP v.4) of the network layer. The roles of the network layer are shown in the IP header. This IP header consists of the following elements.

- *VER (VERsion)*: This 4-bit field indicates the version of the IP system. In the case of IP v.4, VER is 4.
- *IHL (Internet header length)*: This 4-bit field indicates the length of the IP header and is expressed as a multiple of 4 octets. The value written into IHL is typically "5," which corresponds to an IP header length of 20 octets (= 4 octets × 5).
- *TOS (type of service)*: This 8-bit field indicates the degree of importance of the IP packet or the quality of communication and service. The meanings of bits 0 through 7 are listed in Figure 2.14.

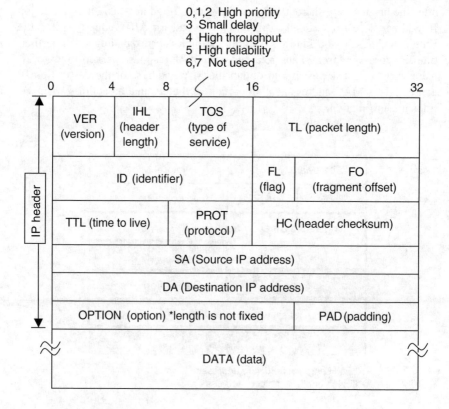

FIGURE 2.14 IP packet format (IP v.4).

Basic Mechanisms of Connection-Oriented Network 23

- *TL (total length of packet)*: This 16-bit field indicates the length of the packet, including the header, expressed as octets.
- *ID (IDentifier)*: This 16-bit field is used to combine fragmented data when transferring data to the upper layer.
- *FL (FLag)*: This 3-bit field is used to control packet division.
  Bit 0: unused (fixed to 0)
  Bit 1:
      0 = packet is dividable
      1 = packet is not dividable
  Bit 2:
      0 = divided final packet
      1 = divided top or intermediate packet
- *FO (fragment offset)*: This 13-bit field indicates which part of the divided original data was stored by 8-octets units. Because the maximum amount of original data that can be expressed is 64 koctets, this is the maximum amount of data that can be transferred as a packet.
- *TTL (time to live)*: This 8-bit field indicates the maximum time during which the relevant packet can exist in the network. The TTL value is reduced by 1 each time the packet passes through the router, and the packet is disposed of when the TTL becomes 0. This strategy makes it possible to automatically detect and dispose of packets that create an endless loop, which otherwise would allow them to exist infinitely.
- *PROT (PROTocol)*: This 8-bit field indicates the upper-layer protocol (e.g., 1 = ICMP, 6 = TCP, and 17 = UDP).
- *HC (header checksum)*: This 16-bit field is used to check the normality of the header.
- *SA (source IP address)*: This 32-bit field indicates a source IP address.
- *DA (destination IP address)*: This 32-bit field indicates a destination IP address.
- *OPTION (option)*: This field has a variable bit length and is available for use in such options as security and time stamping.
- *PAD (padding)*: PAD executes padding to 32 bits in option, etc.
- *DATA (data)*: All of the signaling information is stored in the DATA area.

These packet-formatting functions are utilized when packets are sent using IP. Figure 2.15 shows the structure of the packet format, including the data-link layer, the network layer, the transport layer, and the user data. User data is encapsulated by layer.

Now we turn our focus to the transport layer, taking the TCP protocol as an example. Figure 2.16 shows the structure of the TCP segment header.

- *SRC PORT (source port number)*: This 16-bit field identifies the source port number.
- *DEST PORT (destination port number)*: This 16-bit field identifies the destination port number.

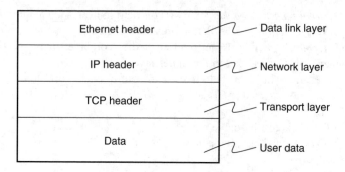

FIGURE 2.15 Packet format using Ethernet header and TCP-IP header.

- *SEQ (sequence number)*: This 32-bit field indicates the sequence number with the number of bytes for securing the normality of data, i.e., it shows the number of bytes from the top that are being transferred.
- *ACK (acknowledgment number)*: This 32-bit field confirms the acceptance of a packet by sending back the SEQ value.
- *HLEN (header length)*: This 4-bit field indicates the length of the TCP header and is expressed as a multiple of 4-byte units. The value written into HLEN is typically "5," which corresponds to a TCP header length of 20 bytes ($= 4 \times 5$).

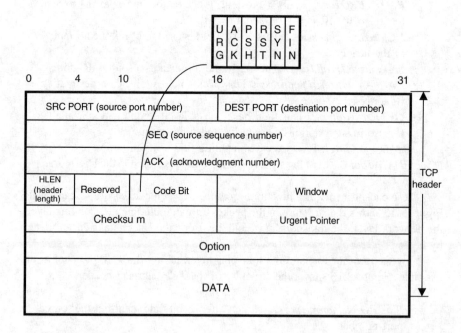

FIGURE 2.16 Structure of TCP header.

Basic Mechanisms of Connection-Oriented Network 25

- *Reserved*: This 6-bit field is reserved for expansion.
- *Code bit*: This 6-bit bitmap is a control flag. These flags are as follows (from the left-hand side of Figure 2.16):
  URG (urgent flag): urgent processing required (1 = urgent)
  ACK (acknowledgment flag):
    1 = acknowledgment required
    0 = acknowledgment not required
  PSH (push flag):
    1 = immediately transfer data to upper-layer application
    0 = timing of data transfer at the receive side is not yet determined
  RST (reset flag): TCP connection is forced to disconnect by setting this flag bit to 1
  SYN (synchronize flag): establishes the virtual circuit; sequence number is initialized by setting flag bit to 1
  FIN (final flag): end of data to be transferred is signaled by setting this flag bit to 1
- *Window*: This 16-bit field indicates how much data can be received after confirming acknowledgment of the sequence number.
- *Checksum*: This 16-bit field checks the normality of the header.
- *Urgent pointer*: This 16-bit field points to the location of data that must be processed urgently when the URG code bit is 1.
- *Option*: This field is unused at present.

Now we are ready to describe in detail the operation of TCP-IP using these fields. Figure 2.17 shows the method of transmission and reception of data between Node A and Node B. In this case, Node A sends a packet with sequential number 1 to Node B, and when Node B receives this packet, it writes 1 into the ack (acknowledgment) packet and sends the ack back to Node A. That is, communication is executed through a combination of sequential packet numbers and acknowledgment. The source Node A does not send packets 2 and 3 unless it receives the ack packet for packet 1 from Node B, thus providing a secure method for transmitting and receiving data.

Consider the case when the ack packet is not sent back. As shown in Figure 2.18, it can be surmised that the ack packet was not sent back because either the transmitted packet 1 was lost or the ack packet itself was lost. In this case, if the ack packet is not received within a specified time, the source Node A judges it as a time-out and resends packet 1. This resending process ensures a highly reliable packet transmission.

Related to the resending process at Node A, there is a possibility that the return of the ack packet was delayed, for example because of a long round-trip time, etc., with the result that double packets arrive at Node B. In such a case, Node B checks the sequential number and discards the duplicate packet and does not return an ack packet for the discarded packet as shown in Fig. 2.19. In this way, it is possible to remove a duplicate packet while continuing to execute

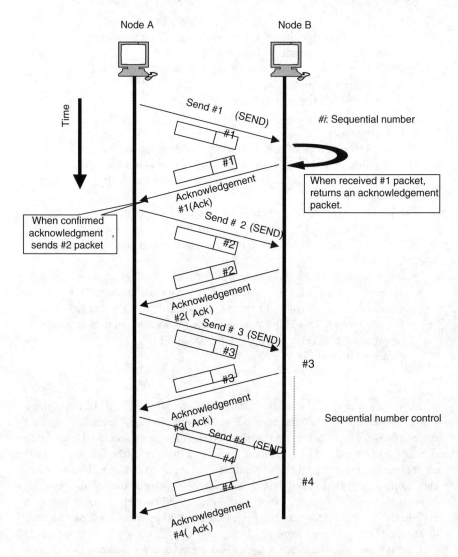

FIGURE 2.17 Acknowledgment and sequential number.

reliable communication. As described previously, methods to ensure reliability include such measures as using a sequential number, packet resending based on judgment of time-out, checking of double sending, and so on.

Next, we describe various measures for improving the performance of communication. In the case of a primitive acknowledgment method, data transmission becomes ineffective when the round-trip time is too long and the source

Basic Mechanisms of Connection-Oriented Network

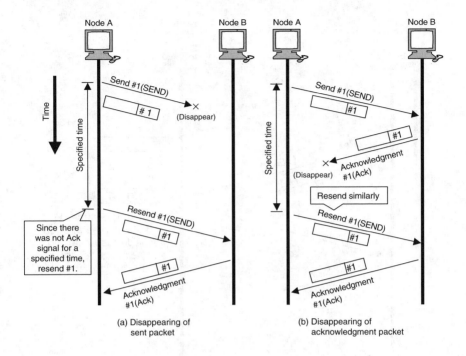

FIGURE 2.18 Improvement of packet transfer reliability by resending process.

node has to wait for the ack packet. Therefore, in TCP, the window-controlling method is utilized, as shown in Figure 2.20. In this method, if we assume that the window includes two packets, source Node A can send two packets without confirming the ack packet. The process at the destination Node B is the same as that of the primitive acknowledgment method, that is, it sends back the ack packet 1 when it receives packet 1, the ack packet 2 for the received packet 2, etc. In this way, it is possible to receive many packets in a short time. But, because the destination Node B must return an ack packet for each received packet 1, 2, ..., n, many useless ack packets are sent back. Transmission overhead can be reduced using the method shown in Figure 2.21, where the destination Node B sends back a single ack packet in response to receiving packet 8. Destination Node B does not send an ack packet if any of packets 1 through 8 are missing. Thus, the ack for packet 8 represents the normal arrival of packets 1 through 8. This method greatly reduces the number of ack packets returned.

To execute packet transmission within the source node more securely, and to transmit an appropriate amount of packets, there is a flow-controlling method (flow-volume control). The destination node generally has a receive buffer to

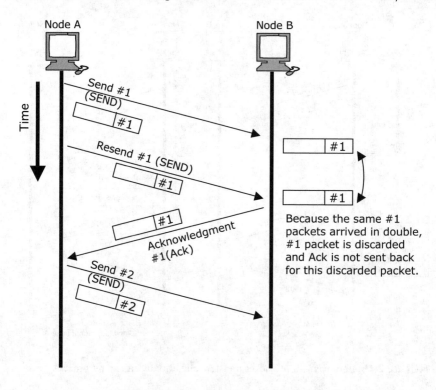

FIGURE 2.19 Discarding the packet when duplicate the same packets arrived in double.

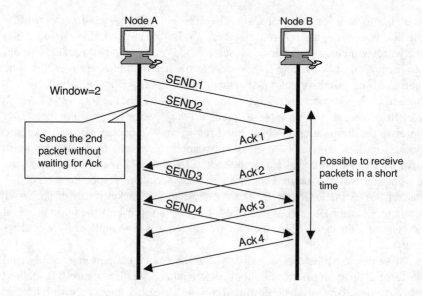

FIGURE 2.20 Window controlling.

Basic Mechanisms of Connection-Oriented Network

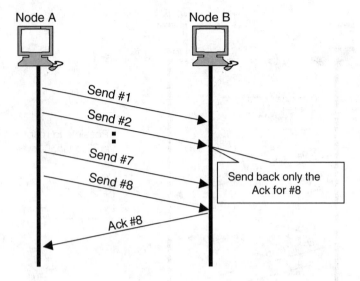

FIGURE 2.21 Saving the ack process by window process.

store the received packets and execute sequential processing. The ack packet that is sent from the destination node specifies the number of bytes acceptable (500 bytes in the example of Figure 2.22) as well as the number of bytes corresponding to packets 0 through 100 that have already been received. That is, an ack message of "packet 101 +500 bytes are acceptable" is returned to the source Node A. From this ack packet, Node A knows that up to 100 packets have been received normally and that the destination Node B is available to receive another 500 bytes. So, Node A transmits Packet 101 +500 bytes. Because the receive buffer of the destination Node B becomes full when it receives this 500 bytes, the next ack packet states that Node B received up to 600 bytes and 0 bytes is acceptable. In response to this ack message, Node A stops sending the packets. When a vacancy (in this case, 200 bytes) is generated in the receive buffer of destination Node B, Node B sends another ack packet stating that Node B has received normally up to 600 bytes and that Node A can send 200 bytes to Node B. When Node A receives this ack packet, it sends 200 bytes of data. In this manner, it is possible to receive securely all of the packets at a rate that does not exceed the system's ability to receive the packets.

From the previous description, it is evident that TCP is an excellent transport-layer protocol capable of providing a secure, reliable, and efficient means of data transmission. Moreover, TCP is an adaptive protocol that can adjust the rate of data transmission to match the system's data-carrying capacity.

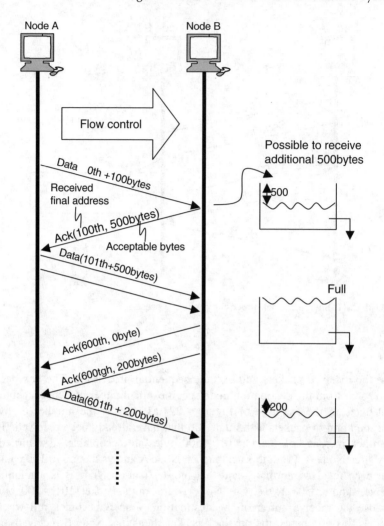

FIGURE 2.22 Flow control between Node A and Node B (control of flow volume).

REFERENCES

1. De Prycker, M., *Asynchronous Transfer Mode, Solution for Broadband ISDN*, 2nd ed., Ellis Horwood, New York, July 1991.
2. Chen, T.M. and Liu, S.S., *ATM Switching System*, Artech House, Norwood, MA, March 1995.
3. Kano, S. and Kuribayashi, S., *Easy to Understand ATM Network Signaling System New Signaling System for Multimedia Communications,* Telecommunications Assoc. and Ohm-sha (in Japanese), Tokyo, May 1996.
4. Stevens, W.R., *TCP/IP Illustrated*, Vol. 1, The Protocol, Addison-Wesley, Reading, MA, Dec 1993.

5. Stevens, W.R., *TCP/IP Illustrated,* Vol. 3, TCP for Transactions, HTTP, NNTP, and the UNIX Domain Protocols, Addison-Wesley, Reading, MA, Jan. 1996.
6. Comer, D.E., *Internetworking with TCP/IP,* Vol. 1, Principles, Protocols, and Architecture, Prentice Hall, New York, Jan. 2000 (4th Ed.).
7. Halabi, S., *Internet Routing Architectures,* 2nd ed., Cisco Press, Indianapolis, IN, Jan. 2000.

3 Connection-Oriented Communications and ATM

3.1 TRANSMISSION METHOD IN ATM

In Chapter 2, we described the basics of the connectionless communication system, especially that of the TCP-IP system. MPLS executes IP forwarding as the initial transmission method, but if MPLS is "cut through," then it operates based on a connection-oriented transmission system. In this chapter, we describe the ATM transmission system, which is one of the transport protocols after cut-through in MPLS and is the most typical method of connection-oriented communications as well as its traffic-control technology and OAM (operation and maintenance) technology [1, 2].

Figure 3.1 shows the basic mechanism of an ATM (Asynchronous Transfer Mode) communication system [3]. ATM communication establishes a network connection prior to starting communication. This virtual connection transfers packets by relaying a VCI/VPI label (virtual-channel identifier/virtual-path identifier) in the ATM header from end to end of the communication path. Actual data transmission is executed through this virtual connection. As a measure of transmission, fixed-length packets (cells) of 53 bytes are utilized. Each node (ATM switch) executes a high-speed routing process with hardware according to the header's information. The route of transmission is determined based on the header's VCI/VPI information. Label relay (relay of the VCI/VPI of the virtual connection) and label swapping (conversion) are executed in each node.

One feature of an ATM communication system is that the packets always pass through the same route during communication. It is also possible to alter transmission speed by the number of fixed-length packets, and high-speed transmission can be executed by hardware routing. When establishing the virtual connection, an ATM system also has the unique ability to specify the route, to judge whether the new connection should be accepted or not, and to control traffic.

ATM network protocol is composed of a physical layer, a data-link layer, and a network layer, as shown in Figure 3.2. The control plane (C plane) consists of a user plane (U plane) and a management plane (M plane). The C plane controls setup/release of the connection, and the U plane transmits and receives the user information. The M plane executes maintenance/recovery processing or collects the traffic or error information as a system management function. As described in Chapter 2, the C plane executes communication using an ATM cell on AAL

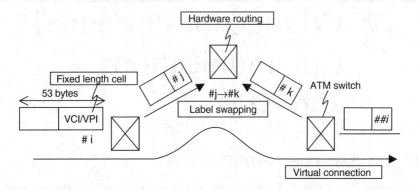

FIGURE 3.1 Basic concept of ATM communication.

(ATM Adaptation Layer), SSCOP (Service-Specific Connection-Oriented Protocol) on layer 2, and Q.2931 on layer 3. On the other hand, in the U plane, multiple AALs have been prepared as the layer 2. AAL1 executes fixed-rate communication, AAL2 executes voice communication, and AAL3/4 executes data communication. AAL5 also executes data communication, but in a simple method that will be described later in Section 3.2.4.

To help in understanding how ATM functions as a connection-oriented communication system, we will compare it with other protocols, using packet communication (x.25) and frame-relay communication as examples (Figure 3.3). Consider a case where user A and user B communicate with each other. The network is composed of nodes (switches) and transmission lines. We define the interface between user and network as UNI (user network interface) and the interface between switches as NNI (network node interface). The protocol for UNI is different from the one for NNI.

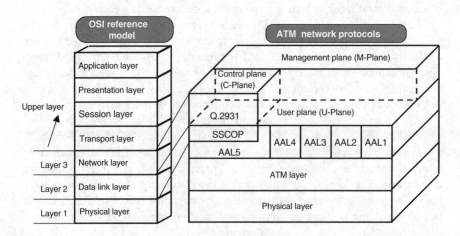

FIGURE 3.2 Protocol structure of ATM network.

Connection-Oriented Communications and ATM

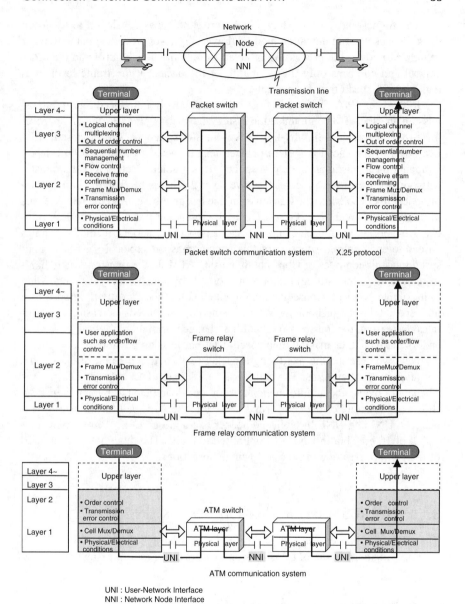

FIGURE 3.3 Protocol stack of ATM, frame-relay, and packet communication systems.

In the case of x.25 packet protocol in Figure 3.3(a), each packet-transfer switch terminates layer 3. That is, each transfer switch checks the normality of all the packets link by link and restructures the packet. In the case of x.25, because layer 3's packet processing is done by software, although secure transmission is possible, throughput is not high.

The frame-relay system shown in Figure 3.3(b) was developed to speed up the x.25 packet communication method. Layer 3 is never processed in the transfer switch. Moreover, the processing in layer 2 omits the flow control and the order control and confirms only the normality of the packet at the frame level, thus making high-speed transfer possible.

In the later 1980s, as optical fiber cable was widely introduced into the transmission line, the ATM communication system shown in Figure 3.3(c) — an improvement over the previous link-by-link packet-transfer system — was proposed. ATM is a novel communication system of fixed-length packets (cell) based on a label-swapping mechanism, as already described. It is a unique feature that the transfer network never executes processing above layer 2. This is because the introduction of optical fiber cable into the transmission line has greatly reduced the rate of errors in transmission. With no need to execute the resending process link by link or to check the normality of each claim, it became possible to adopt a method that relies on end-to-end control. Adoption of this simple transport protocol has made high-speed transmission rates of Gbps possible using ATM. In the transfer switch, hardware switching (forwarding) has become possible by simple label swapping owing to its simple protocol. Consequently, this method has been greatly developed as a high-speed transport technology. With transmission line speeds increasing and error rates decreasing, the transport protocol has changed from a link-by-link type to an end-to-end type, thus making high-speed data transmission possible.

Next we describe the fixed-length packet (cell) that is the unit of data transportation in ATM systems. The cell has a total length of 53 bytes and comprises a header (5 bytes) and an information area or a payload field (48 bytes). Figure 3.4 shows the structure of the cell format. Note that the ATM cell format is different between UNI and NNI. In NNI, the header is composed of a VPI of 12 bits and a VCI of 16 bits, but in UNI, VPI is limited to 8 bits. The functions of each field of the cell are described in the following subsections.

3.1.1 GFC (Generic Flow Control) *Only in UNI

The GFC field is used for flow control between user and network. GFC is used for notifying permission of transmission to users or for multiplexing subscriber.

3.1.2 VCI/VPI (Virtual Path Virtual/Identifier Channel Identifier)

VCI/VPI are used for recognizing a user in combination. VPI is an identifier of virtual path and defines the path as a pipe connecting two locations, such as Tokyo and Nagoya, for example. VCI is attached independently to each user connection. Thus, No. 6 between Tokyo and Nagoya can be distinguished from No. 6 between Tokyo and Osaka. And, because it is possible to manage the fault

Connection-Oriented Communications and ATM

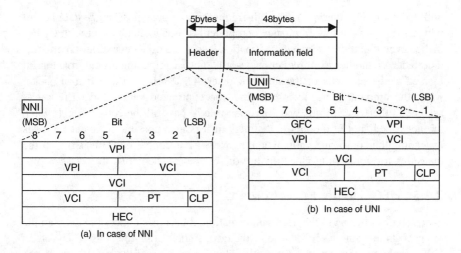

FIGURE 3.4 ATM cell format.

processing by route using a VPI when a fault occurs, reallocation is easily accomplished. Because VCI/VPI are used for routing in the transfer switch, these bits are called "routing bits."

3.1.3 PT (Payload Type)

PT expresses the content of the payload (OAM cell, congestion cell, for example) with 3 bits. The possible combinations are:

000: Usual user cell (noncongestion, intermediate cell of packets)
001: Usual user cell (noncongestion, final cell of packets)
010: Congested user cell (intermediate cell of packets)
011: Congested user cell (final cell of packets)
100: VCI-level OAM cell (segment)
101: VCI-level OAM cell (end-end)
110: OAM cell for resource management
111: reserved

3.1.4 CLP (Cell Loss Priority)

Because ATM is a packet-switching solution, there is a possibility of cell loss occurring in the transfer switch on the path of communication as a result of congestion, etc. A cell with CLP = 1 is discarded with higher priority than a cell

38 GMPLS Technologies: Broadband Backbone Networks and Systems

with CLP = 0. For example, if a high-frequency component of image data is sent with CLP = 1 and a low-frequency component of image data is sent with CLP = 0, then even if part of the image information is lost, the user's eye cannot recognize its effect. As another example, consider a UPC (usage parameter control) circuit located at the entrance of the network where CLP = 0 is assigned to a packet when the amount of traffic is less than the contracted value, and CLP = 1 is assigned otherwise. In this way, it is possible to pass all the cells through when the network is not in a congested state, but it is also possible to reject the cells exceeding the contracted value when the network is congested. This is one of the measures for controlling QoS (quality of service) of the network.

3.1.5 HEC (HEADER ERROR CONTROL)

When an error occurs in the header information, it becomes impossible to assemble the packets in frame level because of the error in delivering the cells. Therefore, a CRC (cyclic redundancy check) is calculated for each bit of the top 4 bytes of the header, and the result is written into a fifth byte field (HEC). By protecting the header information using this HEC, ATM ensures a highly reliable data transmission.

3.2 ATM ADAPTATION LAYER

In this section, we describe the AAL (ATM Adaptation Layer). ATM is a universal protocol targeted for application in multimedia communications. For this purpose, long packets must be divided into short packets (segmentation). To execute adaptation of these divided packets with an upper-layer protocol, five types of AAL protocol layers (AAL1 through AAL5) have been developed (Figure 3.5).

AAL1 is an AAL for fixed-rate transmission and exists to execute emulation of a circuit-switching service. A circuit-switching service communicates using periodic frames, and this requires a fixed-rate communication. But because ATM is based on a packet-switching system, fluctuation or loss of packets can occur unless the timing at both ends is adjusted.

AAL2 is an adaptation targeted for low-bit-rate voice applications. In the case of low-bit-rate voice, it takes a long time to fill the cell's 48-byte payload with a single user's signal. Because voice communication is bidirectional and operates in real time, it is necessary to shorten the time required to build up the cell. Thus, because there is no harm if the 48 bytes are filled with information from multiple channels, it is possible to shorten the time to construct the cell by using an adaptation layer that multiplexes multiple users to a single VCI.

AAL3/4 are adaptation layers targeted for data applications. AAL3/4 divide a long packet into multiple segments (segmentation) and guarantee the assembly of a frame when these segments arrive.

AAL5 is an adaptation layer that executes transmission of the long packet in a simple manner by specifying the top, middle, or end of the packet.

Connection-Oriented Communications and ATM

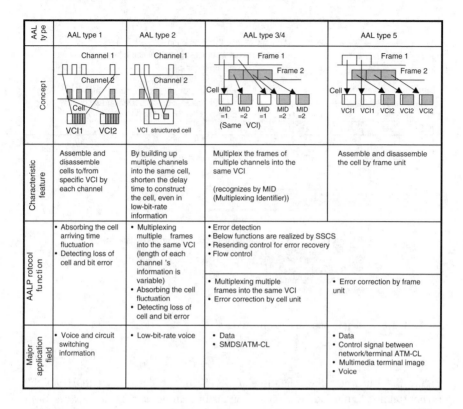

FIGURE 3.5 Function outline of AAL.

3.2.1 AAL Type 1

AAL type 1 is an ATM adaptation layer for executing circuit emulation of fixed-rate communications. It guarantees the removal of fluctuations in cell arrival time and in cell disposition that can occur in ATM transmission.

In AAL1, as shown in Figure 3.6, 1 byte of the 48-byte payload is used as a header of SAR-PDU. The SAR-PDU header consists of CSI (1 bit), SC (3 bits), CRC (3 bits), and parity (1 bit). SN has the role of notifying whether all the cells have arrived or not. That is, the protocol detects cell loss and inserts dummy cells when cells are lost. SN has a mechanism to adjust the timing of sending/receiving using an upper 1 bit and a lower 3 bits.

Most terminals have a self-running clock and execute encoding operations, etc., based on this clock. On the receiver side, it is necessary to regenerate the source-side clock. To regenerate the clock using this SN (sequential number), an SRTS (synchronous residual time stamp) system is utilized. When the lower 3 bits of SN (SC) indicate an odd number, the top 1 bit (CSI) is used to adjust the shift of timing to transmit.

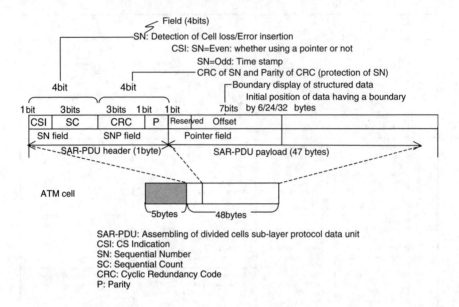

FIGURE 3.6 Structure and function of AAL1.

Figure 3.7 illustrates the operation of SRTS. In the case of an SDH/SONET (synchronous digital hierarchy/synchronous optical network), the network has a stable master clock. Here we define the difference of counter values between the counter based on the master clock (CTR-I) and the counter (CTR-S) based on the transmission-side clock that is used for coding, etc. as α. The receive side also has two counters, and we define the counter based on the receive-side clock that is used for decoding, etc. as β. The receive-side clock frequency is adjusted to become α = β by monitoring the β value.

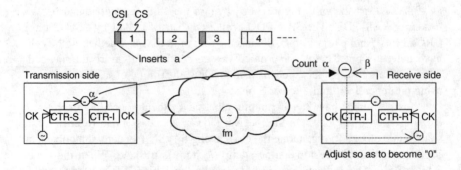

FIGURE 3.7 Operating principle of SRTS.

Connection-Oriented Communications and ATM 41

The difference value between the master counter and the transmission-side counter is communicated using CSI, the top 1 bit of SN, as described previously. When SC is an odd number, the difference α between the counter value of the master counter and the transmission-side counter is added as the CSI bit and can be sent by using multiple cells as shown in Figure 3.7.

3.2.2 AAL Type 2

AAL type 2, which was standardized in 1997 for use in low-speed voice communication, is an ATM adaptation layer dedicated for voice communications that require real-time treatment with variable transmission speed. As an ATM adaptation layer, AAL2 was initially assumed to be suitable for moving-picture communications with a variable bit rate using a differential encoding method. However, because a moving picture has strict requirements for the quality of communication, it is necessary to explicitly specify the traffic characteristics to guarantee the required quality of communication when the peak rate is high. In the case of moving-picture communications, where the bit rate changes depending on the movement of an object on screen, it has been difficult to develop specifications for AAL2 to meet the requirements for quality. Consequently, standardization of AAL2 has been withheld for a long time.

As further background on which this AAL2 standard was established in 1997, when it was assumed that its main application would be low-speed voice communication, it is worth mentioning that voice-encoding technology has greatly advanced since then, and application of ATM to mobile communications including cellular phones was intensively discussed at that time. In fact, low-speed voice communication systems (16 kbps to 4 kbps) have been utilized in mobile communication systems, where effective use of bandwidth is especially required, in compression technology for Internet telephony, and in private networks. If we transmit such a low-speed voice signal after filling the 47-byte field in AAL1, the delay time for constructing the cell becomes long. For example, although the usual voice codec operates with a bit rate of 64 kbps and its delay time for establishing a cell is 6 msec, if the voice is compressed to 8 kbps (as it is for cellular phones), the delay time for constructing the cell becomes 48 msec, which is eight times longer than for a conventional telephone. To shorten this delay time, a method has been proposed that, when the delay time reaches a certain value (20 msec, for example), completes cell construction without waiting until the entire 47 bytes are filled. However, this method is inconsistent with methods that could increase the network usage efficiency by improving the encoding efficiency.

Given this background, and taking into consideration the variable-speed voice-encoding system, an AAL2 protocol that multiplexes packets from multiple channels into a single cell by dividing the voice data into variable-length short packets has been established. AAL2 is different in its sublayer structure from the other types of AAL, which do not multiplex multiple channels into the same cell, especially in that it includes a common-part sublayer (CPS), as shown in Figure 3.8. CPS has such functions as dividing voice data into short packets and mapping voice packets from multiple channels to cells. SSCS (service-specific convergence sublayer) has been

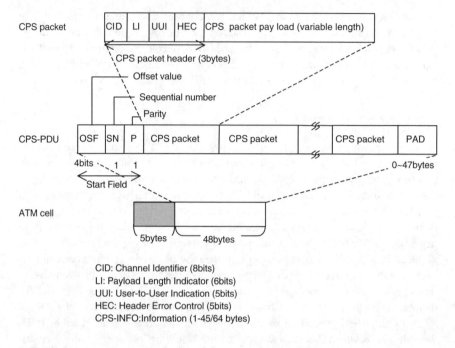

FIGURE 3.8 Structure and function of AAL2.

defined as the future uncompressed voice transmission system, but the details have not yet been specified. In the next subsection, we describe the main features of CPS.

3.2.2.1 Packetizing the CPS

Compressed voice data is packetized into short packets (CPS-packet). This packet includes a header of 3 bytes, which includes HEC (header error controlling information), and a payload of variable length. The header consists of a channel identifier (CID) that indicates the channel to which the CPS packet belongs, a length indicator (LI) that indicates the length of the CPS packet payload, and a user-to-user indication (UUI) that shows the content of the CPS packet payload. To this information, HEC is attached, just as it is in the ATM layer header.

3.2.2.2 Multiplexing the CPS Packets

Because the length of a CPS packet is variable, it does not necessarily fit the expected payload of an ATM cell. Therefore, there is a function that can transmit a CPS packet crossing over two ATM cells. To do this, a start field (STF) of 1 byte is defined as the CPS-PDU header to indicate the crossover part of the CPS packet as an offset value (OSF: offset field). That is, the part that was offset by the OSF field is the top of the first CPS packet within this CPS-PDU. This STF field also includes a sequence number (SN) to detect cell loss and a parity check bit (P).

3.2.2.3 Timer for Cell-Construction Delay

The transmission side of a cell has a function that transfers the residual part of the cell by padding it when the delay time for cell construction exceeds a predefined time. This function, in combination with the previously described function that multiplexes multiple channels into a single cell, prevents the cell delay from becoming too long when the number of multiplexed channels is small or when soundless parts are continued, etc.

3.2.3 AAL Types 3/4

AAL types 3/4 are used for data communication. As originally conceived, AAL3 was discussed as a connection-oriented-type layer, and AAL4 was discussed as a connectionless-type layer. However, because they share common functions up to the CPCS (common-part CS) function, they have been combined as AAL3/4. The main feature of this AAL3/4 layer is that it can multiplex multiple frames into one ATM connection using the multiplexing function of MID (multiplexing identification). Figure 3.9 shows the structure and function of AAL3/4. We describe the features of AAL3/4 in the following subsections.

3.2.3.1 Frame Assembly and MID Multiplexing Functions

In AAL3/4, it is possible to multiplex multiple frames into one ATM connection by using the indication of the segment type (ST), the multiplexing identification (MID) on the SAR header, and the sequence number (SN) of modulo 16 for each MID. The upper-layer data is divided into multiple segments of 44 bytes, each of which constructs an SAR-PDU packet of 48 bytes by adding 4 bytes to itself. SAR-PDU is distinguished by an ST bit that indicates whether it is a beginning cell or end cell of CPCS and is identified by MID. That is, the top of the frame is recognized by ST = 10, and the SAR-PDU of succeeding cells having the same MID value are assembled up to the end cell of ST = 01. The value of SN (sequential number) is used to identify the cell constructing the frame by giving a sequential number, and it can be defined independently for each frame.

3.2.3.2 Error Detection

In each cell that has been multiplexed by MID, cell loss and error insertion are checked by sequential number (SN). Bit errors in each cell are detected by adding a CRC (cyclic redundancy check) code of 10 bits to the SAR trailer.

In cases where cell loss has occurred successively over 16 cells and the value of SN could not be trusted, the CPCS-PDU (CPCS-packet data unit) is assembled, and the indication of length of the CPCS-PDU payload within the CSCS trailer is checked to see whether or not it coincides with the length of the received frame.

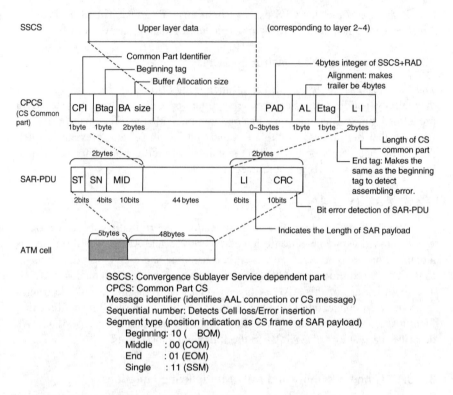

FIGURE 3.9 Structure and function of AAL3/4.

Further, to avoid assembling the wrong CPCS-PDU when a certain frame's last half cells and the next frame's front half cells were lost, different values (Btag = Etag within the same frame) are written into the beginning tag (Btag) of the CPCS header and to the end tag of the CPCS trailer, and these values are checked at the receive side to see if they are the same as that of the transmission side.

SSCS is used for frame error control and resending control corresponding to that of layer 2, but the details of SSCS have not yet been established as a standard specification. As an AAL for connection-oriented communication, AAL5 (described in the following subsection) is considered as the mainstream. This is because the amount of information that can be carried by one cell in AAL3/4 is only 44 bytes owing to the overhead of its SAR header and trailer. Moreover, the complexity of high-speed SAR functions imposes a huge demand on the hardware.

3.2.4 AAL Type 5

At present, AAL type 5 is the preferred layer for data communication because of its low overhead. Figure 3.10 shows the structure and functions of the AAL5

Connection-Oriented Communications and ATM

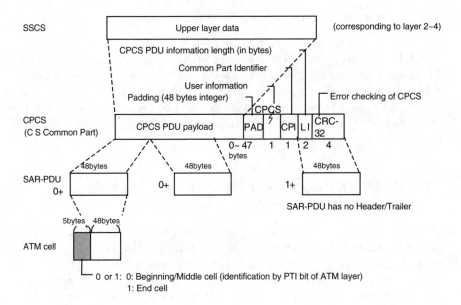

FIGURE 3.10 Structure and function of AAL5.

cell format. AAL3/4 error checks the payload in the SAR sublayer, but in AAL5, the SAR sublayer only assembles the payload, and the error-detection process is executed in CPCS. AAL5 also simplifies processing in that there is no multiplexing function using MID on ATM connection, and error detection is done simply by only CRC checking after assembling the frame without using the cell's sequential number. In addition, the frame's field structure is based on a 32-bit unit in consideration of protocol process by software.

The functions of AAL5 are discussed in the following subsections.

3.2.4.1 Frame Assembly

In AAL5, there is no header and trailer in the SAR sublayer, unlike the situation for AAL3/4. The frame boundary is recognized by the payload type (PT) of the ATM layer's cell header. That is, when the value of user-to-user information indicator is 0, it indicates that the cell succeeds the cell constructing the same frame, and when it is 1, it indicates that this cell is the end cell.

CPCS is composed of user information, padding (PAD) that arbitrates the length of CPCS-PDU to an integral multiple of 48 bytes, a CPCS user-to-user indicator (CPCS-UU) that transmits the user-to-user information of CPCS, a common-part identifier (CPI), a payload length indicator (LI) of CPCS-PDU, and a 32-bit trailer to execute error detection of the whole CPCS frame.

3.2.4.2 Error Detection

Because there is no sequential number (SN) used in AAL5, as there is in AAL3/4, error detection is executed by CRC checking after assembling the CPCS, including checking of cell loss and error insertion. When the end cell is lost, because the error is detected after assembling CPCS crossing over the succeeding frame, the information corresponding to two frames is discarded. In this case, the length information of LI is notified to the upper layer as CPCS-PDU after CRC checking.

3.3 PERMANENT CONNECTION AND SWITCHED CONNECTION

There are two connection types in connection-oriented ATM: a permanent virtual connection (PVC) that is fixed semipermanently, and a switched virtual connection (SVC) that is established call by call (Figure 3.11). PVC is a service in which traffic is always flowing, such as a private line. ATM can be used as a backbone for various services, and it is often used as a pipe for frame-relay communication, for example. This PVC service is set up by an operator in

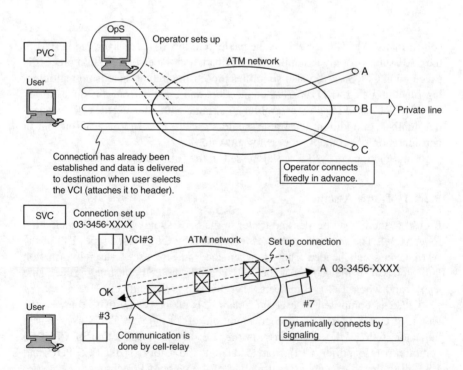

FIGURE 3.11 PVC (permanent virtual connection) and SVC (switched virtual connection).

advance. The granularity of the bandwidth can be set freely, which is a characteristic feature of ATM, making it possible to set up the PVC connection with free bandwidth such as 52.4 Mbps, for example. Because there is no operation of setting up the connection in PVC service, it is possible to provide a communication pipe freely with any amount of bandwidth, and it is also possible to offer multiple QoS on these pipes. On the other hand, SVC can set up the connection by signaling, thus making it possible to provide users with more-dynamic services. In SVC, the connection is set up only when communication is required, and it is released upon completion of communication. However, if there is no vacancy in the resource, the request for connection is refused, an event that is known as a "call loss." Different from telephony, in the case of ATM, it is possible not only to establish the connection, but also to set up the connection with free bandwidth. For example, it is possible to establish a connection at 15 Mbps from point A to point B and a connection at 30 Mbps from point A to point C at the same time. Of course, in SVC it is also possible to set up the requirement for quality in parallel by introducing a concept of QoS.

3.4 TRAFFIC ENGINEERING IN ATM

A main characteristic feature of ATM communication is its excellent traffic engineering technology [4, 5]. Figure 3.12 shows the traffic control technologies used in an ATM node. There are five major functions in traffic controlling.

1. *Connection admission control (CAC)*: Determines whether or not to allow or not to allow for setting up a new connection and guarantees QoS
2. *Usage parameter control (UPC)*: Monitors and controls how much traffic is flowing based upon the amount of traffic that the user has initially specified
3. *Priority control*: Controls the traffic according to the order of priority of connection and attaches an order of priority at the cell level based on a PT (payload type) bit
4. *Shaping control*: Smoothens the burstlike traffic and reshapes the traffic to match bandwidth requirements
5. *Routing control*: Detours traffic away from congested routes

Network resources are managed according to such information as the current state of traffic (estimated amount of traffic by monitoring the buffer in the switching system), route congestion (obtained from operating system, etc.), QoS information (estimated quality of connection based on the cell-loss rate or delay time), and fault information (notice of fault by OAM cell). Based on this traffic information and using traffic-control functions 1 through 5, network resources are controlled to obtain maximum efficiency. The following subsections describe these traffic-engineering control functions in greater detail.

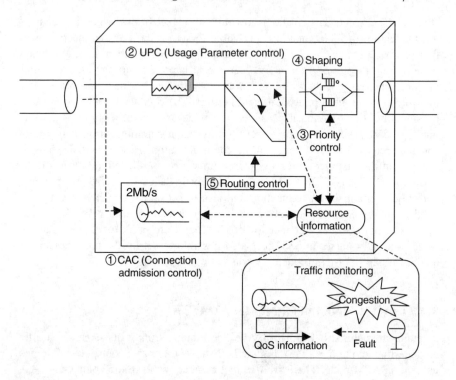

FIGURE 3.12 Traffic control technology in ATM node.

3.4.1 CONNECTION ADMISSION CONTROL (CAC)

Connection admission control (CAC) is the most basic of the ATM traffic controls, and it is used only in connection-oriented communications that securely reserve network resources. Figure 3.13 shows the mechanism of CAC. Here we assume that the current capacity of the transmission line is 150 Mbps, and connection 1 (50 Mbps), connection 2 (30 Mbps), and connection 3 (40 Mbps) have already been established. If a new request for a connection of 50 Mbps came in, it could not be accepted because the total bandwidth would exceed the transmission line's capacity of 150 Mbps. However, if a new request was for a connection at 20 Mbps, it would be accepted. Although this control is based on a simple calculation that sums up the bandwidth requirements, a more dynamic CAC that takes the statistical and multiplexing effects of traffic into consideration is now under development. For example, because the traffic volume is controlled by UPC (usage parameter control), traffic volume that is actually flowing is always less than the specified value when the connection was established and never exceeds that value. In other words, when control is executed based on the reported (specified) traffic volume by the user, there is a possibility that network resources are actually being wasted. Therefore, there is another method of connection control in which the residual bandwidth is estimated by monitoring the actual traffic volume. However, this method is effective

Connection-Oriented Communications and ATM

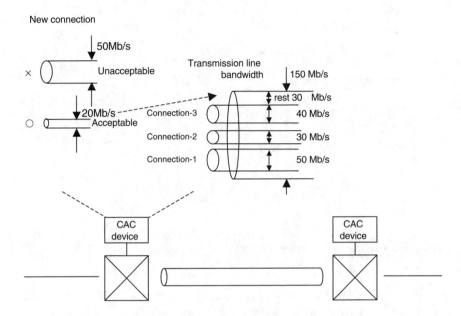

FIGURE 3.13 Mechanism of Call Admission Control (CAC).

only when the traffic volume of each connection is slim and the capacity of the transmission line is large enough to allow for statistical multiplexing. It is also possible to reserve bandwidth exceeding the transmission line's capacity, based empirically on the capacity of the transmission line and the number of connections, much like an airline that overbooks tickets to accommodate predicted cancellations. The effectiveness of CAC can be increased by multiplexing the various connections that differ in quality of service and changing the bandwidth of the connection in the case of congestion or overbooking, etc. However, in most cases a network is operated by combining the methods that enable users to receive services at low cost, that reroute the connection, etc.

3.4.2 Usage Parameter Control (UPC)

Usage parameter control (UPC) continually monitors whether the user whose connection was admitted by CAC is executing communication within the contracted limit of traffic volume in an ATM network [6]. Figure 3.14 shows the concept of UPC. The fundamental purpose of UPC is to prevent QoS from deteriorating for normal connections when any ATM network resources are damaged by malicious intent or through a faulty terminal. Therefore, the controlling ability and accuracy of UPC should be high. However, the idea of quality assurance can change, depending on the design and margin of the network. In addition, because the need for accuracy, control, and the amount of required hardware fluctuates, it is best to fix UPC settings within a legitimate range.

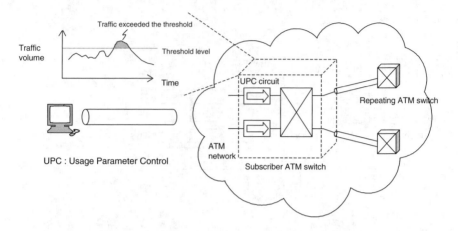

FIGURE 3.14 Operation of Usage Parameter Control (UPC).

Figure 3.15 shows the three typical methods for setting up UPC. The most representative method is the "leaky bucket" algorithm. In this algorithm (also referred to as "bucket with a hole"), arriving cells flow into the bucket while cells are also flowing out of the bucket at a constant rate independent of the depth or number of stored cells. Although the counter value is incremented each time a cell arrives, the counter value is decremented at a constant rate as the cells flow out.

UPC method	Mechanism	Time chart	Operation and feature
(1) Leaky bucket method	Arrived cell, Violation if overflowed, Leaks at constant rate	Depth	• Count up when cell arrived • Countdown at constant rate • Hardware is simple • Controls burst traffic by depth
(2) Credit window algorithm	Reset, Counter r, T, Specified value	T, T	• Count up arrived cells in T hours • UPC monitoring miss may exist • Hardware is simple
(3) Sliding window algorithm (DB system)	Cells go through, Time, T, Number of cells on bridge is specified value.	T, T, T	• Count up the number of cells at all time phase • Accurate monitoring of traffic volume is possible • Hardware is complex

FIGURE 3.15 UPC methods and operation mechanism.

Connection-Oriented Communications and ATM

The threshold value of the counter corresponds to the depth of the counter, and when the counter value exceeds the threshold value, it is interpreted as a violation. In the case where many cells burst into the bucket in a short time, even if the traffic has the same long-time average rate as the one with the nonburst traffic, it is assumed as a violation because too many cells are temporarily accumulated in the bucket. Thus, it is possible to police both average-rate traffic and burstlike traffic using a single bucket.

A problem in circuit design is that the counter value must be decremented at a constant rate. In the case where policing is executed on multiple VCIs, a cell arriving at one VCI is incremented only at that VCI counter, while decrementing must be periodically executed on multiple VCIs at the same time, which has been a constraint in implementing the circuit. To resolve this problem, a measure has been proposed whereby the circuit increments and decrements the counter when a relevant cell arrives and records the time t. In this way, by decrementing the counter by the product of elapsed time Δt and the rate of decrement (leak rate) when a cell arrives, it becomes possible to process only one VCI at a certain timing.

The second method is the "credit window" algorithm, also known as the "jumping-window algorithm," the "fixed-window algorithm," and the "T-X algorithm." (These are essentially the same algorithm.) The credit-window algorithm increments the counter when a cell arrives and resets the counter periodically with time interval T. The hardware structure of this method is the simplest, and it is easy to implement. However, there is a possibility that policing accuracy may suffer if the arriving cells cross over the reset timing and exceed the specified value, leading to an out-of-phase interpretation of time T. For example, assuming that the specified value is X cells, a maximum of $2X$ cells might be allowed to arrive within time T, which would result in a policing error.

The third method is the "sliding window" algorithm, also known as the "dangerous bridge" algorithm, which guarantees the passage of X cells of the maximum number of cells for time T of all phases. Using this method, it is possible to guarantee that less than X cells have arrived within the past T times by memorizing the cell arrival information with a T-bits shift register. However, this method requires a shift register, which can be problematic because hardware requirements become large when the time interval T is long.

Most of the cells deemed to be violators by these UPC circuits are discarded or tagged. If the tagged cells encounter congestion, they are dropped by priority. However, in cases where the network load is light, the cells can be transmitted to the destination user.

3.4.3 Priority Control

Because the ATM network is a statistically multiplexed network based on packet transfer, the loss of cells caused by buffer overflows will occur with a certain probability. To avoid cell loss, the network can allocate the maximum bandwidth to each connection if the capacity of the transmission line is greater than the sum of the bandwidth of multiple connections. However, implementation of such a method can reduce the network's efficiency. This section describes how a network can efficiently process buffer overflows and cell disposal.

Consider the example of an MPEG picture, where the image data is represented by low-frequency components corresponding to the image's outline and high-frequency components corresponding to the fine unevenness in the image. The low-frequency components are assumed to be high-priority information, and the high-frequency components are assumed to be low-priority information. The same approach can be applied to the violation tag in UPC described in the previous section. In an ATM system, the loss of high-priority cells is minimized by executing a priority-control scheme. Figure 3.16 shows the two current methods for priority control.

Threshold control is a method that inhibits the writing of nonpriority cells into the buffer. Using this approach, the buffer exclusively stores only the priority cells, at least from the threshold value to the physical upper limit of the buffer, thus guaranteeing the upper limit of scrappage rate or delay time for the priority cells. However, if we want to shorten the delay time, we must lower the threshold value or reduce the size of the buffer, which makes it difficult to increase the network's efficiency.

Separate-queue control is a method that separates priority cells and nonpriority cells when the cells are written into the queue. By reading out from this queue, it is possible to read out only the priority cells. From the nonpriority queue, it is possible read out the nonpriority cells only when there are no cells in the priority queue. When focusing on the reading operation, the priority cells can fully monopolize the speed of the transmission line. In some cases, however, the nonpriority cells cannot be read out at all. That is, the nonpriority cells are not guaranteed either a maximum scrappage rate or a maximum delay time.

By combining these methods, it becomes possible to protect the priority traffic during congestion and to increase the network's usage efficiency in the normal state.

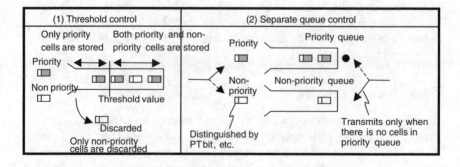

FIGURE 3.16 Mechanism of priority control.

3.4.4 Traffic-Shaping Control

In the case of ATM, the buffer load varies greatly, depending on the traffic's degree of burstness when cells arrive. When many cells arrive at the same time or successively, the buffer load becomes heavy, which can cause cell disposition or overflow. The traffic-shaping control shown in Figure 3.17 is used to smooth the burstlike or violated traffic. The degree of burstness is defined as the ratio of dispersion (moment of the second order or above) to the average (the first-order moment) of traffic volume, and "shaping" means the operation to minimize the moment of the second order or above. The minimum interval between cells is also an important factor, and it is possible to suppress generation of congestion in the subsequent stage of the network by broadening this minimum interval between cells. The minimum interval between cells corresponds to the peak cell rate that was agreed upon in the contract between a user and a service provider. The service provider can use this shaping control to maintain the traffic within the specified contractual level, even if a user transmits an excessive amount of cells into the network.

As a traffic shaper, there is a VC shaper that executes shaping control in each VC. Using this approach, it is possible to read the cells at a rate specified for each VC and to output them to the transmission line. That is, a VC shaper guarantees the bandwidth for each user by modifying the user's traffic to a volume that is within the contractual traffic parameters, or by outputting the cells with a certain rule (e.g., ratio) among multiple users, or by preventing the fairness of each user's traffic from being reduced as the result of mutual interactions between user traffic. When such a VC shaper is utilized, even if a user outputs a high volume of traffic exceeding the specified value, no other users' connections are influenced by it.

3.4.5 Packet Throughput and Discarding Process

When we send data such as a TCP-IP packet, it is divided into multiple ATM cells throughout the ATM network. If any of the packet's cells are discarded, that packet is assumed to be invalid and a resending process is executed. Figure 3.18 shows how the packets are transferred in an ATM network. Any packet that loses even just one

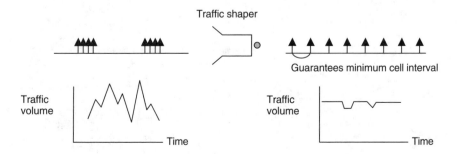

FIGURE 3.17 Principle of traffic shaper.

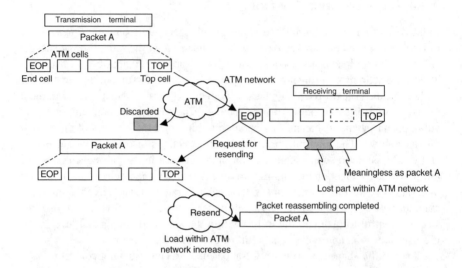

FIGURE 3.18 Packet-level transfer in ATM network.

cell is transferred again by a resending function at the upper layer (TCP layer). Resending may cause a cascade of consecutive resends. These retransferred packets increase the load on the ATM network, possibly causing congestion on the network.

Two methods have been proposed as processes for discarding invalid packets, with both methods focusing on the packet-level performance. These methods are called EPD (early packet discard) and PPD (partial packet discard).

Figure 3.19 shows the mechanism of discarding cells by the EPD method. When the buffer of an ATM switch is filled halfway under the threshold level (TH),

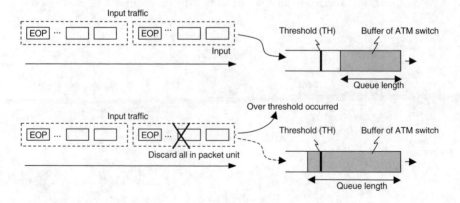

FIGURE 3.19 Mechanism of EPD (early packet discard).

Connection-Oriented Communications and ATM

all of the cells composing a single packet can be stored into the buffer. However, when the buffer of an ATM switch is filled beyond the threshold level (TH), the cells arriving after that are discarded by packet unit from the top to the end cell (EOP, end of packet). As noted in the discussion of the AAL5 protocol layer (Section 3.2.4), the top and the end cells of the several cells that compose the packet can be distinguished. So, in the case when some of the cells from the same packet have already been stored, all of the cells of this packet are stored into the buffer. In this manner, the cells arriving at the terminal that belong to the same packet are either all discarded or they are all stored by packet unit. Figure 3.20 shows the behavior of the throughput of an ATM network (transferring throughput as packet units), comparing the cases with and without EPD control. It is shown that when the load exceeds the threshold level and cell loss occurs in the ATM network, the load becomes increasingly heavy owing to negative feedback by the resending of packets. In the case where EPD control is applied, the decrease in throughput is not so abrupt, showing that the conventional problem of a decrease in packet-transferring capability in the presence of congestion has been resolved.

In the case of PPD, the buffer has no threshold level. When cells overflow from the buffer, all of the cells composing the relevant packet except for the last cell (EOP cell) are discarded. This PPD control can also improve the throughput of the network at the packet level.

3.4.6 CONGESTION CONTROL IN ATM

In a telephone network, congestion is often generated by an abrupt increase of traffic owing to events or disasters. In a packet network such as an ATM network, congestion leads to discarding of cells, and the quality of communication is greatly diminished. On the other hand, because an ATM network is connection oriented, it is possible to reroute the communication or to control the network load. In an

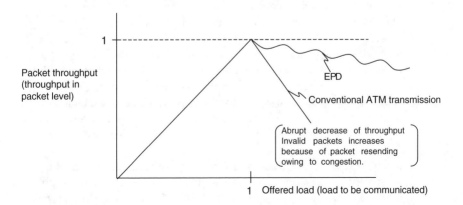

FIGURE 3.20 Packet throughput control by EPD.

56 GMPLS Technologies: Broadband Backbone Networks and Systems

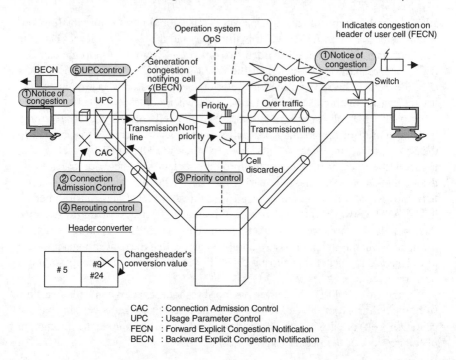

FIGURE 3.21 Operation of ATM network when congestion occurs.

ATM network, high-level traffic-control techniques have been combined to avoid congestion. Figure 3.21 shows an example of a control method that is executed when congestion occurs in an ATM network. The network attempts to recover from congestion using measures 1 through 5 shown in Figure 3.21.

1. ATM network notifies users that congestion has occurred in network by setting the congestion indicator bit in the ATM cell header to 1 (FECN, forward explicit congestion notification). Users who were notified of the network's congestion would then reduce their transmission rates. Although this may be effective in nature, it is ineffective when users do not cooperate by reducing the transmission rates of their terminals.
2. Network stops admitting new connections by CAC and attempts to recover from congestion by disconnecting current users who are in the process of communication. Gradual recovery is expected by this control.
3. In the congestion state, high-priority traffic is reserved as much as possible through priority control. If enough bandwidth has been assigned to priority connection when the system was designed, it is possible to transfer the priority traffic without diminishing the quality of communication.

4. If required, routing control is executed to detour the current connection to another route. This option is possible only because ATM networks are connection oriented. It is also possible to detour the route away from the hot points and to improve the network's overall efficiency.
5. Compulsive control by UPC (usage parameter control) executes restriction of call or input into the network. This is the final control procedure when there is no other way to recover from congestion.

By combining these congestion-control techniques, network congestion can be avoided or the network can recover to a stable state.

3.5 BEARER CLASS AND SERVICES

ATM is a universal transfer protocol for multimedia traffic supporting various services and multiple bearer classes. Table 3.1 shows the bearer classes and the features of various ATM services.

3.5.1 Constant Bit Rate (CBR)

CBR is a pseudocircuit-switching class in which a peak cell rate has been determined. This service class was developed to achieve STM (Synchronous Transfer Mode) services on an ATM network and is called "circuit emulation." In an ATM network, it is possible to reserve the required network resources at peak rate and to execute services that maintain the quality of communication. It guarantees a small CTD (cell-transfer delay) and CDV (cell-delay variation) as well as a small percentage of cell loss. However, it is a prerequisite that the nature of the traffic be known in advance based on the user's report. Maximum rate is determined by the speed of the transmission line. This service class can be applied for a service such as circuit emulation that transfers traffic with ATM.

3.5.2 Variable Bit Rate (VBR)

This is a service class that specifies the traffic pattern with an average cell rate and a maximum burst length besides the peak cell rate. This class has been developed for voice/image communications using a variable-rate encoding method as the target of services. For example, there is a distinct difference in the nature of traffic between the frame period of image and the period of scene change. A peak rate determined by codec hardware is also one of the parameters. VBR is categorized by two types: rt-VBR (real-time VBR) for real-time communication and nrt-VBR (non-real-time VBR) for accumulated communication. Both CBR and VBR are premised on the fact that the nature of traffic is known in advance. As for communication quality, nrt-VBR guarantees a small cell loss rate below some specified maximum, while rt-VBR also guarantees small CTD (cell-transfer delay) and CDV (cell-delay variation).

TABLE 3.1
ATM Services Category

	CBR	VBR		UBR	UBR+	ABR	ABT	
		rt-VBR	nrt-VBR				ABT-IT	ABT-DT
Traffic parameter	Peak cell rate	Peak cell rate Average cell rate Maximum burst rate		None		Peak cell rate Minimum cell rate	Peak cell rate	
QoS parameter (quality assurance)	Transmission delay Maximum delay Variation Cell loss rate	Transmission delay Maximum delay Variation Cell loss rate	Cell loss rate	None	Lowest packet rate	Transmission rate is controlled according to state of network	Transmission delay Block ratio Cell loss rate (when route/bandwidth were established)	
Object of services	Circuit switching data	Real time Variable encoding voice/image	Non-real time	Data transfer (especially TCP/IP)		Data transfer within/between LANs	Burstlike data transfer Contents transfer	
Feature	Assumed that user knows the nature of traffic					Supposed as an ATM-layer service to transfer data		

3.5.3 Unspecified Bit Rare (UBR)

Data-transfer protocols such as TCP-IP include such functions as rate control or resending data when information is lost. Therefore, in the ATM layer, a UBR service was defined to expedite sending out as much information as possible whenever a transmission line is available.

Because UBR is a service that does not guarantee a quality of communication at all, if a connection is once completely cut off, it becomes impossible to measure normally the RTT (round-trip time) of TCP-IP in the upper layer and to calculate an exact time-out time to execute resending, which is a problem. Thus, a UBR+ has been proposed to address this problem. UBR+ guarantees a minimum bandwidth at the packet level. To do this, as shown in Figure 3.22, the traffic exceeding the minimum bandwidth is tagged, and this tagged traffic is selectively discarded at the repeater node when congestion occurs. In contrast, the traffic under the minimum bandwidth is guaranteed its QoS because it is possible to measure normally the RTT of TCP-IP. Besides this method, there is also a method that, instead of tagging traffic by UPC, executes shaping based on the minimum bandwidth to guarantee QoS and allocates the residual bandwidth fairly between connections.

3.5.4 Available Bit Rate (ABR)

ABR is a service that maximizes the use of the available bandwidth of current repeater links as effectively and fairly as possible. In ABR service, the network sends information about the available bandwidth to each user, and based on this information, each user controls the transmission rate.

For example, if there is an available bandwidth of 10 Mbps on the 30-Mbps link and there are four connections that are currently using this link (active connections), ABR tries to maintain fairness and guarantee QoS by allowing each connection to use 2.5 Mbps to maximize the usage of bandwidth. Ideally, loss of cells does not occur within the network. Figure 3.23 shows a mechanism of

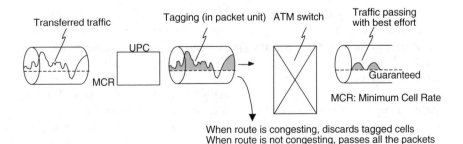

FIGURE 3.22 Minimum bandwidth guarantee type best effort communication by UBR+.

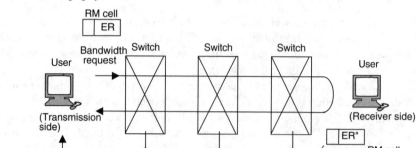

FIGURE 3.23 Bandwidth changing by ABR.

bandwidth changing in ABR service. User sends the data attaching the maximum possible rate to transfer the resource-management cell (RM cell) with a constant period (for example, one time per 32 user cells). The repeater node notifies an available rate (ER: explicit rate) that was calculated taking the current usage condition and available bandwidth into consideration so as to share fairly between connections to the transmission-side terminal by attaching to this RM cell. It is possible to obtain the information on minimum available bandwidth on all the routes by rewriting the ER value only when the available bandwidth of the relevant node is less than the threshold value, depending on the ER value that has been written into the RM cell.

3.5.5 ATM Burst Transfer (ABT)

As a method to transfer burstlike data effectively, there is an ABT (ATM burst transfer) service. This ABT service is categorized into two types: ABT-IT (ABT-immediate transmission), which attempts to transfer burst data without reserving the bandwidth in advance, and ABT-DT (ABT-delayed transmission), which reserves the required bandwidth on the route in advance and transfers burst data.

Figure 3.24 shows the procedure of ABT-IT transfer. In this ABT-IT, information is transmitted at the same time when the bandwidth request is sent. If it is impossible to reserve enough resources at the halfway point of the route, the connection is blocked once and is retried. In the case of this example, the connection is retried by setting the bandwidth to one-half the value of the rate, and resending is executed starting from the top cell (packet). Although ABT-IT

Connection-Oriented Communications and ATM

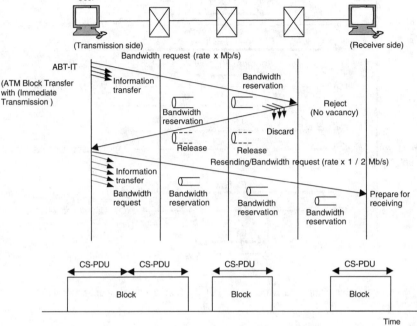

FIGURE 3.24 Mechanism of ABT and image of cell transferring by ABT.

may be superior to ABT-DT in setup time, when not enough bandwidth has been reserved, ABT-IT may require more time to transfer the data because resending might be required many times. On the other hand, ABT-DT can surely transfer the data, but when the amount of data is small, it is wasteful because the resource is occupied without operation until connection is established. Therefore, ABT-DT is suitable for burstlike data transfer or content delivery.

3.6 OAM FUNCTION IN ATM

Maintenance, fault detection, and quality measurement are all key factors in network operation, so it is indispensable to assign an operating system (OpS) to manage and support these functions on the communication equipment side. The OAM (operation and maintenance) function is a generic designation of various operation/maintenance functions to support the high quality and stable transmission of information. The OAM function consists of (1) a performance-monitoring function, (2) a fault-detection function, (3) a fault-notification function, (4) a system-protection function, and (5) a point-of-fault identification function.

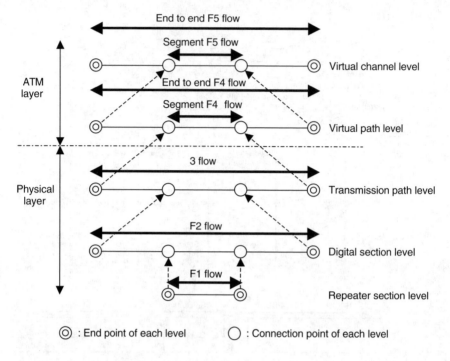

FIGURE 3.25 OAM level and flow.

The OAM function is defined for each notifying function on the following five levels and takes a hierarchical structure as shown in Figure 3.25:

Repeater section level
Digital section level
Transmission path level
Virtual path (VP) level
Virtual channel (VC) level

In each hierarchical level, the exchange of OAM information is executed independently of the other levels. The reason for making the OAM structure hierarchical is so that each layer can be developed independently, as in the OSI layer model.

The OAM function when ATM communication is executed on SDH (synchronous digital hierarchy) is illustrated in Figure 3.26. In the physical layer, which is located under the ATM layer, there are three layers. From low to high layer, these are F1 flow (repeater section level), F2 flow (digital section level), and F3 flow (transmission path level). In the ATM layer, there are two layers, F4 flow (VP level) and F5 flow (VC level). Each of these layers executes fault detection independently of the others.

Connection-Oriented Communications and ATM

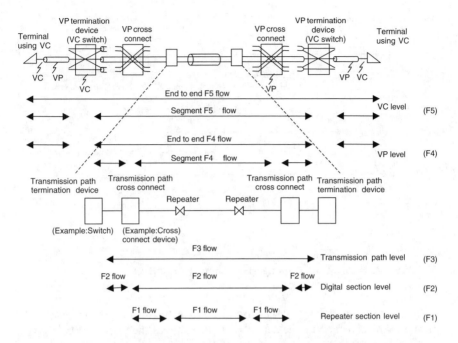

FIGURE 3.26 Relationship between OAM flow and equipment in physical layer and ATM layer.

Flows from F1 to F5 are called "layer escalation," and as shown in Figure 3.27, the upper flow expands and notifies the fault detected by the lower flow. Figure 3.27 illustrates the case in which a fault occurs in a transmission line between repeaters, making it impossible to communicate. When the point of fault notifies that a fault has occurred by sending an AIS (alarm indication signal) and this signal is detected at the endpoint, it generates an RAI (remote alarm indication) signal in the reverse direction and an AIS corresponding to the next higher flow layer in the forward direction. In this way, it is possible to detect all of the faults that occur in this repeater section at both endpoints of the transmission path.

Let us look at the OAM function of the ATM layer in greater detail. ATM has excellent OAM and traffic engineering functions for connection-oriented communications. Figure 3.28 shows the outline of operation of the alarm-transmission function in the ATM layer. The node that detects a fault sends forward an AIS (VP-AIS) cell to notify that a fault was detected, and the endpoint sends back an RAI (the same way as in SDH) so that the originating point of VP can know that a fault has occurred. This triggers a switching of the transmission path.

OAM cells are transferred together with user cells of VP or VC, so a method is required to distinguish them from user information. In the OAM cell for VP, a VPI value of the relevant VP is given, and a special VCI value (for segment:

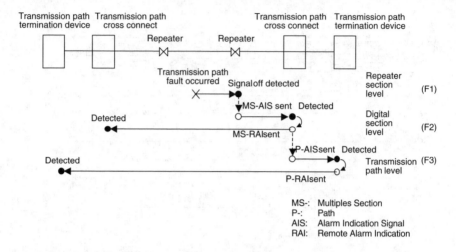

FIGURE 3.27 Example of alarm transmission in physical layer.

0003H; for end-to-end: 0004H) is also given to distinguish it from the user cells. In the OAM cell for VC, the same VCI/VPI value as the relevant VC is given, and a special PTI value (for segment: 100B; for end-to-end: 101B) is also given to distinguish it from the user cells.

The top byte of the payload of an OAM cell is used to identify which OAM function the OAM cell will use. For example, in the case where an OAM cell notifies the alarm in a forward direction (as described in the next section), 0010000B is inserted. And into the remaining payload area, information defined individually by each OAM function and an error-detection bit is inserted. An AIS OAM cell is sent out every 1 sec, and if an AIS is not detected after 2.5 sec by soft state or timeout mechanism, the endpoint assumes that the fault has been recovered.

Because ATM executes asynchronous transfer of cells, cells do not arrive when there is no information or when the transmission line is cut off by a fault. Therefore,

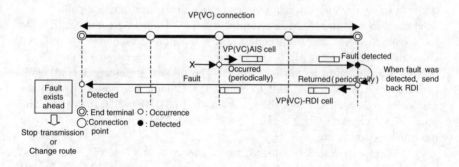

FIGURE 3.28 Outline of operation of alarm-transmission function in ATM layer.

Connection-Oriented Communications and ATM

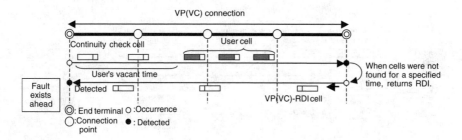

FIGURE 3.29 Outline of operation of continuity-check function.

it is always necessary to check whether a normal connection is being established or not. This is accomplished by executing a continuity check, a function that monitors whether a VP or VC connection is being established normally or not.

Figure 3.29 shows the outline of operation of the continuity-check function. The transmission-side endpoint sends the OAM cell for continuity checking in the event that no user cells are transferred for a specified time period (1 sec). The receive-side endpoint judges that a fault or a LOS (loss of continuity) has occurred in the transmission path when it fails to receive the user cells of relevant VP (or VC) or continuity-check cells for a specified time period (about 3.5 sec). When LOS is detected, the endpoint terminal periodically sends the VP-RAI cell (or VC-RAI cell), which indicates that VP (or VC) in the reverse direction is out of order to the opposite-side endpoint. Recovery from the LOS state is judged when a user cell or a continuity-check cell is received.

Next we describe the loop-back function as a measure to isolate the point of fault. The loop-back function is designed to confirm continuity in the specified section in VP or VC without interrupting the services. Figure 3.30 shows the outline of operation of this loop-back function. An OAM cell for loop back is sent out from the device at the end of the section where continuity needs to be confirmed. In this loop-back cell, an ID number of the device to which the loop-back cell should be returned is attached. When the device that has been specified

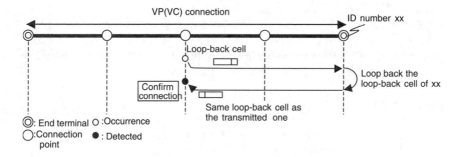

FIGURE 3.30 Outline of operation of loop-back function.

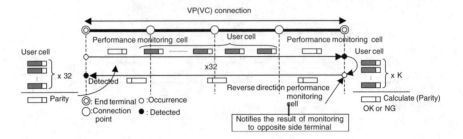

FIGURE 3.31 Outline of operation of performance-monitoring function.

as the loop-back point receives this loop-back cell, the device sends back this cell in the reverse direction. When the device that originally sent the loop-back cell receives the returned loop-back cell, the continuity of the specified section is confirmed. By repeating this procedure, we can easily locate the point of fault.

To check cell loss and quality degradation in ATM communication, there is a performance-monitoring function as a high-level OAM function. Performance monitoring is a function that measures the quality of transmission such as loss of cell in each VP or VC and notifies the measurement result to the opposite-side endpoint. Figure 3.31 outlines the operation of the performance-monitoring function. The transmission-side endpoint sends an OAM cell for performance monitoring in every specified number of user cells (called a "block"). This performance-monitoring cell contains information such as a parity for all the user cells included within the block, the number of user cells actually transmitted, etc. The receive-side endpoint compares the content of the received block with the content of the performance-monitoring cell and measures the number of lost cells and the number of error codes, etc. Even if just one cell was lost, it can be detected as loss-of-cell, and a bit error within the payload can also be detected by parity check. In the performance-monitoring cell, an area to notify the measurement result to the opposite-side endpoint has been defined. When the receive-side endpoint notifies the measurement result obtained at the receive side to the transmission-side endpoint, this information is inserted into the notifying area in a reverse-direction performance-monitoring cell.

As seen here, ATM can execute high-level operation and maintenance functions using various OAM functions. In this chapter, we have described the basic principles of ATM. Readers who want more details about connection-oriented communication technology can consult the literature [7–10].

REFERENCES

1. Cuthbert, L.G. and Sapanel, J.-C., ATM, the broadband telecommunications solution, *IEE Telecommun. Serv.*, 29, London, UK, Jan. 1993.
2. Saito, H., *Teletraffic Technologies in ATM Networks*, Artech House, Norwood, MA, Feb. 1994.

3. Onvural, R.O., *Asynchronous Transfer Mode Networks: Performance Issues,* 2nd Ed., Artech House, Norwood, MA, Oct. 1995.
4. Saito, H., *Teletraffic Technologies in ATM Networks,* Artech House.
5. Sato, K., *Advances in Transport Network Technologies,* Artech House, Norwood, MA, Sept. 1996.
6. Chen, T.M. and Liu, S.S., *ATM Switching Systems,* Artech House, Norwood, MA.
7. Coover, E.R., *ATM Switches,* Artech House, Norwood, MA, March 1995.
8. De Pryker, M., *Asynchronous Transfer Mode: Solution for Broadband ISDN,* 2nd ed., Ellis Horwood, New York, July 1991.
9. Rahmar, M.A., *Guide to ATM Systems and Technology,* Artech House, Norwood, MA, Sept. 1998.

4 Internet Protocol (IP)

Internet protocol (IP) is the third-layer protocol, and it consists of an IP forwarding protocol and an IP routing protocol. The routing protocol creates a routing table, and the forwarding protocol transmits the IP packets according to the routing table. In this chapter, we will first describe the IP forwarding technology and then discuss the IP routing technology.

4.1 IP FORWARDING

In the IP layer, a node within a network is called a router. IP packets are delivered to the destination host after being transferred between routers hop by hop, much like a bucket brigade. The router does not need to know the complete route information to the destination address. It is enough to know just which neighboring router is the nearest to reach the destination address and to transfer the IP packets to the neighboring router as the next hop of the trip. In this way, in the global-scale Internet, IP packets are transferred to the desired destination address.

The router determines the destination of packets according to a routing table that includes information about the relationship between the destination address and the next hop. In this process, called IP forwarding, information in the IP header and the address play an important role. Section 4.1.1 describes the structure of the IP header and each of its fields. Section 4.1.2 describes the address to be used for searching the routing table. Section 4.1.3 describes the routing table.

4.1.1 IP Header

The IP header includes a destination address. The router determines the next hop using this IP header (Figure 4.1) The IP header also includes such information as a TTL (time to live) and a header checksum. TTL is used to prevent IP packets from existing infinitely when a routing loop is generated. Specifically, the TTL of an IP packet is reduced by 1 at each hop and discarded when it becomes 0. The header checksum is used to detect the bit error of the header. Header error is fatal in packet-transmission networks where the transmission process inside the network is determined based on the information in the packet header. If a bit-error occurs in a packet header during transmission, the packet cannot be transferred to the right destination. Bit-error is detected by calculating the checksum of the packet header. At the transmitter side, a checksum is calculated assuming the IP header as a sequence of 16-bit words in a form of 1's complement and is put into the header checksum field. At the receiver side, the sum of the received IP header (including the checksum) is calculated in a form of 1's complement

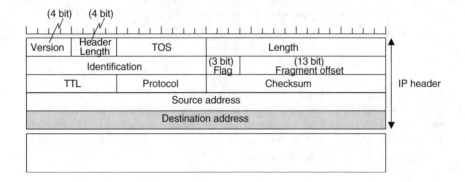

FIGURE 4.1 Structure of IP header.

assuming it as a sequence of 16-bit words. If the result of the calculation at the receiver side is all 1's, the receiver side judges that there is no error and that the received packet is normal. If the result is not all 1's, the receiver judges that there is an error and discards the received packet. Calculation of the header checksum is repeated every time the header is changed. For example, when the TTL counter is reduced by 1, the header checksum is recalculated.

The other fields in the IP header (Figure 4.1) include a version, a header length, a TOS (type of service), a datagram length, a datagram identifier, a flag, a fragment offset, a protocol identifier, and an originating IP address. The length of a standard header is 20 bytes. The flag (3 bits) and fragmentation offset (13 bits) are related to fragmentation. Fragmentation means that the IP packet is divided when the IP packet length exceeds the MTU (maximum transmission unit) of the link layer. In that event, the original IP packet is divided into multiple pieces. One of the three bits of flag is used as a "more fragments" bit. When this bit is set, it indicates that the IP packet is an intermediate one, and when this bit is not set, it indicates that the IP packet is the last one of the fragment. Because the IP packet is transferred hop by hop, when a route is changed during transmission, the order of arrival of the IP packets at the receiver side may change. Therefore, to restore the original IP packet in the correct sequence, a fragment offset (13 bits) is added to the fragmented IP packet to indicate where the received IP packet belongs. The fragment offset shows the position from the top of the payload of the original IP packet in a unit of 8-byte. One of the three bits of flag is used as a "don't fragment" bit. When this bit is set in the original IP packet, this IP packet cannot be fragmented. It is also possible to set other optional fields, such as a "record route," a "timestamp," or a "source routing." For more details, refer to the literature [1, 2].

4.1.2 IP Address

In IP v.4, a 32-bit address is utilized. The IP address is delimited by an 8-bit unit, and each divided part is expressed in decimal system, e.g., 129.60.83.222. By using this IP address, all of the hosts connected to the global Internet are uniquely

Internet Protocol (IP)

distinguished. The IP address consists of a network block and a host block and has been classified according to the number of hosts that the network can accommodate. For unicasting, classes A, B, and C are available. In classes A, B, and C, the first 8 bits, 16 bits, and 24 bits are used, respectively, as network blocks. Identification of class A, B, or C is executed by using the top few bits. If the first bit is 0, it is class A. If the top two bits are 10, it is class B. And if the top three bits are 110, it is class C. Therefore, in classes A, B, and C, the number of networks that can be defined is limited to the range that can be expressed by 7 bits, 14 bits, and 21 bits, respectively, in a binary system. The number of hosts that can be accommodated in each network class is the maximum number (minus 2) that can be expressed by 24 bits, 16 bits, and 8-bits, respectively. (Two cases are excluded to accommodate the naming convention that all 1's of the host block indicate a broadcasting address and that all 0's of the host block means that it is a network itself.)

For example, a class B network can accommodate 65,534 hosts ($= 2^{16} - 2$) per network. It is rare that one network accommodates such a great number of hosts. To effectively accommodate host addresses in a class B network, a technology called "subnetting" is utilized. Subnetting is a technique to further subdivide the host block's address to allocate the address. For example, if we divide the host block of class B address into two parts and define a subnet and host block using the first 4 bits and the next 12 bits, respectively, the network can accommodate 4,094 ($= 2^{12} - 2$) hosts, and it is possible to create 16 subnets from the network address of a single class B network.

4.1.3 Forwarding Table (Routing Table)

Each router executes hop-by-hop routing using a forwarding table. The forwarding table has such field as (1) destination address, (2) next hop's IP address, (3) network interface number, etc., corresponding to each entry. To the destination address (1), either a host address, a network address, or a default address is written in. In the case where the host address is written in, it means that this destination address is a host that can be reached directly from any of the network's routers. When a router receives the IP packet with this destination address, the router transfers this packet directly to this host via a network interface. In the case where the network address is written in, there are two possibilities: either this address exists within the network to which the router is connected, or it does not. In the former case, the IP packet that goes to this destination address is broadcast to all the hosts in this network. In the latter case, the IP address is transferred to the next hop's router. If there is nothing that corresponds to the destination address, it is possible to use the default setting. The next hop's address (2) is either the IP address of the network interface of the router itself or the IP address of the next hop's router. In the former case, the destination address (1) can be reached directly from the network interface of the network to which the next router is directly connected. In the latter case, the destination address (1) cannot be reached directly. In the case where the network interface number (3) is written in, it shows the number of the network interface to which the IP packet is transferred.

The forwarding table is searched by using the destination address (1) as a key. If the host can be reached directly from the next router, the host address is used as the destination address. Otherwise, the network address is used as the destination address.

As previously described, there are three classes — A, B, and C — of network address. Among these, the class B network can create only 16,384 (= 2^{14}) Internet addresses. Consequently, there is concern that if the number of hosts connected to the Internet continues to increase, there will be a shortage of class B addresses. If and when that happens, we will have to build the network by combining multiple class C addresses with class B addresses. Although this approach solves the problem of a shortage of class B addresses, it creates a new problem in that, by using multiple class C addresses to build one network, the number of entries in the routing table increases unnecessarily. To resolve both problems at the same time, a technology known as CIDR (classless interdomain routing) has been proposed [3, 4].

In this CIDR technology, the multiple IP addresses are integrated into a single routing table as follows:

1. The addresses share the same bits from the top of the IP address with the same bit pattern (this is called a "prefix").
2. The routing table and the routing algorithm are expanded so that they can use the 32-bit address and the 32-bit mask.
3. The routing protocol is expanded so that it can handle the 32-bit addresses and 32-bit masks that have been introduced.

Figure 4.2 shows a comparison between addressing methods by CIDR and by class A, B, and C. In this figure, the address range that can be expressed

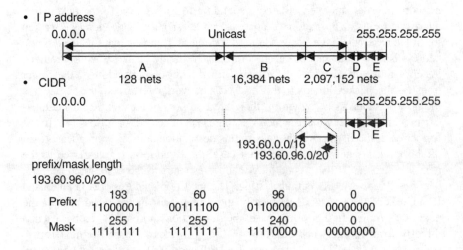

FIGURE 4.2 IP address and CIDR.

Internet Protocol (IP)

by 32 bits, from 0.0.0.0 to 255.255.255.255 is given by a straight line. The class A address can define only 128 networks, and the class B address can define up to 16,384 addresses, and the class C address can define up to 2,097,152 addresses. The class D address is used for multicasting, and the class E address is used for experiments. In CIDR technology, a network address is expressed by a 32-bit IP v.4 address and mask length. For example, the address 193.60.96.0/20 is expressed by a 32-bit IP v.4 address of 11000001 00111100 01100000 00000000 and a 32-bit mask of 11111111 11111111 11110000 00000000. As seen from the CIDR straight-line expression, this address corresponds to a part of the block of the network address of 193.60.0.0/16.

In searching the routing table using the CIDR method, a table-searching technique called a "longest prefix matching" search is utilized. When the destination address is the network address, in CIDR, the destination address is expressed by using a 32-bit IP address and a 32-bit mask.

When an IP packet arrives, the destination address of the IP packet is extracted from its header and is used as a search key in the routing table. Because a mask bit is "don't care," it may match multiple entries. In this case, the entry whose bits coincide with the destination address used as the search key over the longest span of bits is selected as a search result. This is known as "longest-prefix matching." Many algorithms have been proposed for this longest-prefix-matching search [5–8]. Because the part beyond the prefix length is not considered, the destination address of an IP packet may match multiple entries in the forwarding table. In this case, the longest matched entry is again selected as the search result. Figure 4.3 shows a concept of a longest-prefix-matching search. In this case, when an IP packet with a destination address of 148.32.96.4 arrives, even though two entries of 148.32.0.0/16 and 148.32.96.0/24 partly coincide with the arriving IP address, the latter entry of 148.32.96.0/24 is selected as the search result because it shares more bits with the IP address than the other entry of 148.32.0.0/16.

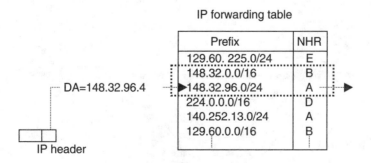

FIGURE 4.3 Concept of "longest prefix matching" search and the forwarding table.

4.2 IP ROUTING

IP routing is a technology for creating a routing table. This technology is important because it enables the Internet to operate autonomously and distributedly and to be developed on a massive scale.

Because the Internet is such a huge network, it is not feasible to create a routing table by centrally managing the destination addresses. For this reason, a routing protocol was developed as a mechanism to create a routing table autonomously and distributedly. The routing protocol creates a routing table while making the routers communicate with each other and exchange information, and based on this routing table, it creates a forwarding table to transfer the IP packets. The routing table is dynamically changed corresponding to the change of network status or operating policy of the network.

We can view the IP routing protocol of the Internet from two perspectives: as a hierarchy to which the IP routing protocol is applied, and as an algorithm that the IP routing protocol uses [9]. In this section, we first describe the hierarchical classification that is applied to the IP routing protocol, and then we describe the algorithm classification that is used by the IP routing protocol. Finally, we describe OSPF (Open Shortest Path First) and BGP-4 (Border Gateway Protocol) as a typical example of IP routing protocols.

4.2.1 HIERARCHY OF ROUTING PROTOCOLS

The Internet has been developed by mutually connecting multiple networks. An individual network within the Internet is called an autonomous system (AS). Operation inside the AS can be determined independently of any other AS. An enterprise network, an in-campus network, or an ISP's network are all examples of an AS.

Routing in the Internet is divided into two hierarchical layers: inside-AS and inter-AS. In each of these hierarchical layers, the purpose of the routing protocol and the environment used are different. The routing protocol used within an AS is referred to as IGP (Interior Gateway Protocol), and the protocol for communication from one AS to another is called EGP (Exterior Gateway Protocol). Figure 4.4 shows how IGP and EGP are applied in networks. EGP is used between

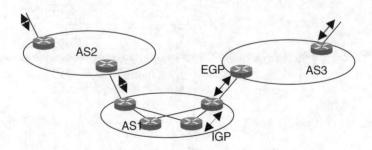

FIGURE 4.4 IGP and EGP (inside-AS routing and inter-AS routing).

ASs. Border routers of neighboring ASs mutually exchange the route information using EGP. Routers inside an AS exchange route information via IGP.

IGP is a routing protocol that is used inside AS and creates a routing table so that an IP packet can be transferred to the destination address along the shortest path. Typical routing protocols include RIP, IGRP, OSPF, IS-IS (Intermediate System to Intermediate System), and so on. EGP is an inter-AS routing protocol, and thus it must be capable of handling a huge amount of routing information. It is also required to create the routing table while reflecting the effect of each AS's operating policy on the table. A typical and popular routing protocol is BGP-4.

4.2.2 Categorization of Routing Protocol by Algorithm

Routing protocols are categorized into various types based on their operating principles. There are distance-vector types, link-state types, and a path-vector type [10]. Table 4.1 provides a brief description of the operating principles for each type of routing protocol. In the distance-vector type, each router exchanges a distance vector that consists of a destination address, a distance to the destination address, and a next-hop router's address; updates its own distance-vector table according to the distance vector of each destination address from neighboring routers; and calculates the next hop of each destination address. In this distance-vector type, as described later, network size is restricted owing to the problem of looped routing. Typical examples of distance-vector-type protocols are RIP and IGRP.

The path-vector-type routing protocol exchanges a path vector instead of a distance vector. The advantage of this approach is that it does not restrict the network size. A typical example of a path-vector-type routing protocol is BGP. The link-state type is a protocol in which the same network topology is shared by all of the routers by exchanging the link-state information among the routers, with each router calculating the shortest route from itself to the destination router on this topology. Typical examples of this link-state-type protocol are OSPF and IS-IS.

TABLE 4.1
Operating Principles of Routing Protocols

Type	Outline	Example
Distance vector	Seeks the next hop in the shortest route by exchanging a distance-vector table (includes destination address information, etc.); Bellman-Ford algorithm is used	RIP IGRP
Path vector	Selects the route with the shortest path length that is possible to avoid a routing loop by exchanging a path vector (includes through-node information to destination address, etc.)	BGP
Link state	Each node has topology information by exchanging a link state and calculates the shortest route by using topology information	OSPF IS-IS

TABLE 4.2
Distance-Vector Table

Destination	Next Hop	Distance
129.60.225.0/24	E	3
148.32.0.0/16	B	32
148.32.96.0/24	A	7
224.0.0.0/16	D	29
140.252.13.0/24	A	103
129.60.0.0/16	B	50

4.2.2.1 Distance-Vector-Type Protocol

The distance-vector-type protocol creates the routing table by exchanging a distance-vector table (a table of distance to each destination address and the next router) with neighboring routers. Table 4.2 shows an example of a distance-vector table. In this distance-vector table, pairs of distances from a router to various destination addresses and the links to the neighboring routers used at that time are recorded.

The distance-vector-type protocol has been designed based on the Bellman-Ford algorithm [11]. Initially, this protocol operates assuming that each router knows the "costs" of the links to the neighboring routers. That is, only the information of a neighboring router and the cost of the link to it are recorded into the distance vector. Each router exchanges the distance-vector table with neighboring routers, and if the router judges that the distance vector obtained from certain neighboring routers could reach the next hop in a shorter distance than its own distance vector, it updates its own distance-vector table with the obtained distance vector. Whenever a router updates its distance-vector table, it notifies the neighboring routers, and they update their tables. If each router continues to update the distance table autonomously and distributedly, the contents of the distance-vector table eventually converge. The route according to this converged distance-vector table becomes the shortest route. Figure 4.5 shows the operating concept of a distance-vector-type routing protocol. In this example, router X obtains a distance-vector table from router B (destination = A, next hop = B, distance = 50) and from router C (destination = A, next hop = C, distance = 100). From this, router X knows that it is possible to reach router A at a cost of 52 (= 50 + 2) via router B and at a cost of 101 (= 100 + 1) via router C. Therefore, router X decides to adopt router B as the next hop to go to router A.

In the Bellman-Ford algorithm, each node updates the distance-vector table based on Bellman's formula shown below. Assuming that the link cost between neighboring nodes i and j is $d(i,j)$ and that the shortest distance from node i to node j at time t is $D(t,i)$, when node i obtains the distance vector from node j, the shortest distance at the next instance of time $D(t+\delta,i)$ is given by Bellman's formula:

$$D(t+\delta,i) = \min[D(t,i), d(i,j) + D(t,j)]$$

Internet Protocol (IP)

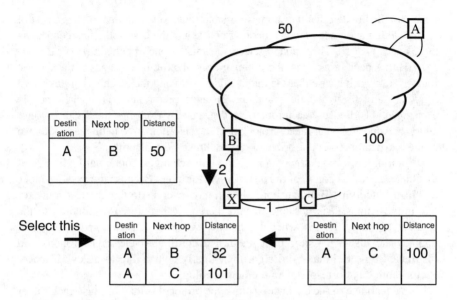

FIGURE 4.5 Operating concept of distance-vector-type routing protocol.

The next hop to the destination node is replaced with the distance vector that was selected by $\min[D(t,i), d(i,j) + D(t,j)]$.

In the example of Figure 4.6, the source node seeks to find the shortest route to the destination. Here we assume that the source node i at time t can reach the destination with a cost of the shortest distance $D(t,i)$. We also assume that the neighboring

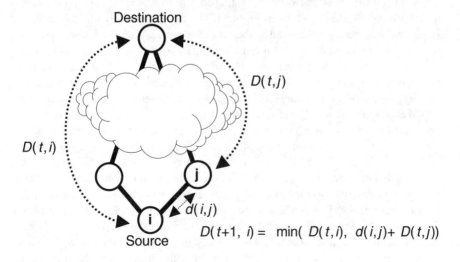

FIGURE 4.6 Bellman's formula.

node j notified node i that the shortest possible route to the destination has a cost of $D(t,j)$. Node i and node j are neighboring nodes, and the cost of the link connecting these nodes is $d(i,j)$. Then, node i selects the smaller one of either $D(t,i)$ or $d(i,j) + D(t,j)$. It has been proved that the Bellman-Ford algorithm converges to the shortest route even in such a case that the initial value $D(0,i)$ and the neighboring link's cost $d(i,j)$ are any positive values and the distance-vector table is asynchronously exchanged [11]. Because each router can distributedly exchange the distance-vector table, each updates the distance-vector table. The algorithm used in this case is known as the "asynchronous and distributed-type Bellman-Ford algorithm."

If a link has a breakdown and the topology changes, routers at both ends of this link detect the change of the distance-vector table. Because the asynchronous and distributed-type Bellman-Ford algorithm converges to the shortest route for any initial values, it is possible to find the shortest route corresponding to the changed topology. That is, when the distance-vector table is changed, each router updates the distance-vector table in turn according to the asynchronous and distributed-type Bellman-Ford algorithm, finally converging to a new distance-vector table that corresponds to the new topology.

An advantage of the distance-vector-type protocol is that, because each router can operate asynchronously, autonomously, and distributedly, it is enough to know only the connection relationship with the neighboring routers. Except for the distance-vector table, each router is not required to hold a high volume of data. (In the case of a link-state-type protocol described later, each router is required to hold a huge volume of topology information.)

A disadvantage of the distance-vector-type protocol is that, when a link fault occurs and a route to the special destination is completely shut down, it is not possible to recognize that the route is blocked. In such a case, the distance of the distance vector that is addressed to that destination will continue to increase infinitely. This problem is known as "counting to infinity." Several countermeasures have been proposed to resolve this problem. The most commonly utilized method is the one that sets an upper limit to the cost of a route within the network. When the cost of a route exceeds this limit, the network assumes that the cost is infinite and that the route to the destination is completely broken. Therefore, the distance-vector-type protocol is limited in that it cannot be applied to a large-scale network.

Figure 4.7 shows the change of distance to a destination address and the next hop of each source router against time. For example, the source router A has distance vectors (0, A), (1, B), and (2, B) for destinations A, B, and C, respectively, at time 0. Note that the terms in parentheses mean (distance, next hop) to the destination address. In Figure 4.7, we assume that the cost of all the links is 1. When the link AB goes down, router B assumes that there is a route to router A just after the fault, but, because router C notifies that it is possible to go to router A with a cost of 2, router B judges that it is possible to go to router A via router C with a cost of 3. Thus, at time 2, the distance vector of router B to router A becomes (3, C). Next, because router B notifies router C that it is possible to go to router A with a cost of 3, router C judges that it is possible to go to router A via router B with a cost of 4. Thus, at time 3, the distance vector of router C to router A becomes (4, B).

Internet Protocol (IP)

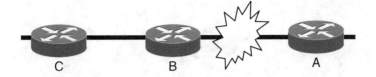

Time Src.	Dst.	0	1	2	3	4	5	6
A	A	(0,A)	(0,A)	(0,A)	(0,A)	(0,A)	(0,A)	(0,A)
	B	(1,B)	**(Inf,B)**	(Inf,B)	(Inf,B)	(Inf,B)	(Inf,B)	(Inf,B)
	C	(2,B)	**(Inf,B)**	(Inf,B)	(Inf,B)	(Inf,B)	(Inf,B)	(Inf,B)
B	A	(1,A)	**(Inf,A)**	**(3,C)**	(3,C)	**(5,C)**	(5,C)	**(7,C)**
	B	(0,B)	(0,B)	(0,B)	(0,B)	(0,B)	(0,B)	(0,B)
	C	(1,C)	(1,C)	(1,C)	(1,C)	(1,C)	(1,C)	(1,C)
C	A	(2,B)	(2,B)	(2,B)	**(4,B)**	(4,B)	**(6,B)**	(6,B)
	B	(1,B)	(1,B)	(1,B)	(1,B)	(1,B)	(1,B)	(1,B)
	C	(0,C)	(0,C)	(0,C)	(0,C)	(0,C)	(0,C)	(0,C)

(dist,nh)

Detect link down

FIGURE 4.7 "Counting to infinity" problem.

Following this pattern, the distance-vector table is exchanged repeatedly between router B and router C, with an ever-increasing cost to get to router A.

Because of the "counting to infinity" problem, the distance-vector-type protocol judges that, when a distance to the destination exceeds a certain fixed value, the distance is infinite and a route to that destination does not exist. If the distance to be assumed as infinite is increased, the time that the distance-vector table takes to converge becomes long. If the distance to be assumed as infinite is decreased, it limits the size of the network. Various methods have been proposed to shorten the time to converge a distance-vector table, including split horizon/poison reverse, hold down, dual-cost metrics, triggered update, and DUAL [10].

4.2.2.1.1 Split Horizon/Poison Reverse

"Split horizon" is a method that does not notify a distance vector to a destination address for a neighboring node that becomes the next hop for a certain destination. "Poison reverse" is a method that notifies a distance vector that has an infinite distance. Using these methods, the distance-vector table converges more promptly (see Figure 4.8). In the example of Figure 4.8, after the link AB goes down at time 1, router C does not notify the distance vector (2, B) to router A as is. The reason is because router B is the next hop of this distance vector. Instead, it notifies the distance

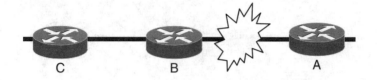

Src.	Dst.	Time 0	1	2 Detect link down	3	4	5	6
A	A	(0,A)	(0,A)	(0,A)	(0,A)	(0,A)	(0,A)	(0,A)
A	B	(1,B)	**(Inf,B)**	(Inf,B)	(Inf,B)	(Inf,B)	(Inf,B)	(Inf,B)
A	C	(2,B)	**(Inf,B)**	(Inf,B)	(Inf,B)	(Inf,B)	(Inf,B)	(Inf,B)
B	A	(1,A)	**(Inf,A)**	**(Inf,C)**	(Inf,C)	(Inf,C)	(Inf,C)	(Inf,C)
B	B	(0,B)	(0,B)	(0,B)	(0,B)	(0,B)	(0,B)	(0,B)
B	C	(1,C)	(1,C)	(1,C)	(1,C)	(1,C)	(1,C)	(1,C)
C	A	(2,B)	(2,**B**)	(2,B)	**(Inf,B)**	(Inf,B)	(Inf,B)	(Inf,B)
C	B	(1,B)	(1,B)	(1,B)	(1,B)	(1,B)	(1,B)	(1,B)
C	C	(0,C)	(0,C)	(0,C)	(0,C)	(0,C)	(0,C)	(0,C)

(dist,nh)

FIGURE 4.8 Split-horizon/poison-reverse method.

vector (Inf (infinite), B) to router B. From this, at time 2, the router knows that it is impossible to reach router A and changes the distance vector to router A to (Inf, B).

However, even if the split-horizon/poison-reverse method is used, it is not possible to resolve completely the counting-to-infinity problem, as seen in Figure 4.9. When using the distance-vector-type protocol, the network still must assume that, when the distance exceeds a certain value, the distance is infinite. In the example of Figure 4.9 at time 0, the distance vectors of routers B, C, and D to router A are (1, A), (2, B), and (2, B), respectively. Here, if we assume that link AB went down at time 1, then router B notifies routers C and D that the distance vector to router A became (Inf, A). Either router C or D, whichever one received this distance vector earlier than the other, changes the distance vector to router A. Here, we assume that router C received the distance vector earlier than router D. (If router D received it earlier than router C, the counting-to-infinity problem still occurs in the same way.) At time 2, router C changes the distance vector to router A to (3, D). That is, router C compares the link state (Inf, A) from router B to router A and the link state (2, B) from router D to router A, and it judges that it is possible to reach from router D to router A with two hops. This is because the link state (Inf, A) from router B to router A has not yet been notified to router D, so router D continues to notify the distance vector to router A as (2, B). Next, at time 3,

Internet Protocol (IP)

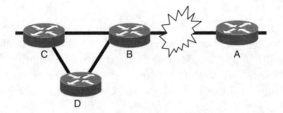

Time		0	1	2	3	4	5	6	7	8	9
Src.	Dst.			Detect link down							
A	A	(0,A)	(0,A)	(0,A)	(0,A)	(0,A)	(0,A)	(0,A)	(0,A)	(0,A)	(0,A)
	B	(1,B)	(Inf,B)	(Inf,B)	(Inf,B)	(Inf,B)	(Inf,B)	(Inf,B)	(Inf,B)	(Inf,B)	(Inf,B)
	C	(2,B)	(Inf,B)	(Inf,B)	(Inf,B)	(Inf,B)	(Inf,B)	(Inf,B)	(Inf,B)	(Inf,B)	(Inf,B)
	D	(2,B)	(Inf,B)	(Inf,B)	(Inf,B)	(Inf,B)	(Inf,B)	(Inf,B)	(Inf,B)	(Inf,B)	(Inf,B)
B	A	(1,A)	(Inf,A)	(Inf,A)	(Inf,A)	(4,C)	(4,C)	(4,C)	(7,C)	(7,C)	(7,C)
	B	(0,B)	(0,B)	(0,B)	(0,B)	(0,B)	(0,B)	(0,B)	(0,B)	(0,B)	(0,B)
	C	(1,C)	(1,C)	(1,C)	(1,C)	(1,C)	(1,C)	(1,C)	(1,C)	(1,C)	(1,C)
	D	(1,D)	(1,C)	(1,C)	(1,C)	(1,C)	(1,C)	(1,C)	(1,C)	(1,C)	(1,C)
C	A	(2,B)	(2,B)	(3,D)	(3,D)	(3,D)	(3,D)	(6,D)	(6,D)	(6,D)	(9,D)
	B	(1,B)	(1,B)	(1,B)	(1,B)	(1,B)	(1,B)	(1,B)	(1,B)	(1,B)	(1,B)
	C	(0,C)	(0,C)	(0,C)	(0,C)	(0,C)	(0,C)	(0,C)	(0,C)	(0,C)	(0,C)
	D	(1,D)	(1,C)	(1,C)	(1,C)	(1,C)	(1,C)	(1,C)	(1,C)	(1,C)	(1,C)
D	A	(2,B)	(2,B)	(2,B)	(Inf,B)	(Inf,B)	(5,B)	(5,B)	(5,B)	(8,B)	(8,B)
	B	(1,B)	(1,B)	(1,B)	(1,B)	(1,B)	(1,B)	(1,B)	(1,B)	(1,B)	(1,B)
	C	(1,C)	(1,C)	(1,C)	(1,C)	(1,C)	(1,C)	(1,C)	(1,C)	(1,C)	(1,C)
	D	(0,D)	(0,D)	(0,D)	(0,D)	(0,D)	(0,D)	(0,D)	(0,D)	(0,D)	(0,D)

(dist,nh)

FIGURE 4.9 Problem of split-horizon/poison-reverse method.

router D changes the distance vector to router A to (Inf, B). That is, router D compares the distance vector from router B to router A (Inf, A) and the distance vector from router C to router A (3, D). At this time, because the split-horizon/poison-reverse method operates, router D does not accept the distance vector from router C, but it does accept the distance vector from router B (Inf, A). Then, at time 4, router B changes the distance vector from router C to (4, C). That is, router B compares the distance vector from router C (3, D) and the distance vector from router D (Inf, B). Again, because the split-horizon/poison-reverse method is operating, router B does not accept the distance vector from router D, but it does accept the distance vector from router C (3, D). Further, at time 5, router D changes the distance vector to router A to (5, B). That is, router D compares the distance vector from router B (4, C) and the distance vector from router C (3, D). Here again, because the split-horizon/poison-reverse method is operating, router D does not accept the distance vector from router C, but it does accept the distance vector from router B (4, C). After this, the distance vector is repeatedly exchanged among routers B, C, and D, and the distance to router A continues to increase.

4.2.2.1.2 Hold Down

"Hold down" is a method that, when a downed link is detected, stops to switch the route for a certain time period. When a link that is directed to the destination goes down, the distance to the destination is advertised as infinite for a certain time period. That is, this method attempts to resolve the counting-to-infinity problem by waiting until this information reaches throughout the network.

4.2.2.1.3 Dual-Cost Metrics

"Dual-cost metrics" is a method to shorten the converging time when the link cost changes greatly, depending on the link. In the examples discussed so far, the link cost was assumed to be the same for all the links, and the number of hops was considered as the distance to the destination. But there is also a case where the link cost changes greatly, depending on the link. In such a case, it becomes a problem to determine what distance should be set as the threshold for assuming the cost to be infinite. To resolve this problem, a link cost that is used only for resolving the counting-to-infinity problem is introduced. A link cost for obtaining the shortest route is also used in parallel. By using the former link cost, it becomes possible to identify the unattainable destination.

4.2.2.1.4 Triggered Update and Periodic Notification

In the distance-vector-type protocol, the time to converge a distance-vector table differs greatly depending on whether a node updates its distance-vector table at constant intervals or, in addition, executes notification if it detects a faulty link to the neighboring nodes. The latter case can converge the distance-vector table faster than the former case. This method is called "triggered update."

4.2.2.1.5 DUAL

When a link to a neighboring node goes down, the DUAL method searches for another neighboring node that is closer to the destination node than the source node requesting a connection. In the example of Figure 4.10, it is assumed that the lowest-cost node from the source can reach the destination with a cost of 10 via node *j*. When a fault occurs in the lowest-cost route, the route must be switched. At this point, DUAL considers a detour to the neighboring node with a cost of 9, which is less than 10. This newly identified neighboring node is called a "feasible successor." If a feasible successor cannot be found quickly, the neighboring node searches further to seek other feasible successors.

4.2.2.2 Path-Vector-Type Protocol

The path-vector-type protocol creates a routing table by exchanging a path-vector table that includes the path information of each destination between each router and the neighboring router. Because the path information of each destination is written in the path-vector table, it is possible to prevent routing loops from occurring. In the case of Figure 4.11, there are three routes to destination A as viewed from node F: D-B-A, E-B-A, and E-C-A. These routes are called path vectors. By advertising the path vector and the destination in pairs, it is possible to detect whether a loop has

Internet Protocol (IP)

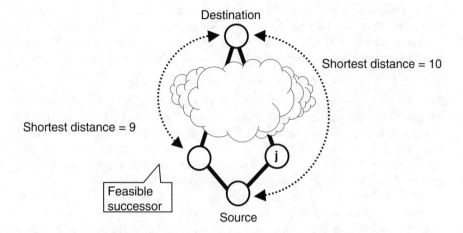

FIGURE 4.10 DUAL's feasible successor.

been created or not on the route. That is, it is possible to prevent routing loops from occurring by not advertising the route information to the node that is on the path vector.

4.2.2.3 Link-State-Type Protocol

The link-state-type protocol creates a routing table by exchanging each router's link state (link information of each router) to obtain the network topology. It then calculates the shortest route for each destination based on this information.

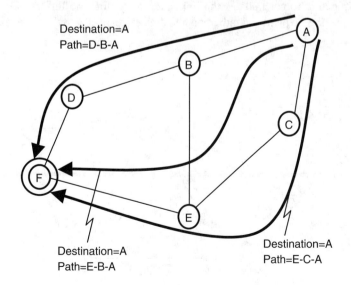

FIGURE 4.11 Path-vector-type routing protocol.

The important mechanisms in this link-state-type protocol are synchronization of the link states and an algorithm to calculate the shortest route. In this section, we first describe the synchronization of the link states in detail and then describe the "Dijkstra method," an algorithm for calculating the shortest route.

4.2.2.3.1 Synchronization of Link States via "Flooding"

The link state represents the relationship of connection between neighboring routers. Each router places the connection information with neighboring routers into a link-state packet and sends this to all of the routers within the network. At this point, because the routing table has not yet been created, the link-state packet cannot be transferred to the other routers using the conventional routing mechanism of IP packets. Thus, a method called "flooding" is used.

During flooding, each router resends the link-state packets received from a neighboring router to all of its neighboring routers. In this way, it is possible to send the link-state packets to the whole network, even though the routing table has not yet been created. To keep the number of link-state packets to a manageable level, not all of the packets received from the neighboring routers are "flooded" to the network; only the link-state packets that the router itself has newly taken in are "flooded" to neighboring routers. That is, the link states that have been received in duplicate are not taken in and are not executed as "flooding" (see Figure 4.12).

When a link state is changed, the originating router of the link state puts this fact into a link-state packet and executes "flooding" to all other routers within the network. For example, if a link goes down, the router puts this information into a link packet and executes "flooding." However, because the link state changes with time, there must be some way to distinguish the most recent link state. This is accomplished by assigning a sequence number to each link state. The originating router updates the link state every time the link state changes, increments the sequence number on the link-state packet by 1, and executes

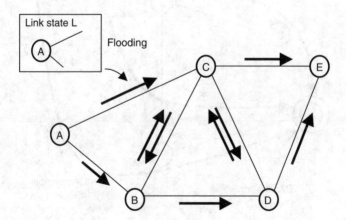

FIGURE 4.12 Concept of link state and "flooding."

Internet Protocol (IP)

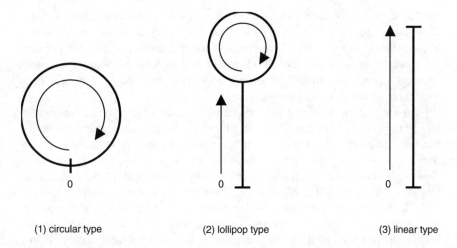

(1) circular type   (2) lollipop type   (3) linear type

FIGURE 4.13 Three types of sequence-number space for link state.

"flooding." A router that receives the link-state packet accepts the one whose sequence number is greater than the others as the latest link-state packet.

The design of the sequence-number space of the link state greatly affects the robustness of the link-state protocol. The sequence number may be destroyed while it is stored in the memory of the router or while being transferred on the link during the process of "flooding" to the whole network. If routing is continued without noticing that the link state has been destroyed, a routing loop or a black hole may occur. Therefore, when we use the link-state-type protocol, it is vital that the link state be synchronized perfectly.

The sequence-number space for the link state can be either circular type, lollipop type, or linear type (see Figure 4.13). In the case of a circular type, it is not possible to distinguish between large or small sequence numbers unless all of them exist in the half of the circle. If three sequence numbers are distributed as shown in Figure 4.14, because it is impossible to judge which is

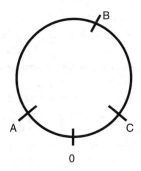

FIGURE 4.14 An example where it is not possible to judge the order of sequence numbers.

the latest link state, each time a router receives a link state it assumes that it is the latest one, accepts all the link states successively, and executes "flooding" to other neighboring routers. This type of "flooding loop" caused a big problem in the early stage of protocol development in ARPANet (Advanced Research Projects Agency Network), and the network could not operate normally until the router software was corrected.

As a countermeasure to this problem, a lollipop-type numbering system has been proposed (Figure 4.13). In this case, there is no problem when the sequence numbers are on the handle part of the lollipop, but when the sequence numbers enter the circle of the lollipop, the system is prone to the same flooding-loop problem as the circular-type numbering system. Such problems can be avoided by using a linear-type numbering system, which is used in most link-state-type protocols. In the sequence-number space of the linear-type numbering method, when the sequence number reaches some specified maximum value, all of the routers discard their link states, and the originating router executes "flooding" of the link state using a sequence number of some specified initial value.

4.2.2.3.2 Dijkstra's Algorithm

Dijkstra's method is an algorithm to calculate the lowest-cost route. All the links have a positive value. It is easy to understand this algorithm when we imagine that the cost tree is extended successively starting from the originating node. The concept of Dijkstra's algorithm is as follows. First, it seeks the node with the lowest-cost route, then it seeks the next node with the lowest-cost route, and again it seeks the next node by repeating these processes until the lowest-cost route to the destination node is selected. That is, at first, a node that has the lowest-cost route among the nodes that can reach the destination with one hop is selected (this is assumed as node A). In this way, node A, which has a minimum lowest-cost route to the destination, is determined. Then, the next-lowest-cost node is selected from the nodes except for node A and the nodes neighboring node A. In this way, it is possible to select the node with the next-lowest-cost route to the destination compared with node A. This procedure is repeated until all the nodes have been tried.

Assuming that D_i has the lowest cost from node i to the destination node, and that Q is a group of nodes of which the lowest-cost route has not yet been determined, Dijkstra's method seeks node i of which D_i is the minimum from group Q. It then excludes this from group Q and updates D_j for another node j belonging to group Q by applying $D_j = \min(D_j, d_{j,i} + D_i)$. This process is repeated until group Q becomes empty. The algorithm operates properly because all of the link costs have a positive value. That is, once the node with the lowest cost has been obtained, it is not possible to find a shorter route by taking another new link. In the example shown in Figure 4.15, the nodes that can reach the destination with 1 hop are A, B, C, and D, and these can be reached with a link cost of 1, 3, 4, and 3, respectively. Therefore, node A is first determined. At this point, the node group Q for which the lowest-cost route has not yet

Internet Protocol (IP)

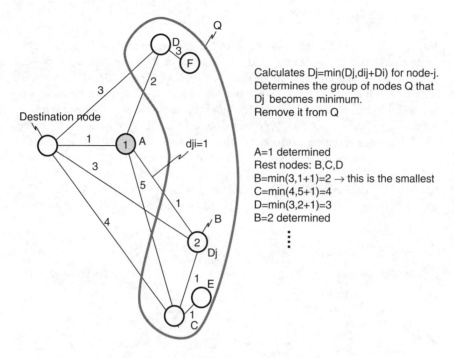

FIGURE 4.15 Example of executing Dijkstra's algorithm.

been determined becomes {B, C, D}. In this group Q, the node with the minimum cost to the destination is selected from the nodes that can reach the destination with 1 hop and the nodes that can reach node A with 1 hop. In this case, the link cost from node A to node B is 1, and the cost from the destination to node B is 2. Because this is smaller than the cost from destination node to the other candidate nodes, node B is determined. This process is repeated until group Q becomes empty.

4.3 EXAMPLE OF ROUTING PROTOCOL

4.3.1 OSPF

4.3.1.1 Principle

OSPF (Open Shortest Path First) [12, 13] is a link-state-type routing protocol that is used as IGP (Interior Gate Protocol). This protocol advertises the link information that each router directly holds to other routers and creates the topology database of the whole network. Further, it calculates the shortest route to the destination using Dijkstra's algorithm based on this topology database.

4.3.1.2 Link State

The link state expresses the information of the link that extends from each router. As an identifier to recognize the link, the identifier of the router that exists on the opposite side of the link is utilized. In the example of Figure 4.16(a), router A is connected to routers B, C, and D by a point-to-point link. Router A expresses each link with identifiers of B, C, and D. Thus, the link state that is advertised by router A becomes {B, C, D}. A link state that advertises in this way is called a "router LSA" (link-state advertisement). On the other hand, in the example of Figure 4.16(b), when we see router A through subnetwork (is the layer-2 network such as Ethernet and token-ring, for example) in the IP layer, the neighboring router to router A is B, C, and D. At this time, the link state of network is advertised as {AS, B, C, D}, and each router advertises router LSA, which shows that the list of links originating

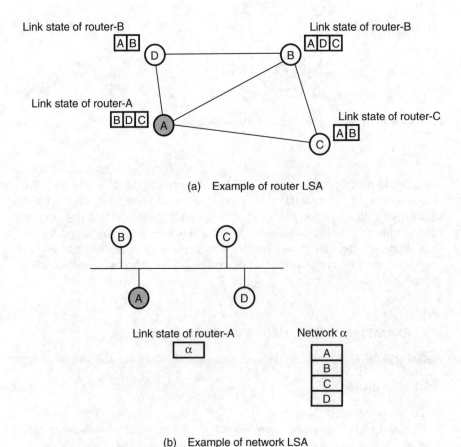

FIGURE 4.16 Example of advertising the link state.

from it. The link state of the network advertised is called a "network LSA," and the router that advertises the network LSA of network is called the "designated router" (DR). In this way, it is possible to reduce the number of link states to be advertised by advertising the network LSA. That is, in the case of Figure 4.16(b), because router A has a neighboring relationship with other routers, if router LSA is advertised by such a method as in Figure 4.16(a), router LSA is advertised as {B, C, D}. In the same way, if other routers also advertise the link state of the routers that have a neighboring relationship, and if the N routers are connected to the network at the same time, it is required to advertise $N(N-1)$ link states. But, as described above, it is possible to reduce the number of advertised link states to as low as $(N+1)$, where $N+1$ means N = router LSA and 1 = network LSA.

By flooding the link state, each router builds a topology database. When the size of the network using OSPF becomes large, problems of scalability make it difficult to manage the topology of the whole network in each router. In this case, the network is divided into multiple areas, and only the topology within each of these areas is managed, with the other areas being managed using only the summarized network information.

4.3.1.3 Scalability and Hierarchization

In a link-state-type protocol, the link information is "flooded" throughout the whole network. Thus, when the network size becomes large, the number of link states to be handled increases. In OSPF, scalability is maintained by introducing two hierarchical layers, and the network is divided into multiple areas. One area — called the "backbone area" — is selected from these divided areas to serve a special role. The backbone area and the other areas are connected in the form of a hub and spoke. That is, all of the other areas are connected to the backbone area with a direct link.

Within each area, each node advertises all of the link states and grasps the complete topology within the area. Advertising of network addresses between areas is executed via the backbone area. At the area border router (ABR) between areas, the link state is bundled for advertising, but the topology information is not advertised outside the area. It is possible to further reduce the number of link states advertised by the ABR by combining addresses.

When the link state is "flooded" over the area, it is flooded according to the idea of the distance-vector-type protocol. A network address that belongs to a certain area is first advertised to the backbone area; then, it is readvertised from the backbone area to the other areas. At this point, because the topology information within each area has been hidden, advertising is executed according to the distance vector. If there are multiple routes for a certain network address, the distances of the various routes to the address are calculated from the sum of distance vectors, and the smallest one is selected (see Figure 4.17).

Figure 4.17 shows how the route from area 0.0.0.1 to node A is calculated. Between area 0.0.0.1 and 0.0.0.0 (backbone area) are nodes B and C (both functioning as ABRs). Within area 0.0.0.0, the lowest-cost route from node A to node B takes cost $c1$, and the route from node A to node C takes cost $c2$. Node B,

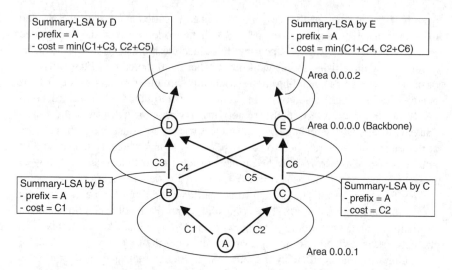

FIGURE 4.17 Flooding between areas.

which is an ABR, uses a summary LSA to advertise that it is possible to reach node A with a cost c1 over the area 0.0.0.0. Node C, also an ABR, uses a summary LSA to advertise that it is possible to reach node A with a cost c2 over the area 0.0.0.0. When this summary LSA reaches the ABR between area 0.0.0.2 and area 0.0.0.0, the ABR readvertises it over the area 0.0.0.2. To node D, which is an ABR, routing is executed within area 0.0.0.0 from nodes B and C with a cost of c3 and c5, respectively. Therefore, when node D readvertises the summary LSA to destination A over the area 0.0.0.2, it advertises by selecting the route with lower cost. In this case, it compares c1+c3 (node B) and c2+c5 (node C). If the former is larger, node D selects the route that node B has advertised; otherwise, it selects the route that node C has advertised. Note that when the summary LSA is readvertised over the area 0.0.0.2, a distance-vector-type route calculation is also executed. The respective topologies of area 0.0.0.0 and area 0.0.0.2 cannot be seen from outside these areas, so the route that is selected is a combination of the shortest route with the lowest cost. This method is the same as that used in the distance-vector-type protocol, which changes its own distance vector to correspond to the lowest-cost distance vector that is transferred from the neighboring router.

The process in the ABR node E is the same as that in node D. To node E, routing is executed within area 0.0.0.0 from nodes B and C with a cost of c4 and c6, respectively. Therefore, when node E readvertises the summary LSA to destination A over the area 0.0.0.2, it advertises by selecting the route with lower cost. In this case, it compares c1+c4 (node B) and c2+c6 (node c). If the former is larger, node E selects the route that node B has advertised; otherwise, it selects the route that node C has advertised.

Internet Protocol (IP)

4.3.1.4 Aging of LSA (Link-State Advertisement)

A link state is distinguished from other link states by the LS-age field of the LSA common header, the LS sequence-number field, and the LS-checksum field. Whenever a new link state is generated from an originating router, the LS sequence-number field is incremented, and an LSA-update packet is advertised. In OSPF, the same link state can exist in the network until it exceeds some predefined time limit. If there is no change in the content of the LSA within this time limit, the LSA is refreshed and advertised again, and the sequence number is incremented by 1. By default, the link state is advertised every 30 min (or some other time as defined by the variable "LS refresh time"). If the link state is not updated after 1 h (or some other time as defined by the variable "max age"), the link state is deleted from the database of each router.

The LS-checksum field includes the checksum of the link state, which is calculated by using Fletcher's algorithm. Once an LSA instance has been created, this value is never changed. Therefore, the LS-age field is excluded from the objective of checksum.

4.3.1.5 Content of LSA

The LSA has a common header (see Figure 4.18) and is distinguished by three fields: advertising router, LS type, and link-state ID. The advertising router identifies the router that has created the relevant LSA. The LS type indicates the type of LSA. For example, type 1 indicates that the LSA is a router LSA, and type 2 indicates that the LSA is a network LSA. The link-state ID is used to

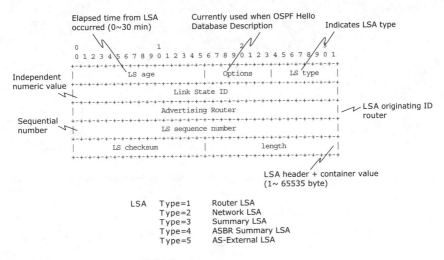

FIGURE 4.18 Structure of LSA header.

distinguish the LSA that has been advertised by the relevant advertising router. The value used for this link-state ID has been fixed for each LS type.

LS age and LS sequence number are used to establish synchronization by judging which of the same LSAs is the most recent. When an advertising router creates the LSA, an initial value 0 is set to LS age, but when the LSA is advertised and delivered to each router, each router executes aging of LS age. The time unit of LS age is seconds. LSA is specified to re-create the advertising router at every 30 min, even if there is no change in its content. This is called a "refresh of LSA." When LS age exceeds the Max Age (maximum age, set to 60 min by default), the router that detected this advertises the LSA as an LSA of Max Age before deleting it from the link-state database. Other routers can also delete this LSA. And, even if the LS age does not actually exceed the Max Age, it is possible to delete the LSA from the network by advertising LSA using Max Age for LS age. This process, called "premature aging," cannot be executed by anyone other than the advertising router.

Type 1 is called a router LSA, which indicates the connection relationship to the neighboring routers (see Figure 4.19). To the link-state ID field, the same value as in the Advertising Router field is entered. To the #links field, the number of links coming

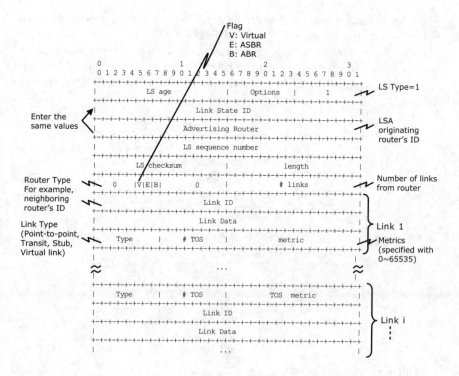

FIGURE 4.19 Structure of router LSA (type 1, for routers connected point-to-point).

Internet Protocol (IP)

from the router is entered. Link information time's number of links are entered to the LSA header. Link information — link ID, link type (point-to-point, transit, stub, virtual link), and metrics (specified with a value of 1 to 65535) — is entered into the router structure. The link-ID field contains information about the node to which this link is connected. Therefore, the link ID differs depending on the link type. When the link type is point-to-point, the router ID of the neighboring router is entered. When the link type is "transit," the router ID of the designated router (DR) is entered. When the link type is "stub," the IP subnet number is entered. And when the link type is "virtual link," the router ID on the other end of the link is entered.

Type 2 is called a network LSA, which indicates the connection relationship to the subnets when multiple routers are connected to a broadcast-type link (e.g., Ethernet) (see Figure 4.20). To the link-state ID field, the interface address of the DR is entered. The network mask and router ID of the router that is connected to network are also recorded in this network LSA.

Type 3 is called a summary LSA and is used when an OSPF-operated network is divided into multiple areas and hierarchized. As the size of the OSPF-operated network increases, a scalability problem arises, as each router manages the whole network topology. To overcome this problem, the network is divided into multiple areas, with each area managing its own topology and advertising only the summarized network information to other areas. It is the summary LSA that uses this summarized network information for advertising (see Figure 4.21). ABR creates this summary LSA. To the link-state ID, the network address is entered. When the address is ASBR, a type-4 summary LSA is used. In this case, the router ID of ASBR is set to the link-state ID.

Type 5 is called an AS-external LSA and is used when packets are injected to OSPF taking the route that is advertised from BGP, etc. as an external route.

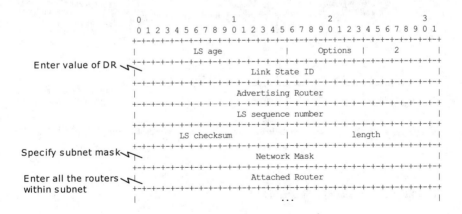

FIGURE 4.20 Structure of network LSA.

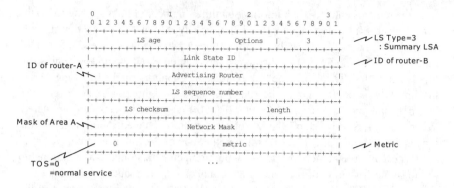

FIGURE 4.21 Summary LSA from router A to arena X.

Figure 4.22 shows the model of AS-external LSA. In this case, router A is an ASBR and advertises 10.0.0.0/8 of BGP into OSPF. In this way, the packets addressed to 10.0.0.0/8 are transferred outside the AS via router A. In this LSA, only the type-5 LSA that is advertised to the entire AS that OSPF is operating is executed as "flooding" (types 1, 2, and 3 execute "flooding" only within the area) (see Figure 4.22). The link-state ID is an IP address of the external route. To indicate which ASBR has injected this external route, an ASBR summary LSA (Figure 4.23) of type 4 is used when information that router A receives from BGP is transferred to area B, as seen in Figure 4.24.

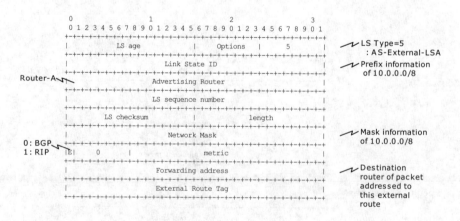

FIGURE 4.22 Format of AS-external LSA.

Internet Protocol (IP)

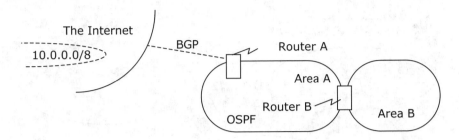

FIGURE 4.23 ASBR summary LSA.

4.3.2 BGP-4

4.3.2.1 Principle

BGP-4 [14–16] is an EGP that is a path-vector-type protocol. It exchanges route information (attainable destination address) assuming AS (autonomous system) as a node. It transfers a list of AS numbers on the way of the route (path list) together with the route information (see Figure 4.25). This is done to avoid the occurrence of routing loops. When an AS receives the route information advertised from a neighboring AS, it accepts the destination information only when its own number does not exist in the list of AS numbers that was advertised together with the route information. In other cases, because a routing loop is generated, it does not accept the route information that was transferred from a neighboring AS. In Figure 4.25, when AS3 is advertised the route information to 10.1.1.1 from AS4, AS3 notifies it to AS1. At the same time, to notify that it is possible to reach 10.1.1.1 via AS3 and AS4, it advertises the path vector AS3-AS4 to AS1 by attaching it to the route information of 10.1.1.1. In the same way, when AS1 advertises this route information to AS2, AS1 advertises the route information 10.1.1.1 attaching the AS1-AS3-AS4 as a path vector.

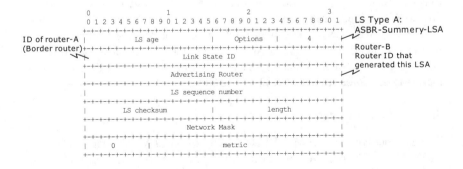

FIGURE 4.24 Data-injection model to OSPF of router A from BGP.

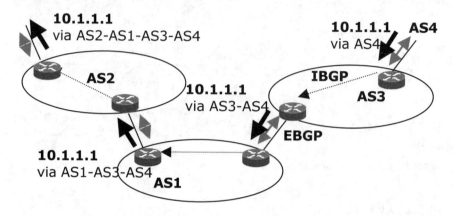

FIGURE 4.25 Concept of operation of BGP-4 (path-vector-type routing protocol and loop detection).

It is also possible to assign attributes to route information to reflect various operation policies. BGP-4 is required to handle a vast amount of route information. In BGP-4, the same route information is exchanged just one time using TCP (Transmission Control Protocol), and is not exchanged repeatedly many times.

Not all of the routers within AS need to speak BGP-4; only the border router located at the border to neighboring ASs is required to speak BGP-4. A TCP session is built between border routers, and route information is exchanged during this session. A border router that receives route information advertised from a neighboring AS advertises this information to all of the border routers within its AS. A session between border routers belonging to different ASs is called an EBGP (Exterior Border Gateway Protocol) session, and a session between border routers belonging to the same AS is called an IBGP (Interior Border Gateway Protocol) session. In Figure 4.26, the border router (BGP speaker) of AS1 has established the EBGP session with the border router of AS3. When this border router of AS1 receives the route information from the border router of AS3 via this EBGP session, it then relays the route information to the other border routers of AS1 via IBGP session. The other border routers of AS1 then notify the route information notified via IBGP session to the border router of neighboring AS. For example, the border router of AS1 neighboring AS2 notifies the route information to the border router of AS2 via EBGP session.

4.3.2.2 BGP Message

BGP has four types of message: UPDATE, OPEN, KEEPALIVE, and NOTIFICATION. The OPEN message is used when BGP peers start to build a session together. OPEN notifies such information as a version number of BGP protocol, the AS number of the peer, and a "hold timer" to the peer to establish the BGP session (see Figure 4.27). It is also possible to execute authentication as an option, for

Internet Protocol (IP)

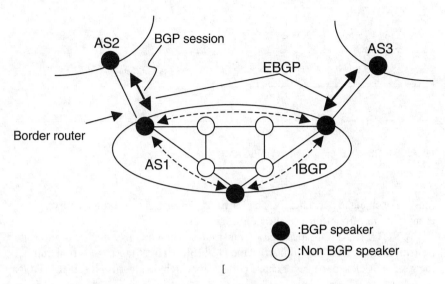

FIGURE 4.26 Border router (BGP speaker) and BGP session.

example, when the BGP version or AS number are invalid, it does not establish a session with the peer. The hold timer uses the minimum of the time values that were exchanged with the peer, and it must be longer than 3 sec. If the holder timer were 0, session management using a KEEPALIVE message would not be executed, and the session would always be considered as being up. If the session is not established, the fact and the cause are notified with a NOTIFICATION message.

KEEPALIVE is used to confirm the mutual connectivity of BGP peers periodically. If no UPDATE message or KEEPALIVE message is exchanged during the time specified by the hold timer, it is assumed that the BGP session has been disconnected. If the BGP session is to be maintained, a KEEPALIVE message

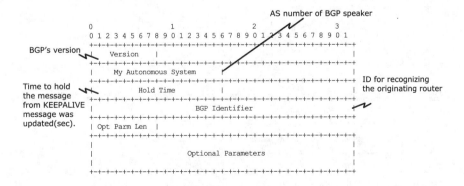

FIGURE 4.27 BGP message (OPEN message).

FIGURE 4.28 BGP message (NOTIFICATION message).

must be exchanged periodically. The KEEPALIVE message is typically exchanged at an interval of one-third of the hold timer value.

A NOTIFICATION message is transmitted when an error occurs in the state transition of the BGP session. After the NOTIFICATION message is transmitted, the session is closed. An error cause code is also included in this NOTIFICATION message (see Figure 4.28).

The UPDATE message is used to exchange the route information. Route information and path attributes of unattainable and attainable routes are included in this message (see Figure 4.29). It is possible to include the information on multiple unattainable routes and attainable routes. The path attribute is applied to all of the attainable routes included in a single UPDATE message.

The BGP speaker executes state management of the BGP session while exchanging BGP messages. The state-transition machine of a BGP session consists of six states: Idle, Connect, Active, OpenSent, OpenConfirm, and Established (see Figure 4.30). The Idle state is the state just after the BGP speaker is started up,

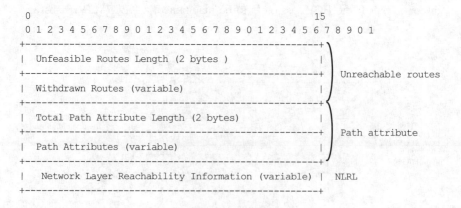

FIGURE 4.29 BGP message (UPDATE message).

Internet Protocol (IP)

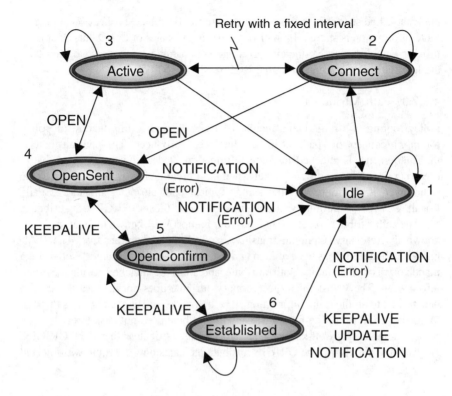

FIGURE 4.30 State transition of BGP session.

which is a state until a TCP session for the BGP session is established between peers. The Idle state transits to the Connect state when the TCP session is established. If a TCP session is not established within a certain time period, it transits to the Active state and retries to establish a TCP session within a certain time interval. When a TCP session is established at the current state or the Active state, it transmits the OPEN message to peers using the TCP session, and at the same time, it transits to the OpenSent state. When it receives an OPEN message from a peer in this state, it transits to the OpenConfirm state. When it judges that it is possible to establish a BGP session by comparing the OPEN message that it has transmitted to a peer with the content of the OPEN message that it received from a peer, the state transits to the Established state. When the state moves to the Established state, exchanging of route information is started. If an error occurs in the states of OpenSent, OpenConfirm, or Established, a Notification message is transmitted, and the state transits to the Idle state after disconnecting the BGP session.

4.3.2.3 Path Attributes

Policy routing is a remarkable feature of BGP. To realize this feature of policy routing, a concept of "path attributes" has been introduced. The path attribute is information that is attached to route information. It determines the route that is used by priority for routing a packet when a different path attribute is given to the route, even if the route was the same. Typical path attributes include ASPath, LocalPref, MED, Community, etc. Figure 4.31 shows the list of BGP path attributes.

The path attribute is categorized into four groups: Well-known mandatory, Well-known discretionary, Optional transitive, and Optional nontransitive. The Well-known path attribute has to be able to understand all the BGP speakers. Well-known mandatory is a mandatory path attribute and must be attached to all the route information. The Well-known discretionary attribute does not necessarily have to be attached to all the route information. The transitive path attribute passes through to another BGP peer. The nontransitive path attribute does not pass through.

As mandatory (Well-known mandatory) path attributes, there are ORIGIN, ASPath, and NextHop. The ORIGIN attribute indicates how the route was injected

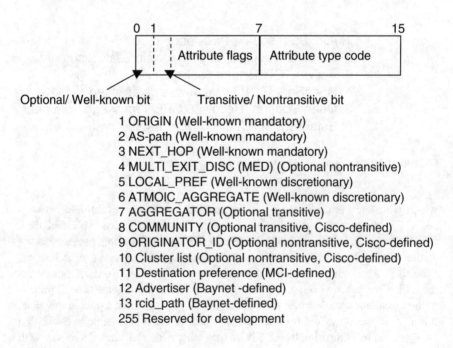

FIGURE 4.31 BGP path attributes.

Internet Protocol (IP)

to BGP at the originating AS of the route, and it takes a value such as IGP, EGP, Incomplete (static route), etc. The ASPath attribute indicates the path vector to the route, and the number of the AS through which the packets pass to reach the route is set as this attribute. When the number of the AS itself is included in the ASPath attribute of route information that was received via the EBGP session, that route is not taken in to avoid generating a routing loop. The NextHop attribute indicates the address of the router to which the packet is transmitted to reach that route.

One of the Well-known discretionary path attributes is LocalPref. The LocalPref attribute controls the outbound traffic, and it attaches a value to the route and executes routing using route information with a higher value of LocalPref for the same route (see Figure 4.32). LocalPref is valid only within the same AS and does not exchange the value between BGP peers of other ASs. Figure 4.32 shows a case where a network address of 128.213.0.0/16 is in AS3. If the border routers R11 and R12 set the LocalPref attribute relating to this route to 200 and 300, respectively, when advertising this route to AS1, the traffic from the other routers within AS1 to the address of 128.213.0.0/16 is routed by using R12 because it has a larger value of the LocalPref attribute. Here, if R12 went down, a route to R11 would be used.

The MED (multiple exit discriminator) attribute is used to control the inbound traffic. When two ASs (AS1 and AS2, for example) are connected at multiple points (BGP speaker), this attribute is used to notify which BGP speaker is to be used when AS1 accepts the inbound traffic to the counterpart AS2. The counterpart AS2 executes routing by using route information having a smaller value of the MED attribute. In Figure 4.33, it is assumed that there are border routers R11 and R12 in AS1 and that the network address of 128.213.0.0/16 is located at a point closer to R11. At this time, each border router R11 and R12 in AS1 advertises the address 128.213.0.0/16, setting the MED value to 120 and 200,

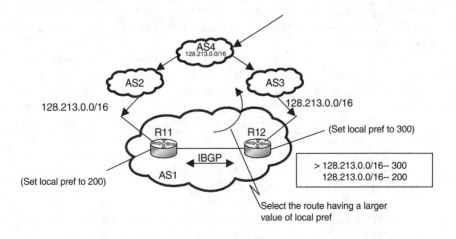

FIGURE 4.32 Routing control of outbound traffic using LocalPref attribute.

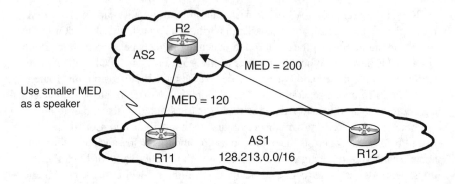

FIGURE 4.33 Routing control of inbound traffic using MED attribute.

respectively. Then, the border router of AS2 handles the route information from R11 by priority. By doing this, AS1 receives traffic addressed to 128.213.0.0/16 from the border router R11, which is closer to that destination address. By using an IGP metric within AS1 of that route as an MED value, it is possible to execute routing using a BGP speaker that is closer to the destination address within AS1 (see Figure 4.33). MED is an Optional nontransitive attribute.

Community is a tag that is attached to route information to execute the routing policy. By using this tag, it is possible to apply the same routing policy to multiple routes. Applying the policy using this tag (Community attribute), rather than manually applying the same policy to individual route information, makes the policy-setting work easier, and it is possible to attach multiple Community attributes to one route. The Community attribute is four octets, and usually the AS number is used for the upper two octets. The Community attribute of which the upper two octets is 00h, FFh has been reserved, and using the No-Export (FFF1h), No-Advertise (FFF2h), or No-Export-Subconfed (FFF3h), etc. is limited as a Well-known Community. Community is an Optional nontransitive attribute.

Figure 4.34(a) shows a case where AS1 is advertising three routes: two routes—129.41.5/17 and 129.41.100/17—with a prefix length of 17 and route 129.41/16, which combines the other two routes with a prefix length of 16. As for the route with a prefix length of 17, AS1 wants to receive the traffic from AS2 using different links (Link 1 and Link 2 in the figure). Such route control can be executed in the event that 129.41.5/17 exists close to Link 1 and 129.41.100/17 exists close to Link 2. However, in this example, there is no need to distinguish the difference of routes with the prefix length of 17 when entering from AS3 to AS2. Therefore, when advertising these routes from AS2 to AS3, it is better not to advertise the route with a prefix length of 17, but to advertise only the route with a prefix length of 16. In such a case, the route with a prefix length of 17 is advertised from AS1 to AS2 attaching a tag called No_Export using the Community attribute. At the same time, the route with a prefix length of 16 is

Internet Protocol (IP)

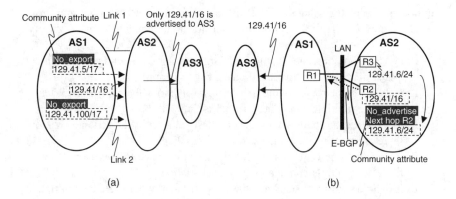

FIGURE 4.34 Application of routing policy using a Community attribute.

advertised from AS1 to AS2 without attaching such a tag. AS2 advertises only the route with a prefix length of 16 and does not advertise the route with a prefix length of 17 to AS3. By doing this, it becomes possible to execute policy route control that, while executing fine route control between AS1 and AS2, reduces the number of routes to be advertised between AS2 and AS3.

In Figure 4.34(b), AS1 and AS2 are connected with a LAN, and only R1 of AS1 and R2 of AS2 are exchanging the route information via EBGP on this LAN. Assume that the R2 is a destination router of 129.41/16 and that R3 is a destination router of 129.41.6/24. In this case, R2 advertises the route 129.41.6/24 to R1 using a NextHop indicating that R1 is the next hop. In this way, R2 can transfer the packets directly to R1 without passing through R2 for route 129.41.6/24. R1 advertises this route to other BGP routers within AS1 using an IBGP session. However, the route 129.41.6/24 that was notified from R2 to R1 using the NextHop is important for R1, but it is not important for other BGP routers within AS1. Therefore, it is better not to notify this route to other BGP routers within AS1. In such a case, a tag called "No_Advertise" is attached to the route 129.41.6/24 advertised by the NextHop using the Community attribute so that R1 does not notify this route to the other border routers within AS1. R1 does not advertise the route to which this tag has been attached to the other BGP routers within the same AS during their IBGP sessions.

The difference between Figure 4.34(a) and Figure 4.34(b) is whether or not the route is advertised within an AS. In discussing the two examples in Figure 4.34, a key point was to demonstrate the ease of realizing a complex policy control by using the Community attribute.

4.3.2.4 Rules of Route Selection

When receiving the route information on the same route from multiple BGP peers using BGP, certain rules are required to select the optimum route from the multiple-route information. The rule of route selection is based on attributes indicating the path attribute of BGP, as described previously. An optimum route is selected according to the following six rules in descending order of priority.

Rule 1: Select the route having the largest value in LocalPref attribute. When there are multiple candidates, rule 2 is applied.
Rule 2: Select the route having the shortest value in ASPath attribute. When there are multiple candidates, rule 3 is applied.
Rule 3: Select the route having the smallest value in MED attribute. When there are multiple candidates, rule 4 is applied.
Rule 4: Select the route that is the closest in distance to NextHop. When there are multiple candidates, rule 5 is applied.
Rule 5: Select the route obtained via EBGP session. When there are multiple candidates, a route with the smallest BGPID is selected. If all of the candidates have been obtained via IBGP session, rule 6 is applied.
Rule 6: If all of the routes have been obtained via IBGP session, the route with the smallest BGPID is selected.

4.3.2.5 IBGP and EBGP

There are two types of BGP sessions: EBGP and IBGP. An EBGP session is a session between BGP speakers in different ASs; an IBGP session is a session between BGP speakers within the same AS. Route information obtained from other ASs via an EBGP session is transferred to other BGP speakers within the same AS via an IBGP session, and that BGP speaker transfers the route information to other ASs via an EBGP session. In this way, route information is transferred over multiple ASs (see Figure 4.35). The rule of advertising the route information via BGP sessions differs between IBGP sessions and EBGP sessions. Route information received via an EBGP session is advertised to other BGP peers via an EBGP session or an IBGP session. Route information received via an IBGP session is advertised to other BGP peers via an EBGP session, but it is never advertised to other BGP peers via multiple IBGP sessions (see Figure 4.36). This rule is results from the fact that the BGP is a path-vector-type routing protocol. That is, when detecting the routing loop using the ASPath attribute, it

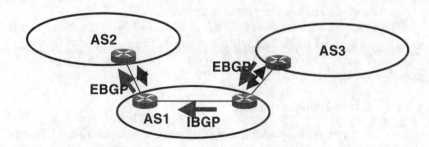

FIGURE 4.35 Concept of transferring route information over multiple ASs.

Internet Protocol (IP)

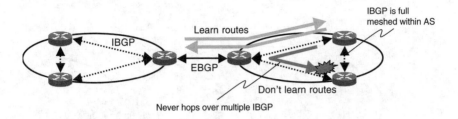

FIGURE 4.36 Rule of advertising via IBGP session and EBGP session.

judges that a routing loop has occurred if the same AS number exists in the ASPath attribute. If the route information received via an IBGP session were to be advertised via an IBGP session, the same AS number would appear in the same ASPath, and it would become impossible to detect a routing loop. For this reason, route information received via an IBGP session is inhibited from being advertised further via other IBGP sessions.

4.3.2.6 Scalability

The rule that inhibits advertising the route information obtained via an IBGP session further via other IBGP sessions means that the IBGP session must be set up in full mesh. That is, the BGP speaker must directly establish the IBGP session with all other BGP speakers within the same AS. If we assume that the number of BGP speakers within the same AS is N, then the total number of IBGP sessions is given as $N(N-1)/2$, and it causes a scalability problem (Figure 4.37). To resolve this problem, two methods have been proposed. The first is called the route-reflector method, and the other is called the confederation method.

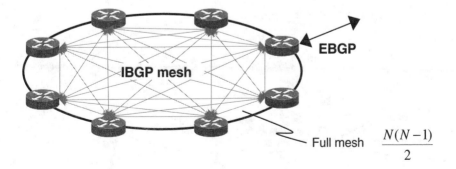

FIGURE 4.37 Scalability problem.

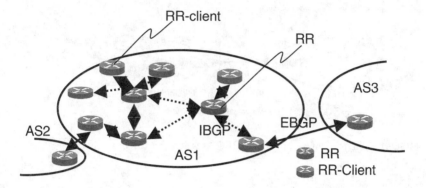

FIGURE 4.38 Route reflector.

4.3.2.6.1 Route Reflector

Route reflector is a method that resolves the scalability problem by introducing the concept of hierarchy in IBGP (see Figure 4.38). BGP speakers are categorized into two types: route reflector (RR) and route-reflector client (RR-Client). RR-Client operates in the same way as a BGP speaker. The RR-Client takes a certain RR as a parent and establishes an IBGP session with only that parent. It depends on the parent RR to advertise the route information that the RR-Client obtains by itself within an AS or to obtain the route information from another BGP speaker within the same AS. That is, route information that the RR-Client itself obtains is advertised to all of the parent RRs, and the route information from other BGP speakers within the same AS is obtained through the parent RR. On the other hand, RR establishes an IBGP session in full mesh, and at the same time, it establishes IBGP sessions between all the RR-Clients whose parent is the RR itself. The RR transfers the route information obtained via IBGP session with the other RRs only to the RR-Client whose parent is the RR itself. And the RR transfers the route information that was obtained from

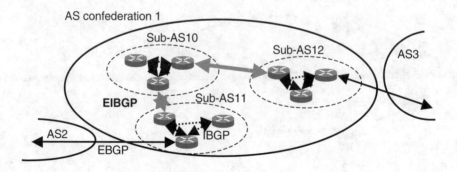

FIGURE 4.39 Confederation.

its RR-Client to other RR-Clients whose parent is the RR itself as well as to all the RRs.

4.3.2.6.2 Confederation

Confederation is a method to resolve the scalability problem by distributedly governing the BGP speakers within an AS (see Figure 4.39). The AS is divided into multiple SubASs, and each SubAS functions like a single AS. Within a SubAS, IBGP speakers are connected in IBGP full mesh, but SubASs are connected mutually with an EIBGP (External Interior Border Gateway Protocol) session. Although the LocalPref and NextHop attributes are Optional nontransitive path attributes, they make it possible to pass the boundary of SubAS. Such attributes as AS-confed-set or AS-confed-sequence are introduced to prevent a routing loop from occurring between SubASs.

REFERENCES

1. Baker, F., Requirements for IP Version 4 Routers, RFC 1812, June 1995; available on-line at http://www.ietf.org/rfc/rfc1812.txt.
2. Stevens, W.R., *TCP/IP Illustrated,* Vol. 1, Addison-Wesley, Reading, MA, 1994.
3. Rekhter, Y. and Li, T., An Architecture for IP Address Allocation with CIDR, RFC 1518, Sep. 1993; available on-line at http://www.ietf.org/rfc/rfc1518.txt.
4. Fuller, V., Li, T., Yu, J., and Varadhan, K., Classless Inter-Domain Routing (CIDR): an Address Assignment and Aggregation Strategy, RFC1519, Sep. 1993. available on-line at http://www.ietf.org/rfc/rfc/1519.txt.
5. Degermark, M., Brdnik, A., Carlsson, S., and Pink, S., Small forwarding tables for fast routing lookups, in *Proc. ACM SIGCOMM î97,* p.p. 3–14, Cannes, France 1997.
6. Waldvogel, M., Varghese, G., Turner, J., and Plattner, B., Scalable high speed IP routing lookups, in *Proc. ACM SIGCOMM î97,* p.p. 25–36, Cannes, France 1997.
7. Gupta, P., Lin, S., and McKeown, N., Routing lookups in hardware at memory access speeds, in *Proc. IEEE INFOCOM î98,* IEEE, p.p. 1240–1247, Piscataway, NJ, 1998.
8. Uga, M. and Shiomoto, K., A fast and compact longest match prefix look-up method using pointer cache for very long network address, in *Proc. IEEE ICCCN î99,* IEEE, p.p. 595–602, Piscataway, NJ, 1999.
9. Huitema, C., *Routing in the Internet,* Prentice Hall, New York, 1995.
10. Perlman, R., *Interconnections: Bridges, Routers, Switches, and Internetworking Protocols,* 2nd ed., Addison-Wesley, Reading, MA, 2000.
11. Bertsekas, D. and Gallager, R., *Data Networks,* 2nd ed., Prentice Hall, New York, 1992.
12. Moy, J., *OSPF: Anatomy of an Internet Routing Protocol,* Addison-Wesley, Reading, MA, 1998.
13. Moy, J., OSPF, Ver. 2, RFC 2328, Apr. 1998; available on-line at http://www.ietf.org/rfc/rfc2328.txt.
14. Halabi, S., *Internet Routing Architectures,* 2nd ed., Cisco Press, IN, USA 2000.
15. Stewart, J.W., III, *BGP4: Inter-Domain Routing in the Internet,* Addison-Wesley, Reading, MA, 1999.
16. Rekhter, Y. and Li, T., A Border Gateway Protocol 4 (BGP-4), RFC 1771, Mar. 1995; available on-line at http://www.ietf.org/rfc/rfc1771.txt.

5 MPLS Basics

5.1 PRINCIPLE (DATAGRAM AND VIRTUAL CIRCUIT)

5.1.1 BOTTLENECK IN SEARCHING IP TABLE

In IP packet transmission, the next hop is determined by searching the routing table using the destination address included in the IP header as an index key. In the network that has introduced CIDR (classless interdomain routing), searching is executed on an IP routing table based on the longest-prefix matching method (see Chapter 4). The longest-prefix matching method checks the matching of the IP address between the destination address of the IP header and the combination of each entry's address and the network mask in the routing table. That is, the method extracts the network address and the network mask of each entry in the routing table. It then calculates a logical AND between the destination address of the arriving IP header and the network mask bit by bit, and it takes the entry that matched over the longest bits with the network address as the next-hop address.

As the number of entries in the routing table increases, execution of the longest-prefix matching search takes more and more time to obtain a result. (Some backbone routers have hundreds of thousands of entries in the routing table.) Therefore, an efficient method for searching the routing table is required. As one of the high-speed searching methods of a routing table, there is a "Patricia tree" method that uses a tree structure for data searching. The Patricia tree is a kind of binary tree, and the method for creating it is as follows.

First, the prefixes in the IP transmission table are expressed in a binary system. The binary prefixes of all entries are then checked, starting from the top bit. When both 0 and 1 occur in the same bit position, a new branch node is created, with the addresses whose bits are 0 and the addresses whose bits are 1 belonging to separate child trees under the newly created branch node. A Patricia tree is created by successively branching off the tree with additional child trees until the search can no longer continue (or until all addresses have been classified).

As we can surmise from the process of creating the Patricia tree, searching is executed sequentially from the top of the tree according to the IP address given as an index key. That is, the search process begins by checking the reference bit position that has been registered to the top node of the Patricia tree. If the bit at the reference bit position is 0, it branches to child 0, and if it is 1, it branches to child 1. At each branch, the Patricia tree is searched in the same way. Whenever an effective node with an entry of the IP transmission table is found during the search, the IP address given as the key is compared with the prefix, and if they coincide, the value is recorded. As the tree search continues, a value is recorded whenever a match is detected. Searching is finished when the prefix of the effective node and the IP address given as a key become mismatched, i.e., when the search attains a node beyond which it is impossible to branch off any more. When the search is finished, the last recorded prefix is taken as the answer.

Figure 5.1 shows an example of a Patricia-tree search. Figure 5.1(b) shows the binary expression for the prefixes of all entries in the IP transmission table of Figure 5.1(a). The prefixes can be divided into two groups: one is a group in which the first bit of the prefix is 0, and the other is a group in which the first bit is 1. As seen in Figure 5.1(b), a new branch is created at the first bit. There are multiple prefixes with a first bit of 1, but these can be further subdivided by

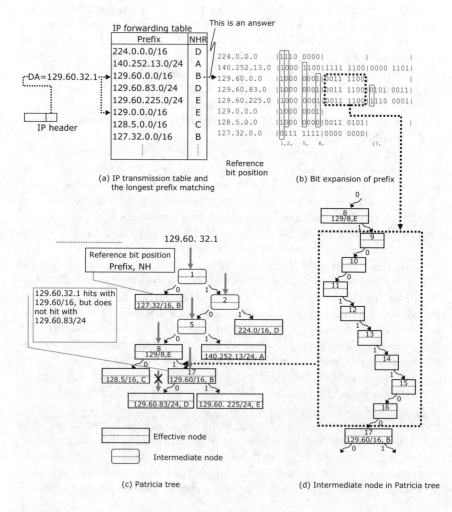

FIGURE 5.1 (a, b) Longest-prefix matching method and (c, d) Patricia tree.

MPLS Basics

looking at the second bit, where the second branch occurs. Next, there are multiple prefixes whose second bit is 0. The third and fourth bits are also 0, but the fifth bit contains both 0 and 1. So, the prefixes can again be subdivided into two groups at the fifth bit, and another branch is placed there. In the same way, another branch occurs at the 8th bit and at the 17th bit. The result is the Patricia tree shown in Figure 5.1(c), created from the IP transmission table give in Figure 5.1(a). Note that a Patricia tree generates two types of nodes: effective and intermediate. The effective node is a node that has an entry corresponding to the IP transmission table, and the intermediate node is a node that exists to counterbalance an effective node. However, because the tree only branches when the bit differs at the reference position, there are no useless intermediate nodes in a Patricia tree.

In Figure 5.1(b), three prefixes — 129.60.0.0/16, 129.60.83.0/24, and 129.60.225.0/24 — have common values of "0011 1100" from the 8th bit to the 16th bit. Note that the Patricia tree does not branch at each bit position; it only branches at bit positions where there is a difference. In contrast, if the tree were to branch at each bit position in the case of a 32-bit IP v.4 address, 32 branches would be required. The value of the Patricia tree is that it reduces the number of branches from a maximum of 32 to a number that depends on both the place and the value of different bits. If we assume that the number of entries of an IP transmission table is N, in the case of well-balanced tree, the number of branches required can be as low as $\log_2(N)$.

In Figure 5.1(c), the IP address 129.60.32.1 is given as an index key. If we search the Patricia tree starting from the top node, the 8th bit is an effective node, coinciding with an entry of the IP transmission table [129/8]. Then, because the 8th bit is 1, the search advances to a child tree branching to the right. The next node (the 17th bit) is also an effective node, coinciding with the entry of the IP transmission table [129.60/16]. Then, because the 17th bit of 129.60.32.1 is 0, the search advances to the left. However, the search does not coincide with the given entry of the IP transmission table [129.60.83/24]. Therefore, in this case, the matched entry is determined as [129.60/16] by falling back to the previously corresponding entry (the 17th bit).

5.1.2 Speeding Up by Label Switching

When the IP routing table is large, the longest-prefix matching method needs a complex method such as the Patricia tree to search the IP routing table. This search method has proved to be a bottleneck in router speed. To resolve this problem, the concept of MPLS (Multiprotocol Label Switching) has been introduced to IP packet transmission [1, 2].

In MPLS, instead of searching a routing table using a destination address in the IP header as an index key, a label is attached to the packet, and the next address is determined by searching the transmission table using this attached label. In contrast to IP, which is connectionless, MPLS is connection oriented. MPLS sets up the virtual connection between the start point and the end point and executes packet transmission using this virtual connection. The label is used to recognize this virtual connection.

112   GMPLS Technologies: Broadband Backbone Networks and Systems

Figure 5.2(a) shows a principle of IP datagram communication (connectionless). The destination address (DA = 140.252.13.34) is written in the packet header. The destination address is a value that is uniquely determined throughout the network, and a different value is given to each host within the network. Each router has a routing table in which a relationship between the destination address and the next hop is recorded. When a packet reaches a router, the router searches the routing table using the destination address that was extracted from the IP header as an index key, and then determines the next hop. By executing the same procedure in each router, the packet is delivered to the host specified by the destination address.

In contrast, Figure 5.2(b) shows a principle of connection-oriented communication by virtual circuit. In this case, a label is attached to the packet header. The label

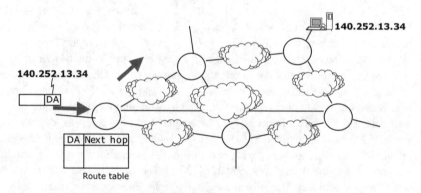

(a) Principle of IP datagram communication (connectionless)

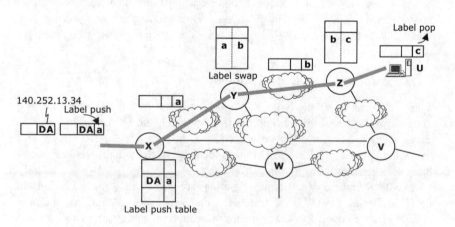

(b) Principle of connection oriented communication by virtual circuit

FIGURE 5.2 (a) Connectionless communication by IP datagram and (b) connection-oriented communication by MPLS virtual connection.

MPLS Basics

does not necessarily need to be unique throughout the network; indeed, it is sufficient to be unique in each link. It should be noted that, in datagram communication, an address is used to recognize the host, but in virtual-circuit communication, a label is used to recognize the flow (virtual circuit). The label is converted link by link. Each switch holds the relationship between the label used at the input side and the label used at the output side in the label table for each flow. (Note that in IP datagram communication, a node is referred to as a router, but in virtual-circuit communication, each node is called a "switch.") The switch searches the label table using the label extracted from the header of packet that arrives at the input-side link as an index key, and determines the label to be used at the output-side link. Then, it replaces the label of the packet header with this label for the output-side link and transfers it to next hop. In virtual-circuit communication, the relationship between the input-side label and the output-side label is set to the label table of each switch along the route through which the packet is transferred in advance.

MPLS is a method in which the concept of connection-oriented communication by virtual circuit is applied to IP datagram communication. When packets enter from an IP network to an MPLS network, a label that is used to transfer the packets within the MPLS network is attached to each packet. When packets exit from an MPLS network and enter into an IP network, that label is removed, and connectionless communication by IP datagram is resumed. In an MPLS network, each node is called a "label-switch router" (LSR). In particular, an LSR located at the edge of an MPLS network is called a "label-edge router" (LER), and a virtual connection (virtual circuit) is called a "label-switched path" (LSP).

An IP address is assigned uniquely to each host and is used to recognize all the hosts in the world connected to the Internet. A 32-bit IP address is used in IP v.4, and a 128-bit IP address is used in IP v.6. The high number of bits is required to uniquely distinguish the vast number of hosts connected to the Internet all over the world. When using a CIDR (classless interdomain routing) method, it is required to use the longest-prefix matching search to search the routing table. On the other hand, in MPLS, because an MPLS label is used to recognize the LSP in each hop, a different value can be used for each hop even if the LSP is the same. An MPLS label is dynamically assigned every time the LSP is set. The label table of each LSP can have just the LSP's information that has been set up to the LSR, and the label to distinguish the LSP can be just a number to recognize the circuits that have been set up in that LSR. Therefore, it becomes possible to search the label table directly using the label as an index key, thus enabling a high-speed searching of the label table (Table 5.1). By introducing the MPLS label, it becomes unnecessary to execute the complex longest-prefix matching search of an IP routing table.

There is an LER on the boundary of an MPLS network. The LER attaches a label to the packet and transfers the packet into the MPLS network. The packets with the same label are handled with the same transmission procedure within an MPLS network. This transmission procedure is called a "forwarding equivalent class" (FEC). As an example of FEC, there is a group of entries transferred to the same LER in the IP routing table (an example of FEC1 is presented in Table 5.2). It is also possible to change the transmission procedure or route according to the

TABLE 5.1
Comparison of Requirements between IP Address and MPLS Label

	Role	Uniqueness of Value
IP address	Recognize uniquely the destination host Route the IP packets Search the IP routing table	Must be unique throughout the Internet
MPLS label	Recognize uniquely the virtual circuit (VC) Recognize uniquely the FEC (forwarding equivalent class) Forward the MPLS packets Search the MPLS label table	Must be unique, but only by the link or by the node

originating address, even if the destination address is same. In such a case, an FEC is created with a pair of the originating address and the destination address (FEC2 in Table 5.2). It is also possible to change the transmission procedure by application, and in such a case, an FEC is defined by using a TCP/UDP port number (FEC3 in Table 5.2). In this way, MPLS can determine not only the destination address, but also the transmission procedure of packets within the network based on a variety of information, which has been very difficult to realize by the conventional IP datagram forwarding based on destination address.

5.2 LSP SETUP TIMING

5.2.1 TRAFFIC DRIVEN

In traffic-driven timing, LSR executes the normal transmission procedure for IP packets without a label. At the same time, it measures the amount of traffic of each flow, and when the amount of traffic exceeds a threshold, it deals the flow as an

TABLE 5.2
Examples of FEC

	SA	DA	SP	DP	PID
FEC 1	—	141.72.168.0/24	—	—	—
FEC 2	192.168.32.6/24	141.72.168.0/24	—	—	—
FEC 3	192.168.32.0/24	141.72.168.0/24	—	80	TCP

MPLS Basics

FEC and sets up an LSP, where the flow is often defined by using a combination of four fields; the destination address, the destination port number, the originating address, and the originating port number. In particular, by dealing the application that transmits the packets continuously in nature as a flow, it becomes possible to carry traffic efficiently using the packets with an MPLS label. In the early stage of development of MPLS, the effectiveness of traffic-driven timing was asserted based on the measured traffic data [3]. That is, from the results of traffic measurement, which found that a portion of the whole traffic flow was occupying a majority of the overall traffic bytes, it was asserted that it is possible to carry traffic effectively by assigning an LSP to a portion of the whole traffic flow.

5.2.2 Topology Driven

In topology-driven timing, a label is assigned to each entry of the routing table. When searching for an entry in a routing table, a prefix is used as an index key, but by using a label instead of a prefix, it becomes possible to speed up the process of searching a routing table.

Topology-driven timing connects all of the MPLS boundary LSRs mutually in full mesh. When a new LSR is connected to an MPLS network, an LSP is set up from the other LSRs to the new LSR. The route of the LSP is changed dynamically corresponding to changes in network topology or routing policy. Basically, an FEC is assigned to each entry of the IP routing table and an LSP is setup. In each repeater LSR, it becomes possible to access directly the contents of the IP routing table of each repeater LSR by notifying what number of entry of the IP routing table corresponds to the relevant FEC as a label. Using this approach eliminates the need to execute the complex longest-prefix-matching search that has been necessary to search the IP routing table [4].

5.3 PROTOCOL (TRANSFER MECHANISM OF INFORMATION)

5.3.1 MPLS Label

An MPLS label is attached to a field called a Shim header [5], a 32-bit header that is attached to an IP packet and consists of a label value (20 bits), an Exp bit (3 bits), a stack-indication bit (1 bit), and a TTL (time to live) (8 bits). The Exp bit is reserved for experimental work and is used for quality-class mapping when determining the quality of service (QoS) of an IP network using MPLS [6]. The stack-indication bit is used to characterize the label stack. It is possible to create a label stack by attaching multiple labels to an IP packet. If this bit is set, it means that the label exists at the bottom of stack. TTL has the same role as the TTL of an IP header, and it is used to prevent a packet from circulating infinitely should a transmission loop be generated. At the place where an IP packet enters an MPLS network, the TTL value of the IP packet is copied to the TTL field of MPLS header and decremented by 1 every time the MPLS packet is transferred within network, and when it becomes 0,

	SA	DA	SP	DP	PID
FEC1→	*	141.72.168.0/24	*	*	*
FEC2→	192.168.32.6/24	141.72.168.0/24	*	*	*
FEC3→	192.168.32.0/24	141.72.168.0/24	*	80	TCP

FIGURE 5.3 MPLS label format.

it is discarded. Upon exit from the MPLS network, the TTL value of the MPLS header is copied to the TTL field of the IP header (Figure 5.3).

When using a virtual-circuit-type protocol in layer 2, it is also possible to use the layer-2 label as the MPLS label. When using ATM (Asynchronous Transfer Mode) on layer 2, it is possible to use VCI/VPI (virtual-channel identifier/virtual-path identifier), and when using FR, it is possible to use DLCI as the MPLS label [7, 8].

5.3.2 LABEL TABLE

There are three types of label table: NHLFE, FTN, and ILM (Figure 5.4). NHLFE (next-hop label-forwarding entry) is a table that describes a procedure relating to the MPLS packet. To each entry of this table, such information as the next

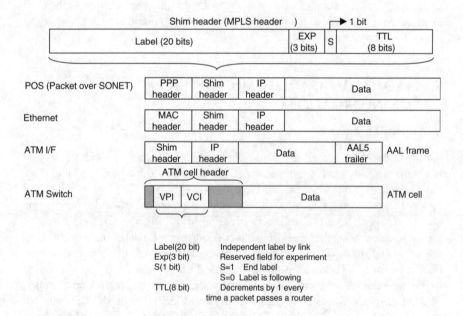

FIGURE 5.4 NHLFE, FTN, ILM.

MPLS Basics

hop and how to process the label is written. The options for how to process the label (Action) are: label exchanging, label hopping, and label pushing after label exchanging. FTN (FEC-to-NHLFE) is a table to relate the FEC to the NHLFE, and it is used to label a packet that has not yet been labeled. ILM (incoming-label mapping) is a table to relate the label to the NHLFE, and it is used in deciding how to process the labeled packet. That is, at the input-side LER, the transfer process of an unlabeled packet is determined by using the FTN, and in the repeater LSR, the transfer process of labeled packets is determined by using the ILM. At the exit-side LER, the label is removed from the labeled packet, and the next hop is determined by using the IP routing table.

The transfer process within an MPLS network in a repeater node is determined based on the label attached to the packet. The LSR searches the label table using the label attached to the packets and determines the LSR of the next hop and the output label. In the LSR's label table, the relationship between the input label and the next hop's LSR and the output label has been recorded in advance. The LSP is determined by setting up the label table of the LSR along the transmission route within the MPLS network. The packets are transferred within the MPLS network while being processed based on a transfer process at each LSR along the LSP, and they finally arrive at the LER on the boundary of the MPLS network. Then, after removing the label, the LER searches the IP routing table using the IP address and determines the next-hop router of the IP network, which is, of course, external to the MPLS network. At the exit-side LER of an MPLS network, the label table is searched by an MPLS label, and the IP routing table is searched by an IP address.

In Figure 5.5, R1 and R7 are LERs, and R4 and R6 are LSRs. H1 transmits an IP packet addressed to R1 and H2. R1 searches FTN and determines the NHLFE entry to which the process content for the label relating to the packet (FEC) addressed to H2 is recorded. To the entry of NHLFE, it is recorded that the next hop (NHOP) is R4, and the label L1 should be attached (Swap&Push (L1)) as the

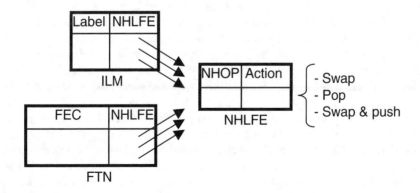

FIGURE 5.5 Principle of packet transfer within MPLS network by label-switching technique.

118   GMPLS Technologies: Broadband Backbone Networks and Systems

content of the process relating to the label. According to this direction, R1 attaches the label L1 to the IP packet and transfers it to R4. This process, which is realized by using two tables of FTN and NHLFE, is as follows: "As for the packet (FEC) addressed to H2, Swap&Push the label L1 and transfer it to the next hop (NHOP) of H4." Then, the MPLS packet to which label L1 was attached arrives at R4, which searches the ILM and determines the NHLFE entry to which the process content for the label of the MPLS packet with the label L1 is recorded. To the entry of NHLFE, it is recorded that the next hop (NHOP) is R6 and that the label L1 should be changed to L2 (Swap (L3)). According to this direction, R4 changes the label of the MPLS packet from L1 to L2 and transfers it to R6. This process, which is realized by using two tables of ILM and NHLFE, is as follows: "As for the packet with attached label L1, switch the label to L2 (Swap) and transfer it to the next hop (NHOP) of R6." In R6, the same process as R4 is executed, and the packet attached to label L3 is transferred to R7. Finally, R7 searches the ILM and determines the NHLFE entry to which the process content for the label of the MPLS packet with the label L3 is recorded. To the entry of NHLFE, it is recorded that the next hop (NHOP) is H2, and the label L3 should be removed (POP) as the content of the process. According to this direction, R7 removes the label L3 of the MPLS packet, makes the IP packet, and then transfers it to H2.

5.3.3 LABEL STACK

In MPLS, it is possible to attach any desired number of labels. The outer label is used for the transfer process, such as determining the next hop to transfer the packet. Label processing executed by a node is one of either Label exchange, Label pop, or Label push after exchange. It is also possible to execute multiple label processes at the same node. If the next hop is the LSR itself, the node executes label pop and executes again the transfer process of that packet using the popped label to outside within the LSR itself (see Figure 5.6).

In the case of Figure 5.6, R1 and R2 are LERs, and R4 and R6 are LSRs. H1 transfers the IP packets addressed to H2. R1 searches the FTN and determines the entry of NHLFE to which the process content for the label relating to the packet (FEC) addressed to H2 is recorded. To the entry of NHLFE, it is recorded that the next hop (NHOP) is R4, and the labels L0 and L1 should be attached (Swap&Push (L1, L0)) as the content of the process relating to the label. According to this direction, R1 attaches the labels L0 and L1 to the IP packet and transfers it to R4. Here, the outer label of the MPLS packet is L1 and inner label is L0. Next, the MPLS packet with the label L1 arrives at R4. R4 searches the ILM and determines the NHLFE entry to which the process content for the label of the MPLS packet with the label L1 is recorded. To the entry of NHLFE, it is recorded that the next hop (NHOP) is R6 and that the label L1 should be changed to L2 (Swap (L3)). According to this direction, R4 changes the label of the MPLS packet from L1 to L2 and transfers it to R6. In R6, the same process as R4 is executed, and the packet with the label L3 is transferred to R7. Finally, R7 searches the ILM and determines the NHLFE entry to which the process content for the label of the MPLS packet

MPLS Basics

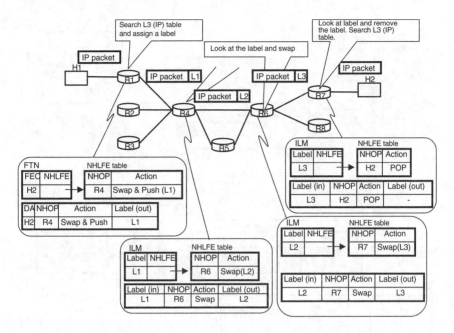

FIGURE 5.6 Processing the label stack processing (Exchange, Pop, Push).

with the label L3 is recorded. To the entry of NHLFE, it is recorded that the next hop (NHOP) is itself (R7) and that the label L3 should be removed (POP) as the content of the process relating to the label. According to this direction, R7 removes the label L3 of the MPLS packet and transfers it to itself. Transferring to itself means to repeat the packet transfer process once again. Here, the packet is still an MPLS packet even after the label L3 was removed, and label L0 is recorded. This L0 is an inner label L0 attached by R1, which is the input-side LER. R7 searches ILM and determines the NHLFE entry to which the process content for the label of the MPLS packet with the label L3 is recorded. To the entry of NHLFE, it is recorded that the next hop (NHOP) is H2 and that the label L3 should be removed (POP) as the content of the process relating to the label. According to this direction, R7 removes the label L3 of the MPLS packet and makes the IP packet and then transfers it to H2.

Thus, as illustrated in Figure 5.7, it is possible to build a hierarchical LSP network by using a label stack.

5.3.4 PHP

In an exit-side LER of an MPLS network, two types of searching are required: label-table searching and IP route-table searching. To speed up the searching process, a method called penultimate-hop popping (PHP) has been proposed (Figure 5.8) for the network (NW). In an LSR that is located one hop before the

120 GMPLS Technologies: Broadband Backbone Networks and Systems

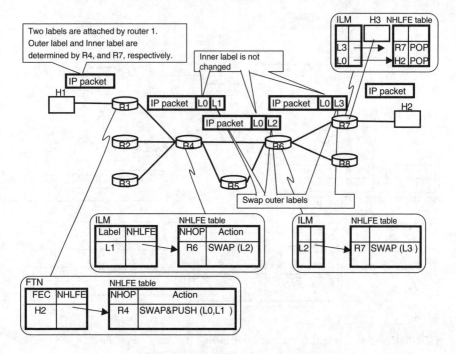

FIGURE 5.7 Hierarchical LSP network.

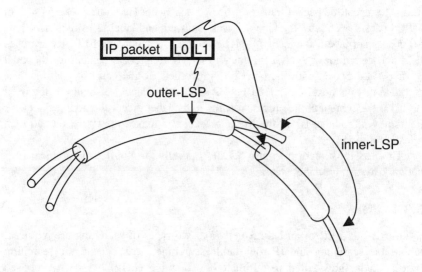

FIGURE 5.8 Penultimate-hop popping (PHP).

MPLS Basics

exit-side LER (penultimate hop), after searching the label table and attaching the output label, the packets are transmitted to the next hop (in this case, LER) from the output interface. However, because the next hop is an LER, it is redundant to use the label attached at this time for searching with label. That is, even if NHLFE is searched by the label in the ILM, because the process content recorded there is POP, it is required to search the IP routing table after popping the MPLS label. Because this is a useless process, in an LSR located one hop before the exit-side LER, the MPLS label is popped and the IP packet is transferred toward the exit-side LER (see Figure 5.8). This lightens the load of the packet transmission process at the exit-side LER.

5.3.5 Label Merge

Label merge is a technique that makes it possible to deal with packets coming from different input port of multiple LSRs as the same FEC. That is, if the packets are dealt with the same transfer process in a downstream LSR as in another LSR, those two FECs can deal with as one FEC by assigning the same label in the downstream LSR (see Figure 5.9). This process, called "label merge," is useful because it reduces the number of labels within the MPLS network.

When executing cell-based transmission, as in ATM, the packet is divided into multiple cells, and the VCI/VPI of each cell's header is used as a label. Therefore, if label merge is executed in a midway LSR, cells from multiple packets are mixed with each other and interleaved, making it impossible to rebuild the packet properly from these cells at the receive side (see Figure 5.10) [4]. Consequently, because

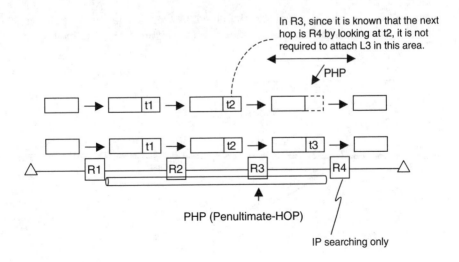

FIGURE 5.9 Label merge.

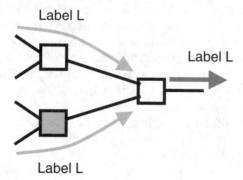

FIGURE 5.10 Problem in VC merge.

the label-merge technique cannot be applied in a midway LSR, different labels must be assigned to each input-side link of LSRs to which multiple packets flow. Further, in such LSRs, the cells must be rebuilt to the packet before they are transferred.

5.4 PROTOCOL (SIGNALING SYSTEM)

To set up an LSP, label information must be exchanged between LSRs using a signaling protocol [9–11]. We can discuss the operation of signaling protocols from three points of view: label assignment, label distribution, and a control system for label assignment/distribution.

5.4.1 LABEL-ASSIGNMENT METHOD (DOWNSTREAM TYPE, UPSTREAM TYPE)

There are two methods for assigning labels used between neighboring LSRs of an LSP: a downstream type and an upstream type (see Figure 5.11). The method to

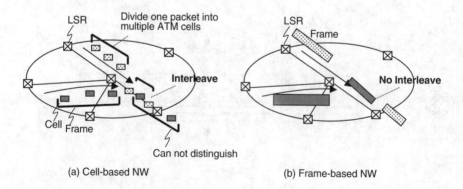

FIGURE 5.11 Label-assignment method.

MPLS Basics

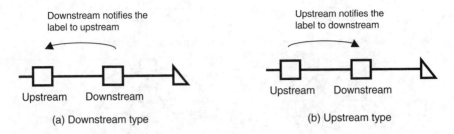

FIGURE 5.12 Label-distribution method.

assign the upstream side is called an upstream type, and the method to assign the downstream side is called a downstream type. In MPLS, the downstream type is commonly utilized. That is, the downstream-side LSR selects an unused label on the link and notifies that it is using that label for the LSP to the upstream-side LSR.

5.4.2 LABEL-DISTRIBUTION METHOD (ON-DEMAND TYPE AND SPONTANEOUS TYPE)

There are two methods to communicate the relationship of the labels used between neighboring nodes (see Figure 5.12). One is called an on-demand type, in which an LSR notifies the correspondence information between the FEC and the label to a neighboring LSR responding to a request for a label assignment from a neighboring LSR. The other method is called a spontaneous type, in which, if there is a relationship between an FCE and a label, an LSR notifies the relationship between the FEC and the label spontaneously to neighboring LSRs.

5.4.3 LABEL-ASSIGNMENT/DISTRIBUTION CONTROL METHOD (ORDERED TYPE/INDEPENDENT TYPE)

There are two methods for controlling label assignment and distribution (see Figure 5.13 and 5.14). One is called an ordered type, which executes assignment of an FEC and a label and distribution of labels that cooperate with each other.

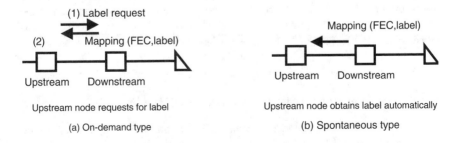

FIGURE 5.13 Label-assignment/distribution control method.

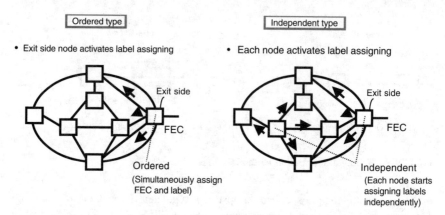

FIGURE 5.14 Two methods for controlling label assignment and distribution.

That is, when an LSR receives information relating to an FEC and label assignment from a neighboring LSR, the LSR assigns the FEC and label and notifies this information to other neighboring LSRs. The other method is an independent type, wherein an LSR assigns a label to FEC and notifies this correspondence information to the neighbor-ing LSRs without receiving a request from these neighbors. In the case of ordered-type control, once an LSR receives the relationship information between the FEC and label from a neighboring LSR and transfers the packet to a neighboring router using that label, then it is guaranteed that the packets are transferred along the LSP. However, this does not work out in the case of the independent-type control. In Figure 5.13, in ordered-type control, a label is assigned to the FEC, and the relationship is notified successively starting from the exit-side LSR toward upstream. In the case of independent-type control, each node independently assigns a label to the FEC and notifies this information in turn to neighboring LSRs on other nodes.

5.4.4 Label-Holding Method (Conservative/Liberal)

As for the holding method of the FEC and the label in each LSR, there are two types: conservative and liberal. The conservative type holds only the assignment information of the FEC and the label of the LSR on the route that the LSP actually passes through. In contrast, the liberal type holds the assignment information of the FEC and the label on the route that the LSP does not pass through. The former has an advantage in that the label-assignment information that each LSR has to hold is small, but the latter has an advantage in that it is possible to recover the LSP on a new route in a relatively short time period when the route of the LSP changes.

5.4.5 Loop-Protection Method (Path Vector/Hop Count)

When a loop occurs on the route of an LSP, it is possible to avoid an infinite turnaround by using TTL in the Shim header. But, separately from this, there are also two methods of preventing a loop in advance when setting up the LSP: vector-type

MPLS Basics

and hop-count-type methods [12]. The vector-type-method notifies the assignment information of the FEC and the label in between LSRs in the same way as a path-vector-type routing protocol, which makes it possible to detect the loop by recording the ID of the LSR itself to the packet. That is, if the ID of an LSR itself is recorded in the control packet that notifies the assignment information of the FEC and label to the other LSRs, it is judged that a loop was generated. On the other hand, in the hop-count-type method, the number of hops that the control packet can transfer is limited. That is, it increments the hop-count number that is recorded on the control packet every time the control packet, which notifies the assignment information of the FEC and label, hops over an LSR. When the counter number exceeds a certain value, the system judges that a loop was generated.

5.4.6 Hop-by-Hop-Type LSP and Explicit-Route-Type LSP

There are two types of routes that can be set up for an LSP: one that is determined by routing protocol and another that is explicitly specified. The former is called a hop-by-hop-type LSP, and the latter is called an explicit-route-type LSP. In the case of a hop-by-hop LSP, when the network topology changes or when a route to a relevant FEC is changed by operating policy, the LSP route can be changed. In an explicit-route-type LSP, the route is specified in advance. Because the input-side LER specifies the LSRs that the LSP must pass through, the route is not changed when the network topology changes. This explicit-route-type LSP is used for policy routing or traffic engineering.

REFERENCES

1. Davie, B. and Rekhter, Y., *MPLS Technology and Applications*, Academic Press, New York, 2000.
2. Rosen, E., Viswanathan, A., and Callon, R., Multiprotocol Label Switching Architecture, RFC 3031, Jan. 2001; available on-line at http://www.ietf.org/rfc/rfc3031.txt.
3. Newman, P., Lyon, T., and Minshall, G., Flow Labelled IP: a Connectionless Approach to ATM, in *Proc. IEEE INFOCOM 96*, IEEE, Piscataway, NJ, 1996.
4. Chandranmenon, G.P. and Varghese, G., Trading packet headers for packet processing, *IEEE/ACM Trans. Networking*, 4, 141–152, 1996.
5. Rosen, E., Tappan, D., Fedorkow, G., Rekhter, Y., Farinacci, D., Li, T., and Conta, A., MPLS Label Stack Encoding, RFC 3032, Jan. 2001; available on-line at http://www.ietf.org/rfc/rfc3032.txt.
6. Le Faucheur, F., Wu, L., Davie, B., Davari, S., Vaananen, P., Krishnan, R., Cheval, P., and Heinanen, J., MPLS Support of Differentiated Services, RFC 3270, May 2002; available on-line at http://www.ietf.org/rfc/rfc3270.txt.
7. Conta, A., Doolan, P., and Malis, A., Use of Label Switching on Frame Relay Networks Specification, RFC 3034, Jan. 2001; available on-line at http://www.ietf.org/rfc/rfc3034.txt.
8. Davie, B., Lawrence, J., McCloghrie, K., Rosen, E., Swallow, G., Rekhter, Y., and Doolan, P., MPLS Using LDP and ATM VC Switching, RFC 3035, Jan. 2001; available on-line at http://www.ietf.org/rfc/rfc3035.txt.

9. Andersson, L., Doolan, P., Feldman, N., Fredette, A., and Thomas, B., LDP Specification, RFC 3036, Jan. 2001; available on-line at http://www.ietf.org/rfc/rfc3036.txt.
10. Jamoussi, B., Andersson, L., Callon, R., Dantu, R., Wu, L., Doolan, P., Worster, T., Feldman, N., Fredette, A., Girish, M., Gray, E., Heinanen, J., Kilty, T., and Malis, A., Constraint-Based LSP Setup Using LDP, RFC 3212, Jan. 2002; available on-line at http://www.ietf.org/rfc/rfc3212.txt.
11. Awduche, D., Berger, L., Gan, D., Li, T., Srinivasan, V., and Swallow, G., RSVP-TE: Extensions to RSVP LSP Tunnels, RFC 3209, Dec. 2001; available on-line at http://www.ietf.org/rfc/rfc3209.txt.
12. Ohba, Y., Katsube, Y., Rosen, E., and Doolan, P., MPLS Loop Prevention Mechanism, RFC 3063, Feb. 2001; available on-line at http://www.ietf.org/rfc/rfc3063.txt.

6 Application of MPLS

MPLS executes transmission of IP packets based on their labels. By using the label, it becomes possible to run various applications that are difficult or impossible to set up under conventional IP datagram transmission, such as traffic engineering or improving the routing efficiency of external routes within an AS (autonomous system) or VPN (virtual private network) [1].

6.1 TRAFFIC ENGINEERING

The purpose of traffic engineering is to optimize the usage efficiency of network resources and network performance [2]. Traffic engineering involves measuring the utilization status of network resources and network performance, changing the parameters as needed to control traffic and routing, and altering the method of treating the network resources.

6.1.1 Problems with IGP

In IGP (Interior Gateway Protocol), such as OSPF (Open Shortest Path First) or IS-IS (Intermediate System to Intermediate System), a route is selected in which the summation of costs allocated to a link is at a minimum. The route is not changed even if the traffic conditions have changed, and the packets are transferred along the route even in the presence of congestion. Figure 6.1(a) illustrates a state where traffic from router 1.1.1.2 to router 1.1.1.6 and traffic from router 1.1.1.4 to router 1.1.1.6 are conflicting in the link in between routers 1.1.1.2 and 1.1.1.4 and in the link between routers 1.1.1.2 and 1.1.1.6, respectively, generating congestion. In this case, it is possible to avoid congestion of these links by bypassing the traffic route from router 1.1.1.1 to router 1.1.1.6 as shown in Figure 6.1(b). However, because routing in IGP is determined by the destination address of packets, it is impossible to detour the packets with the same destination address by the originating address. In IGP, there is no function to change dynamically the route depending on traffic conditions, even if there is congestion in a link on the path, and the packets are simply transferred along the shortest path. Thus, in the conventional IGP, it is impossible to switch the route to take traffic conditions into consideration. The technique of switching the transmission route in response to traffic conditions is called "traffic engineering."

6.1.2 Separation of Forwarding from Routing by MPLS

In MPLS, forwarding of packets is determined by label. Packets are forwarded by using a label table that resides on each LSR (label-switch router) on the route.

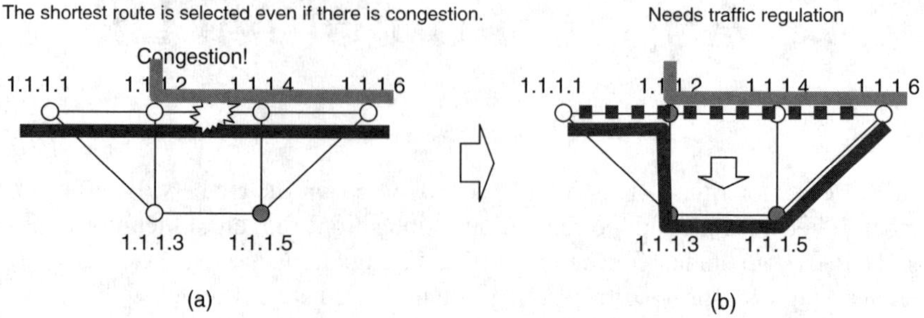

FIGURE 6.1 Problems with IGP.

The label table has an input label, an output label, and information about the next hop's LSR as its entries for each LSP (label-switched path). In MPLS, it is possible to set up the LSP by setting up the label table of each LSR on the route after the route has been determined using a relevant measure. When seen from different angles, it is apparent that transmission of the packets (forwarding) and route control (routing) are separate processes.

On the other hand, in conventional IP networks, the next hop has been determined by the destination address that was attached to the packet, and its relationship has been determined by a routing table of the router. That is, the route to which the packets are transmitted has been determined by the destination address. In this case, it is clear that transmission of the packets (forwarding) and route control (routing) have been combined in a single process.

Figure 6.2(a) shows the basic idea of IP datagram communications (connectionless). The packet header contains the destination address (DA = 140.252.43.4, for example). The destination address is a unique value within the entire network, and different addresses are assigned to all the hosts. Each router has a routing table on which the relationship between the destination address (DA) and the next hop's address is described. When the packet reaches a router, the router searches its routing table using the destination address included in the header as a key and determines the next hop. As the packet moves from router to router, the same procedure is repeated in each router until the packet is successfully transferred to the host specified by the destination address.

In contrast, Figure 6.2(b) illustrates the concept of connection-oriented communications by virtual circuit. In this case, the packet header contains a label instead of an address. The label need not be unique within the entire network, but only within each link. Note that in datagram communications, the address is used to identify the host, but in virtual-circuit communications, the label is used to identify the flow of data (virtual circuit). Label conversion is executed link by link. At each switch, the relationship between the label that is used at the input-side link and the label that is used at the output-side link is held in the label table. (In IP datagram communications, the node is called a router; in virtual-circuit communications, each

Application of MPLS

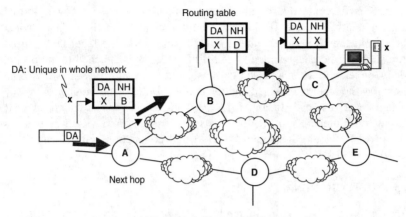

(a) Packet transmission by IP datagram

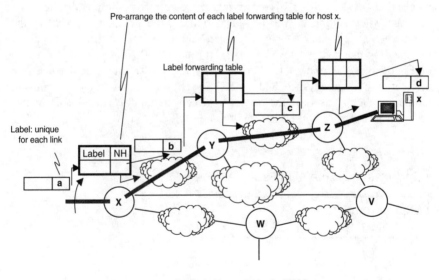

(b) Packet transmission by MPLS

FIGURE 6.2 Concept of separating the routing from forwarding by MPLS.

router is referred to as a switch.) The switch searches the label table using the label contained in the packet header that arrives at the input-side link and determines the label that is to be used in the output-side link. It then replaces the incoming label in the packet header with a new label in the output-side link and transmits the packet to the next hop. In virtual-circuit communications, the relationship between the input-side label and the output-side label is set up in advance in the label table of each switch along the route through which the data flows.

MPLS is a connection-oriented communication method using a virtual circuit, in contrast to the IP datagram communication method. Thus, when packets from the IP network enter the MPLS network, the label that is used for transmitting the packets inside the MPLS network is attached to the packet header. Conversely, when a packet is transferred from an MPLS network to an IP network, the label is removed and the connectionless communication by the IP datagram is resumed. In an MPLS network, each node is called an LSR (label-switch router). The LSR that exists at the edge of an MPLS network is called an LER (label edge router). The virtual connection (virtual circuit) is called an LSP (label-switched path) (Figure 6.2).

6.1.3 Source Routing

Source routing involves setting up the LSP along the specified route in MPLS to avoid congestion. It is possible to set up the LSP along the route by determining the route to the destination node by taking some relevant measure at the originating node of the LSP or by specifying the route information as the signaling message when setting up the LSP. Determining the route of the LSP at the originating node is called "source routing," and the route information specified by signaling message is known as "explicit route" (ER) (Figure 6.3). In the case of Figure 6.3, the route 1.1.1.1 1.1.1.3 1.1.1.5 1.1.1.6 was calculated at the originating node in advance, and the LSP is set up along this route. To transmit the packet along this LSP, the packet to which the label corresponding to this LSP was attached should be inserted to node 1.1.1.1.

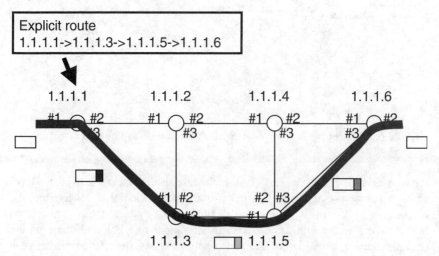

FIGURE 6.3 Source routing and route specifying (explicit routing).

Application of MPLS

An LSP set up in this way is called an ER-LSP. Topology-driven routing is a routing method that assigns the label to each entry of the routing table. Although a prefix is usually used as the key, it is possible to increase the speed of searching the routing table by using the label instead of the prefix as the key. In the case of strict specification depicted in Figure 6.4(a), the route is set up as 0-1(s)-2(s)-3(s)-4. In this method of specification, the letter "s" of "1(s)" means "strict," and the "1" means that the neighboring node must be 1. Therefore, in this case, the route must be 0-1-2-3-4. On the other hand, in the case of loose specification depicted in Figure 6.4(b), the route is set up as 0-2(l)-4. In this method of specification, the "l" of "2(l)" means "loose" and the "2" means that the neighboring node must not be 2. That is, it is okay only if there is a node 2 on the route. Therefore, in this example, it can be either 0-1-2-3-4 or 0-5-6-2-3-7-4.

There are two methods to specify a route: loose specification and strict specification. Strict specification is a method of specifying all of the nodes along the route, and loose specification is a method of specifying selected nodes (or networks or ASs) along the route.

(a) Strict specification

0-1(s)-2(s)-3(s)-4

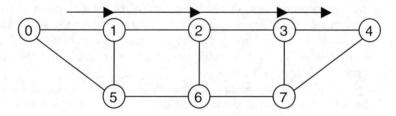

(b) Loose specification

0-2(l)-4

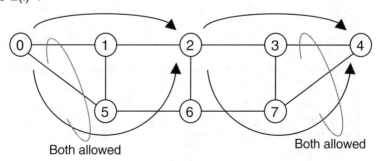

FIGURE 6.4 Concept of strict specification and loose specification.

6.1.4 Traffic Trunk

Traffic trunk is an aggregate of traffic flows of the same class and is carried by an ER-LSP. The attributes that are attached to the traffic link include the following (Table 6.1).

Traffic parameter: an attribute that indicates the traffic performance of traffic trunk such as peak bit rate, average bit rate, or allowable burst size.

Generic path selection and management: an attribute that indicates the rules of path selection and management of traffic trunk. The path-selection rule decides the shortest route that IGP selects and whether route selection should be executed under the restriction of policy or resource condition. The path-management rule decides whether the route should be reconstructed dynamically corresponding to change of network status.

Resource-class affinity: an attribute that indicates which class of resources should be used (or removed). This attribute facilitates flexible application of policy. Classes have been assigned to network resources in advance.

Adaptivity: an attribute that indicates the adaptability to change of network conditions. By using this attribute, it becomes possible to set up an LSP to the optimum route when a new network resource becomes available or a damaged network resource is recovered. An attribute indicating whether reoptimization should be applied or not is also given to the traffic trunk. Adaptivity is absolutely an attribute relating to reoptimization and is not a rule relating to a change of network conditions resulting from a fault or an accident. In the case of a fault, the resilience attribute (described below) is used.

Priority: an attribute that indicates the priority of traffic trunks. It determines the order of route selection when setting up an LSP or when the network recovers from a fault state. It is also used to decide the order of traffic trunks when the preemption attribute (described below) is realized.

Preemption: an attribute relating to preemption (or interruption) of network resources. Specifying the first attribute allows a traffic trunk to become a preemptor. That is, it determines whether or not it is possible to execute preemption on a resource assigned to another traffic trunk. The second attribute determines whether a certain traffic trunk can be preempted or not, that is, whether or not it is allowed that a resource assigned to itself can be preempted from another traffic trunk. As for whether preemption of these resources is possible or not, the priority attribute of a traffic trunk has an influence on it. The case that traffic trunk A can preempt a resource that has been assigned to traffic trunk B is effective only when the following five conditions are approved. (1) A has higher priority than B, (2) A and B are sharing the resource, (3) resource cannot be assigned simultaneously to A and B, (4) A can become a preemptor, and (5) B can be preempted.

Resilience: an attribute that indicates how to respond when a network failure occurs. It specifies whether or not to allow a detour when a network failure

TABLE 6.1
Attributes of Traffic Trunk

Attribute	Description
Traffic parameter	Indicates the traffic characteristics of traffic trunk: bit rate average rate, allowable burst size, etc.
Generic path selection and management	Indicates the rules relating to path selection and path management of traffic trunk. The path-selection rule decides the shortest route that IGP selects and whether route selection should be executed under the restriction of policy or resource condition. The path-management rule decides whether the route should be reconstructed dynamically corresponding to changes in network status.
Resource-class affinity	Indicates which class of resources should be used (or removed). Classes have been assigned to network resources in advance. By using this attribute, flexible application of policy can be easily realized.
Adaptivity	Indicates the adaptability to a change in network conditions. This attribute makes it possible to set up LSP to the optimum route when a new network resource becomes available or when a network resource that has been damaged is recovered. An attribute indicating whether reoptimization should be applied or not is also given to the traffic trunk. (The Adaptivity attribute is related to reoptimization. For a change in network conditions caused by a fault, the Resilience attribute is used.)
Priority	Indicates the priority of traffic trunks. It determines the order of route selection when setting up the LSP or when the network is recovered from a fault state. It is also used to decide the order of traffic trunks when the Preemption attribute is used.
Preemption	Indicates whether a certain traffic trunk can preempt another traffic trunk or whether a certain traffic trunk can be preempted from another traffic trunk.
Resilience	Indicates how to respond when a network failure occurs. It specifies whether or not to allow a detour when a network failure occurs on the route on which the traffic trunk is passing.

occurs on the route on which a traffic trunk is passing. When detouring is allowed, the case is further divided as follows. If there are resources available in a reserved route, detour is executed. Or, detour is executed regardless of whether or not the resources exist in a reserved route. Or, these two cases are combined. Beyond these methods, there may be various other policies to execute detouring when a network failure occurs.

6.1.5 Restricted Route Controlling

When an attribute of traffic trunk, an attribute of resource, and information relating to the topology and status of each link are given, a route that satisfies the conditions of the attribute of traffic trunk is calculated (see Figure 6.5). By giving attributes to a network, high-grade network operation becomes possible. Here, a network resource means a link. A resource attribute is used to color the link. By creating a topology through gathering links with a special color and by routing

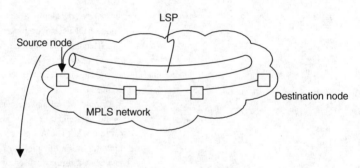

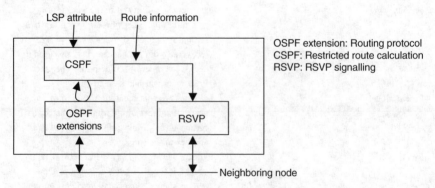

FIGURE 6.5 Concept of CSPF that outputs the route from the input information of traffic trunk attribute, resource attribute, and topology database.

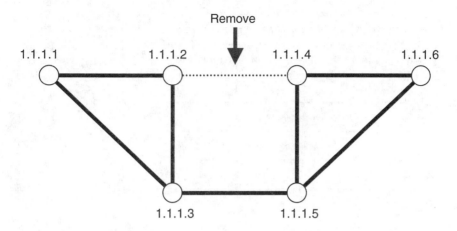

FIGURE 6.6 Concept of CSPF.

using these links, a high-grade policy control can be easily realized. Such a route-computation method for restricted route control is called a Constraint-Based Shortest Path First (CSPF) protocol. In CSPF, a route that satisfies the desired traffic link attributes is calculated by executing a shortest-path computation (such as the Dijkstra algorithm) after removing the links that do not satisfy the desired topology constraints (see Figure 6.6).

CSPF computes the route that satisfies the constraints based on the usage status of each link in the topology. Link-state protocols such as OSPF or IS-IS have been extended so that each node can obtain information about the usage status of each link. Figure 6.7 shows a packet format of opaque LSA that is defined to carry information about the usage status of links in OSPF. Using this extended field, traffic information such as residual resources, maximum bandwidth, etc. is advertised [3].

6.1.6 Setting Up the ER-LSP with RSVP-TE

As signaling protocols used when setting up an LSP, there are CR-LDP [4] or RSVP-TE [5]. When using RSVP-TE, a node number (or network number or AS number) on the way of the route is specified to an object called an ERO (explicit route object). An ERO is carried through the outward route included in the Path message. At the midway nodes, the ERO is transferred to the next hop after extracting its own node number. In Figure 6.8, node 1.1.1.1 creates a Path message addressed to node 1.1.1.6 to set up the ER-LSP to node 1.1.1.6. At this time, node 1.1.1.1 computes the route 1.1.1.11.1.1.31.1.1.51.1.1.6 as the route to node 1.1.1.6, puts it in Path message as an ERO, and then sends it out to the next hop, node 1.1.1.3. When node 1.1.1.3 receives this Path message, it interprets the ERO, removes its own node number, puts it on the ERO, and sends it out to the next hop, node 1.1.1.5. Node 1.1.1.5 executes a similar process and transmits the Path

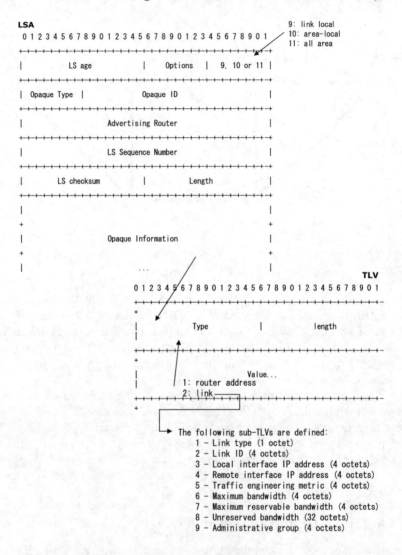

FIGURE 6.7 OSPF-TE opaque LSA format.

message of which the ERO includes node number 1.1.1.6 to the destination node 1.1.1.6. When node 1.1.1.6 knows that the address of the Path message was itself, it sends back the Resv message to the originating node 1.1.1.1. The route through which the Resv message is returned is a reversal of the route over which the Path message has been transmitted.

RSVP-TE is a protocol that extends the RSVP [6], which is a signaling protocol of soft state, for setting up an ER-LSP [7]. RSVP is positioned above the IP layer, and its protocol number is 46. Figure 6.9 shows the packet format

Application of MPLS

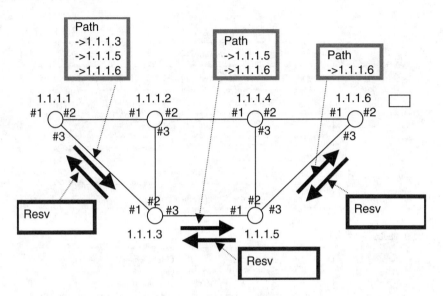

FIGURE 6.8 Image of setting up the ER-LSP with RSVP-TE.

of RSVP-TE. An RSVP packet consists of an RSVP header and an RSVP object. Multiple RSVP objects can be included in a single packet.

An RSVP header includes fields for an RSVP protocol version number (Version), Flag, Message Type, RSVP checksum, TTL, and RSVP packet length (RSVP Length). Currently, the RSVP protocol version number is 1. The Message Types are defined as shown in Table 6.2. The RSVP checksum is calculated as the sum (16 bits) of 1s complement of an RSVP message. TTL (time to live) is decremented by 1 at each hop. The RSVP packet length (16 bits) indicates the total length of an RSVP packet, including the RSVP header, in byte units.

As the RSVP object that is placed following the RSVP header, various types of objects have been defined. The type of object is identified by class number and class type. The format of each object is arranged as, from the front to the end, object length (16 bits), class number (8 bits), class type (8 bits), and object data. There is a correlation between object types and a rule in the order in which the objects are arranged within the packet. The rule is described by BNF (Backus-Naur form) syntax.

Figure 6.10 shows the representation of a Path message and an Resv. message by BNF syntax. Explicit route object (ERO) and record route object (RRO) are newly defined objects to set up the ER-LSP, and they are used, respectively, to specify the nodes to pass and to collect information of the nodes passed. ERO is included in the Path message, and RRO is included in both the Path message and the Resv message. ERO and RRO are composed of multiple subobjects, and node information is included in each subobject. Figure 6.11 shows the format of subobjects used in ERO and RRO. The subobject of ERO consists of a loose bit

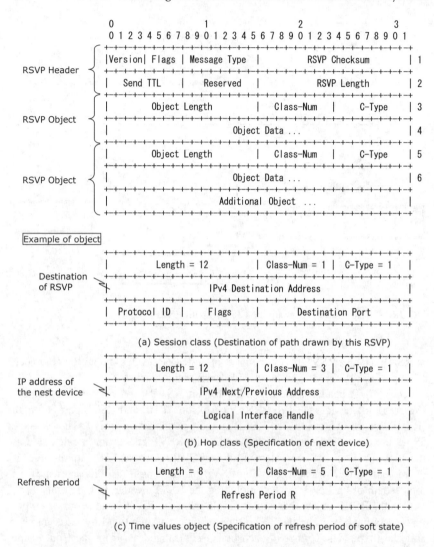

FIGURE 6.9 RSVP message format.

(L), Type, subobject length (Length), and data. The loose bit indicates whether the node is specified to "loose" or "explicit." If this bit is set, it indicates that the node is specified to "loose." The Type field indicates the type of this subobject. Either an IP v.4 node, an IP v.6 node, or an AS number can be selected as the type of node that can be specified by Type. Subobject length indicates the length of this subobject in units of bytes. In the case of Figure 6.11, the example shows that an IP v.4 node was included in the subobject. The subobject of RRO consists of a Type, subobject length (Length), data, and flag (Flags). As for Type, subobject length (Length), and data, these fields are the same as those of ERO. Flags are

**TABLE 6.2
Message Type of RSVP**

Value	Message Type
1	Path
2	Resv
3	PatrErr
4	ResvErr
5	PathTear
6	ResvTear
7	RescConf

```
<Path Message> ::=          <Common Header> [ <INTEGRITY> ]
                            <SESSION> <RSVP_HOP>
                            <TIME_VALUES>
                            [ <EXPLICIT_ROUTE> ]
                            <LABEL_REQUEST>
                            [ <SESSION_ATTRIBUTE> ]
                            [ <POLICY_DATA> ... ]
                            <sender descriptor>
<sender descriptor> ::=     <SENDER_TEMPLATE> <SENDER_TSPEC>
                            [ <ADSPEC>   ]
                            [ <RECORD_ROUTE> ]
<Resv Message> ::=          <Common Header> [ <INTEGRITY> ]
                            <SESSION>  <RSVP_HOP>
                            <TIME_VALUES>
                            [ <RESV_CONFIRM> ] [ <SCOPE> ]
                            [ <POLICY_DATA> ... ]
                            <STYLE> <flow descriptor list>
<flow descriptor list> ::=  <FF flow descriptor list    >
                            | <SE flow descriptor>
<FF flow descriptor list> ::=  <FLOWSPEC> <FILTER_SPEC>
                            <LABEL> [ <RECORD_ROUTE> ]
                            | <FF flow descriptor list>
                            <FF flow descriptor>
<FF flow descriptor> ::=    [ <FLOWSPEC> ] <FILTER_SPEC> <LABEL>
                            [ <RECORD_ROUTE> ]
<SE flow descriptor> ::=    <FLOWSPEC> <SE filter spec list>
<SE filter spec list> ::=    <SE filter spec>
                            | <SE filter spec list> <SE filter spec>
<SE filter spec> ::=        <FILTER_SPEC> <LABEL> [ <RECORD_ROUTE> ]
```

FIGURE 6.10 BNF description of Path/Resv message.

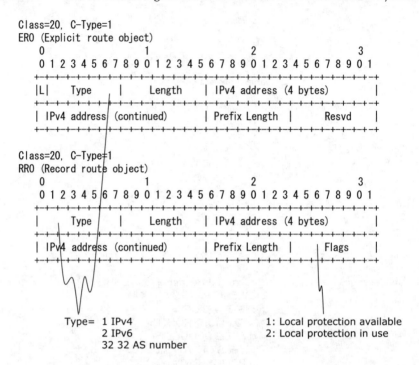

FIGURE 6.11 Format of ERO and RRO.

defined in the case of an IP v.4 node or an IP v.6 node and are used to indicate whether local protection is available or is in use in downstream nodes.

Figure 6.12 shows the concept of processing the ERO object. This is a case where node 0 sets up the ER-LSP addressing to node 4. Here, assume that the route was specified to 1(s)-2(s)-4(l). Where, (s) means strict specification and (l) means loose specification. Node 0 puts this information into ERO and transmits a Path message. Because ERO is specifying the route by setting the next hop as 1(s), the Path message is transferred to node 1. When node 1 received this Path message, it interprets the ERO and confirms that this is addressed to itself. Node 2 removes the subobject relating to itself and transmits the Path message to the next node. Because ERO is loosely specifying the route by setting the next hop as node 4(l), it must select the optimum route to reach node 4. As the route from node 2 to node 4, route 2-3-4 or route 2-6-7-4, etc. can be selected. But because the length of route 2-3-4 is shorter than any other routes, node 2 sends out the Path message to node 3. When node 3 receives the Path message, it interprets the ERO and confirms that the next hop is specified to Loose and node 4. Node 3 does not change the ERO and transfers the Path message to node 4. In this way, the Path message is transferred from node 2 to node 4 according to the route specification of 1(s)-2(s)-4(l).

To record along which route the LSP was set up, an object RRO (record route object) is utilized. This RRO, which is included in the Path message or Resv

Application of MPLS

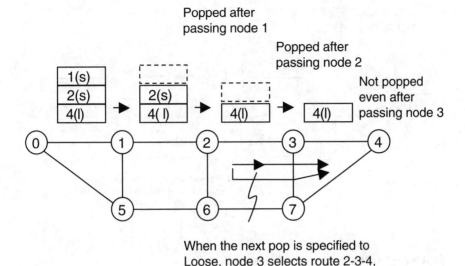

FIGURE 6.12 Processing the ERO and next hop.

message, records all of the node numbers of the nodes to which ER-LSP was set up. The originating node transmits the Path message, which includes the RRO having the IP address of the node itself inside. Intermediate nodes transfer the Path message to the next hop after recording their own IP addresses to the RRO. When the destination node receives the Path message, it extracts the RRO from the Path message, puts it into the Resv message, and sends it back to the preceding hop. This Resv message is transferred in reverse direction to the originating node along the route of ER-LSP. In this return process, intermediate nodes also add their own IP addresses to the RRO. In this way, the nodes on the way of the ER-LSP route can know the ER-LSP route by using an RRO. That is, it is possible to know the upstream route of the ER-LSP by looking at the RRO included in the Path message, and to know the downstream route of the ER-LSP by looking at the RRO included in the Resv message.

The RRO can collect the label information of each hop by setting up a "label recording flag to session attribute" object. Figure 6.13 shows how the route information is collected by using RRO. Let us see how the route information is collected in the case where node 0 set up the LSP from node 0 to node 4. Node 0 sets the "label recording flag to session attribute" object of the Path message and records 0 as its own node number. If the next hop of node 0 is node 1, when node 1 receives the Path message, it records 1 as its own node number to the RRO. Then, if the next hop of node 1 is node 2, when node 2 receives the Path message, it records 2 as its own node number to the RRO. And then, if the next hop of node 2 is node 3, when node 3 receives the Path message, it records 3 as

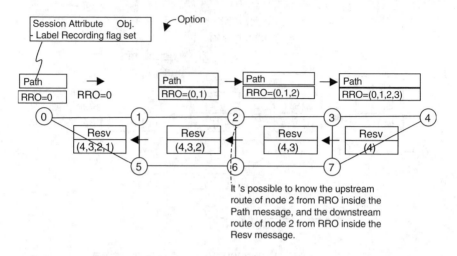

FIGURE 6.13 Processing the RRO.

its own node number to the RRO. If the next hop of node 3 is an end node 4, when node 4 receives the Path message, it sends a Resv message back to the preceding hop. At the same time, node 4 creates a new RRO and puts it into the Resv message. It records 4 as its own node number to the RRO. When node 3 receives the Resv message, it records 3 as its own node number to the RRO and transfers it to the preceding node 2. When node 2 receives the Resv message, it records 2 as its own node number to the RRO and transfers it to the preceding node 1. And when node 1 receives the Resv message, it records 1 as its own node number to RRO and transfers it to the preceding node 0. In this way, the process of collecting the route information is completed. By doing this, it is possible to collect the route information of the LSP.

It should be noted that all of the nodes on the way of the LSP route can collect the route information of the LSP by adding an RRO to the Path message and the Resv message. For example, at node 2, it is possible to know the upstream-side route information by looking at the node number recorded in the RRO inside the Path message, and it is possible to know the downstream-side route information by looking at the node number recorded in the RRO inside the Resv message.

A traffic trunk by an LSP is set up by using RSVP-TE. The traffic trunk is identified by a tunnel identifier (Tunnel-ID). Although the traffic trunk is realized by using an LSP, this LSP can be replaced by another LSP on the way of the route. That is, when it is required to change the attributes of a traffic trunk such as route or bandwidth, etc., it can be done by setting up another LSP with the desired attribute of route or bandwidth, etc., and replacing the old attribute with the new one. The LSP is identified by an LSP identifier (LSP-ID). The Tunnel-ID is recorded in a Session object, and LSP-ID is recorded in a Session_Template object.

Application of MPLS

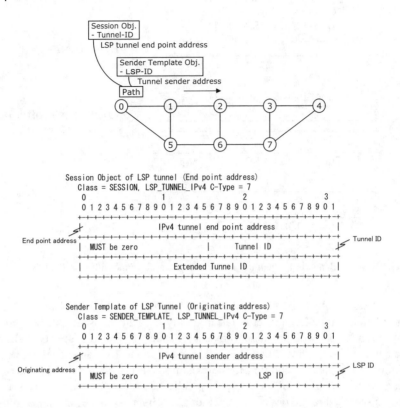

FIGURE 6.14 Session object and Sender_Template of Path message containing Tunnel-ID and LSP-ID, respectively.

Figure 6.14 shows the format of a Session object and a Session_Template object. These objects differ in that either the traffic track is IP v.4 or IP v.6. Figure 6.14 corresponds to the case of IP v.4. To the Session object, an IP v.4 tunnel endpoint address, a Tunnel-ID, and an Extended Tunnel-ID are recorded. The IP v.4 tunnel endpoint address indicates the node address of the receive-side endpoint of the traffic trunk. Tunnel-ID is used to identify the traffic trunk and is not changed from the start to the end of its life. Extended Tunnel-ID is used when it is desired to uniquely identify the traffic trunk between the pair of the originating node and the destination node of the traffic trunk. The endpoint's node address, etc. are recorded in this field. Like the Tunnel-ID, this is not changed from when the traffic trunk is created until it is disconnected. To the Sender_Template object, an IP v.4 tunnel sender address and LSP-ID are recorded. To the IP v.4 tunnel sender address, the originating node's address of the traffic trunk is recorded. An LSP-ID is used to identify the LSP, and this can be changed after the traffic trunk is created until it is disconnected. As described above, if the LSP is replaced when changing the attributes of a traffic trunk, the LSP-ID is also changed.

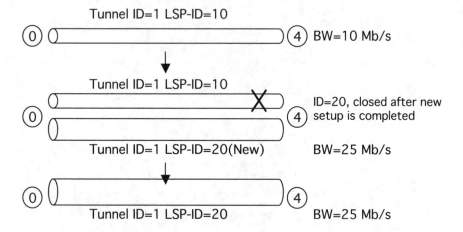

FIGURE 6.15 Bandwidth-changing procedure of traffic trunk with RSVP-TE.

A Session object is put into both the Path message and the Resv message, but a Sender_Template object is put into only the Path message. A Filter_Spec object is recorded with the same format as a Sender_Template. Figure 6.15 shows the bandwidth-changing procedure for the traffic trunk. To change the bandwidth of this LSP tunnel, in addition to the existing LSP (10 Mbps) that has already been set up, another LSP is set up in parallel. Bandwidth of the newly set up LSP is specified to the desired bandwidth (25 Mbps). At this time, bandwidth is reserved with a Shared Explicit style. As shown in Figure 6.16, this Shared Explicit style is a method to specify the bandwidth so that two LSPs can share the same bandwidth. In a section where LSP-ID = 10 and LSP-ID = 20 pass

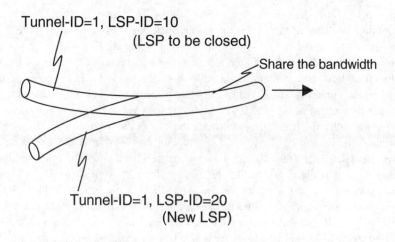

FIGURE 6.16 Shared explicit style.

Application of MPLS

through the same route, it is possible to reduce the bandwidth by sharing the bandwidth. In Figure 6.15, although two LSPs use the same Tunnel-ID, their LSP-IDs are different. That is, Tunnel-ID 1 is used for both LSPs, but LSP-ID = 20 is used in the new LSP while LSP-ID=10 is used in the old LSP (see Figure 6.17). When the new LSP (LSP-ID = 20) is set up, the old LSP (LSP-ID = 10) is disconnected. At this time, the new and old LSPs are switched using a procedure called "make-before-break." That is, switching is executed by first making traffic

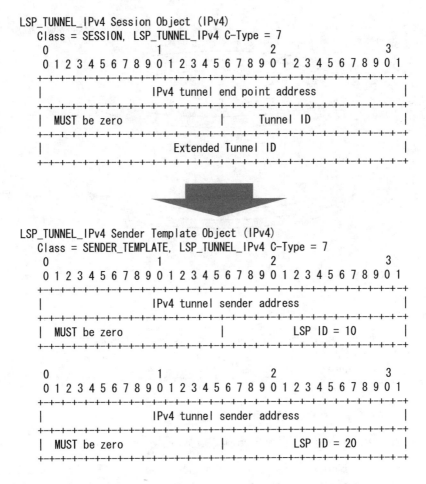

FIGURE 6.17 Procedure for recording Tunnel-ID and LSP-ID to Session object and Sender_Template object of Path message.

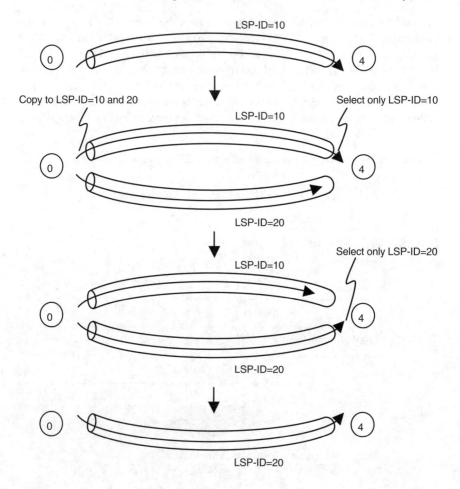

FIGURE 6.18 Switching by "make before break."

flow through both the new LSP and the old LSP, and then by switching the path to the new LSP at the receive end so as to reduce as much as possible the packet loss caused by the switching process (see Figure 6.18).

6.2 ROUTING TO EXTERNAL ROUTE WITHIN AS

6.2.1 Route Exchange by Border Gateway Protocol (BGP)

The Internet works because many autonomously and distributedly managed networks are connected to each other. A network that is autonomously and distributedly managed is called an autonomous system (AS). It is possible to deliver IP

Application of MPLS

packets to all of the hosts connected to the Internet by exchanging the route information between ASs.

For setting the route between ASs, the BGP-4 protocol is widely utilized. BGP-4 is a path-vector type of routing protocol in which the route information and the path information to its route are simultaneously exchanged (see Figure 6.19). The route is expressed by a prefix consisting of a pair of address and network mask, and the path information is expressed by a list of AS numbers through which the IP packets pass to reach the destination address. In the case of Figure 6.19, ISPs A, B, C, and D are exchanging the route information of BGP-4 in full-mesh via NW-a.

When route information between ASs is exchanged by BGP-4, this route information must be transferred to other ASs. An AS is composed of a border router and a core router, and the border router exchanges the route information with the other ASs. When the border router receives the route information from the border router of a neighboring AS, it notifies the received route information to other border routers within the AS. The notified border routers then notify the route information to the border routers of neighboring ASs. In this way, BGP-4 notifies the route information to all of the neighboring ASs. A BGP-4 router sets up an IBGP (Interior Border Gateway Protocol) session against other border routers within the AS itself, and an EBGP (Exterior Border Gateway Protocol) session against the border routers of neighboring ASs, and then notifies the pertinent route information using these sessions.

Figure 6.20 shows a problem of advertising a transit route within an AS. Here, let us assume that the routers 1.1.1.1 and 1.1.1.6 are BGP routers and the others are non-BGP routers. BGP router 1.1.1.6 receives a route 10.1.1.1 from

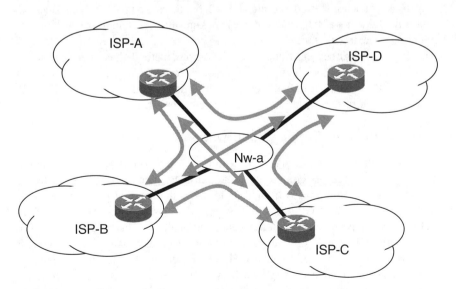

FIGURE 6.19 Exchanging route information via the BGP-4 protocol.

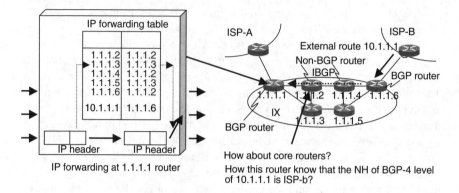

FIGURE 6.20 Notification of route information of BGP-4 within AS.

ISP B. BGP router 1.1.1.6 advertises the route 10.1.1.1 to the BGP router 1.1.1.1 using an IBGP session. Here, it should be noted that the other routers within the AS (from 1.1.1.2 to 1.1.1.5) do not know about the route 10.1.1.1. BGP router 1.1.1.1 advertises the route 10.1.1.1 to ISP A. In this way, the ISP A transfers the packets addressed to the route 10.1.1.1 to the BGP router 1.1.1.1. While the BGP router 1.1.1.1 receives the packets addressed to the route 10.1.1.1 from ISP A, the next hop addressed to the route 10.1.1.1 becomes BGP router 1.1.1.6. However, BGP router 1.1.1.1 is not directly connected to BGP router 1.1.1.6 with a link. If we try to search the route table again, we will see that this is a next hop that connects router 1.1.1.2 directly to BGP router 1.1.1.6 with a link. In this way, packets are transferred to router 1.1.1.2, but there is a problem here. Because router 1.1.1.2 is not a BGP router, it does not know the next hop for route 10.1.1.1, which is an external route. Therefore, the packets addressed to route 10.1.1.1 cannot go forward beyond router 1.1.1.2. Various methods have been proposed to solve this problem, such as making the other routers (from 1.1.1.2 to 1.1.1.5) notify BGP, or using a default route, or inserting the BGP route to the IGP. However, each of these proposed solutions has a problem.

6.2.2 Routing to External Route within AS

To set up routing to an external route within an AS, a core router must route the IP packets toward the external route. Although IGPs such as OSPF and IS-IS are generally used within an AS, a method to insert the external route information that was obtained by BGP-4 into IGP is considered to be effective as a method to realize routing toward the external route within an AS. That is, by having BGP also operate within IX (Internet exchange), it becomes possible to notify BGP information to routers within an AS. However, because the number of external routes is increasing significantly year by year as shown in Figure 6.21, inserting a vast amount of route information may impose a burden on system memory or processors that deal with the routing process. There is another method that sets

Application of MPLS

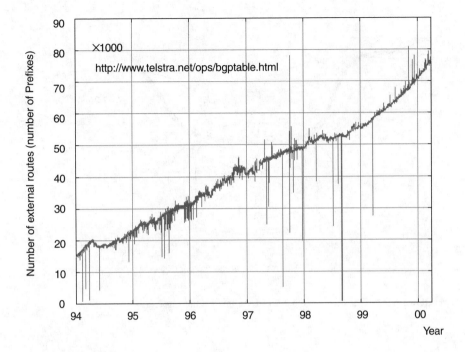

FIGURE 6.21 Trend showing the increasing number of entries in the routing table, 1994–2000.

up a default route to an external route through the core router, but this is problematic in that (a) it is difficult to set up the default route within the core router so as not to generate a routing loop and (b) it may result in limiting the structure of network. There is yet another method that makes BGP-4 also operate in core routers. However, because it is necessary to set up a full-mesh IBGP session between all of the border routers within an AS in BGP-4, it is impossible to apply this method when the number of core routers within an AS becomes great.

6.2.3 Solution by MPLS

By using MPLS, it is possible to set up routing to an external route without placing the brunt of the burden upon core routers [8]. By setting up an LSP tunnel between border routers, incoming IP packets can be transferred from an external source through the AS to an external destination. As seen in Figure 6.22, an LSP is set up corresponding to the IBGP session of a border router of AS. That is, all of the border routers belonging to the same AS are connected in full-mesh by LSP. Transit traffic that comes into AS1 is transferred at border router A by using an LSP addressed to border router B. Here, it should be noted that the traffic from border router A to border router B can be transferred by using the same LSP, regardless

150 GMPLS Technologies: Broadband Backbone Networks and Systems

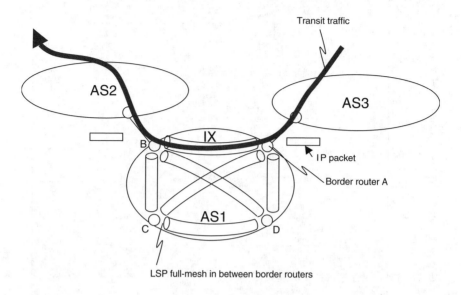

FIGURE 6.22 Transit service achieved via LSP tunneling.

of the route information. In BGP, although information for some tens to hundreds of thousands of routes might be exchanged in between ASs as shown in Figure 6.21, the number of border routers within the same AS is far smaller than the number of routes. Because the transit traffic is transferred along the LSP within AS, the core router within AS need only know the label-transfer information of the LSP, and this number is far smaller than the number of routes. Although the border routers are still required to deal with a lot of BGP route information, the core routers within AS1 are released from the burden of dealing with a vast amount of route information. In this way, core routers are no longer required to be aware of the external routes for transit traffic being transferred from an external AS through the AS itself to another external AS. Moreover, as for the IP packets that are transferred from within the AS toward the external route, once they are transferred to a certain border router from the core router, they can be transferred to the desired border router by using an appropriate LSP tunnel from the border router. A great merit of MPLS is that it facilitates setting up an optimum route and load distribution.

6.3 VIRTUAL PRIVATE NETWORKS (VPN)

In a virtual private network (VPN), a private network is built logically. Depending on the design of a VPN, it might be desirable that different VPNs should be allowed to communicate using the same address, or the goal might be to limit communication between different VPNs. In either case, MPLS can be utilized to satisfy these requirements [1].

Application of MPLS

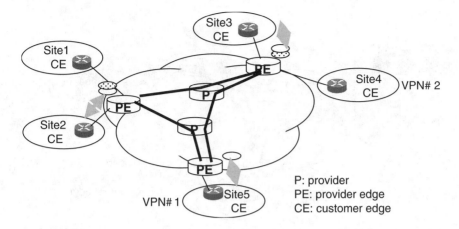

FIGURE 6.23 VPN architecture.

6.3.1 Overlay Model and Peer Model

A VPN is built by connecting multiple nodes through a core network. A VPN is composed of core networks that provide communication through a common carrier for user networks that are connected to the core network at each node. As illustrated in Figure 6.23, the user network consists of multiple sites (Site 1, Site 2, etc.). The core network consists of core routers (P: Provider) and edge routers (PE: Provider Edge). The PE routers of the core network are connected to the edge routers (CE: Customer Edge) of the user network.

There are two methods to connect user networks of different nodes: an overlay model (Figure 6.24) and a peer model (Figure 6.25). In the overlay

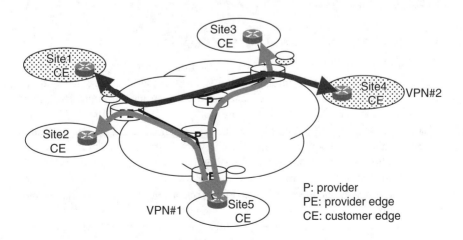

FIGURE 6.24 Overlay model.

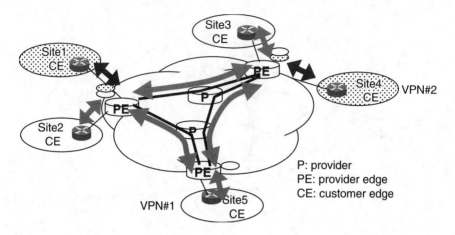

FIGURE 6.25 Peer model (RFC 2547 specification).

model, the CEs of the far-separated nodes are directly connected to each other by an LSP tunnel. Thus, each VPN that is connected through the core network has a different topology. On the other hand, in the peer model, it is the PE routers that are directly connected by an LSP tunnel, and the packets that come from CEs of different VPNs are multiplexed into a single LSP tunnel and transferred to the facing PE router. To distinguish the VPNs included in a single LSP, MPLS labels are utilized. The label attached to identify the VPN is transferred as-is to the destination PE router as the inner label of the label stack. The destination PE router checks the inner label after removing the outer label from the label stack and transfers the IP packets to the CE router of the desired VPN. The VPN architecture of a peer model using an MPLS label stack is described in RFC 2547.

6.3.2 Virtual Routing and Forwarding (VRF)

As a requirement of VPN, there is a restriction relating to communication between different VPNs. Figure 6.26 shows an example of an intranet and an extranet. In the case of an intranet, each entity constructs an exclusive VPN, and in this example, the VPN of the automobile company and the VPN of the bank do not share a VPN mutually. In the case of an extranet, mutual communication is allowed at some of the nodes of the VPN. In the example of Figure 6.26, the New York branch of the automobile company and the London branch of the bank are allowed to communicate mutually, but the Tokyo branch of the automobile company and the Paris branch of the bank are not allowed to communicate via the VPN.

To construct the extranet flexibly in this way, it is better to design the VPN structure by distinguishing the VPN and the Site. A VPN can be constructed to include multiple Sites. The fact that one Site can belong to multiple VPNs makes

Application of MPLS

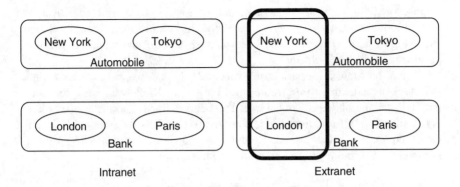

FIGURE 6.26 Example of VPN (intranet and extranet).

it possible to reduce the number of intranets or extranets. Thus, to maximize flexibility, the route information should correspond to each Site, not to each VPN. The route information that corresponds to each Site is called a VRF entity (Virtual Routing and Forwarding entity). In the case of Figure 6.27, Site 3 belongs to VPN A and VPN B, and Site 1 and Site 2 belong to VPN A and VPN B, respectively. The VRF of Site 3 is a sum of VPN A and VPN B, but the VRFs of Site 1 and Site 2 are only VPN A and VPN B, respectively.

6.3.3 MP-BGP

VPN service is realized by exchanging the route information between Sites. A PE router exchanges the route information that belongs to its associated CE with a destination PE, which transfers the route information to the destination Site of the same VPN. The MP-BGP protocol, an extension of BGP-4, is utilized to exchange the route information between PEs (see Figure 6.28). MP-BGP is utilized because (1) it is appropriate for carrying a vast amount of route information; (2) it is possible to exchange route information between PE routers that are not directly

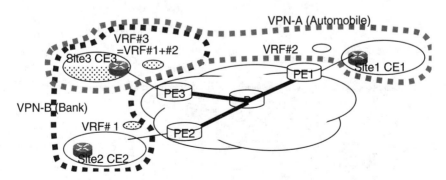

FIGURE 6.27 Flexible service by VRF.

neighboring by using an IBGP session; and (3) it is easy to control the delivery of route information between Sites by assigning attributes to the route information, etc. In Figure 6.28, each PE mutually sets up the MP-iBGP and exchanges the route information between each site.

When the route information of a certain VPN is delivered to related Sites, attributes are attached to the route information. In BGP-4, an extended community attribute called a RouteTarget has been defined, making it possible to notify the route information to a CE related to the VPN by associating it with the RouteTarget as the extended community attribute in each VPN. VPN is implemented through the following three phases,

- VPN-extended community attribute association
- Route advertisement with extended community attribute
- Route selection based on extended community attribute

How the VPN is implemented using extended community attribute is illustrated bolow (see Fig 6.29).

Each VPN is identified by extended community attribute. In Figure 6.29, a value of Automobile is assigned to the extended community attribute RouteTarget in VPN A, and a value of Bank is assigned to the extended community attribute RouteTarget in VPN B. When each PE router receives the route information from its CE router, it attaches the extended community attribute RouteTarget relating to the VPN to which the CE router belongs and notifies this to all of the PE routers.

PE advertises into the core network, the route information with extended community attribute. In the example of Figure 6.29, the PE router 3 thatneighbors CE router 3 of Site 3 attaches both Automobile and Bank to the extended community attribute RouteTarget for the route information from CE router 3 and advertises it. In contrast, the PE router 1 that neighbors CE router 1 of Site 1 attaches only Automobile to the extended community attribute RouteTarget for the route information from CE router 1 and advertises it. And the PE router

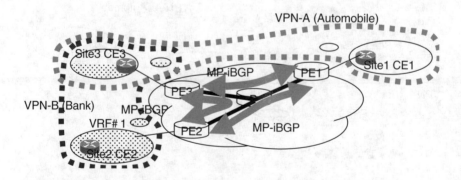

FIGURE 6.28 Advertising of route information by MP-BGP.

Application of MPLS

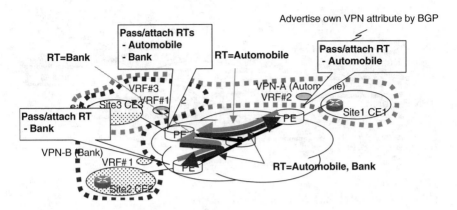

FIGURE 6.29 Advertising the route information of each VPN by the BGP extended community attribute RouteTarget.

2 that neighbors CE router 2 of Site 2 attaches only Bank to the extended community attribute RouteTarget for the route information from CE router 2 and advertises it.

PE receives from the core network, the route information with extended community attribute. The PE router selects only the route information that is related to itself from the route information that is transferred from the other PE routers using MP-iBGP, and then it transfers this information to its associated CE router. That is, in the example of Figure 6.29, the PE router 3 that neighbors CE router 3 of Site 3 selects the extended community attribute RouteTarget to which either Automobile or Bank was attached for the route information notified by the MP-iBGP session with the other PE routers and notifies it to the CE router 3. However, the PE router 1 that neighbors CE router 1 of Site 1 selects only the extended community attribute RouteTarget to which Automobile was attached for the route information notified by the MP-iBGP session with the other PE routers and notifies it to the CE router 1. Similarly, the PE router 2 that neighbors CE router 2 of Site 2 selects only the extended community attribute RouteTarget to which Bank was attached for the route information notified by the MP-iBGP session with the other PE routers and notifies it to the CE router 2. This demonstrates the utility of the extended community attribute RouteTarget. Using this attribute, it is possible to advertise the route information of each VPN while taking extranet flexibility into consideration by associating each Site with a RouteTarget corresponding to the VPN.

In VPN service, it is possible to use the same address between different VPNs. When advertising the route information in between PEs using BGP, uniqueness of address is reserved by attaching an address called a "route distinguisher" (RD) that has been defined to identify the address used in different VPNs. An RD is assigned to each VRF and not to each VPN, i.e., an RD is assigned to each Site. The RD format consists of the AS number (2 octets) of the VPN service provider

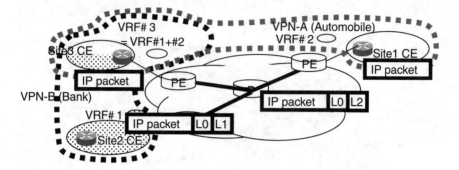

FIGURE 6.30 Label stack (outer label and inner label).

that owns the core network and a type identifier (2 octets) and a number of the type identifier (4 octets).

6.3.4 Notification of Outer Label and Inner Label in VPN

When the route information of each VPN is exchanged, each PE router executes the transfer of IP packets to the route that belongs to the VPN. To transfer IP packets within the core network, a label stack of MPLS is utilized. The outer label is used to transfer to the facing PE router, and the inner label is used to judge to which router the IP packets should be transferred at the facing PE router (see Figure 6.30). That is, an LSP tunnel has been set up between PE routers, and the outer label is used to flow the IP packets through the LSP tunnel. The LSP tunnel between PE routers is set up by using a protocol such as LDP, CR-LDP, or RSVP-TE in the same way as an ordinary LSP tunnel. As seen in Figure 6.31, the inner label is used to identify the CE and is advertised when exchanging the route information by MP-BGP [9].

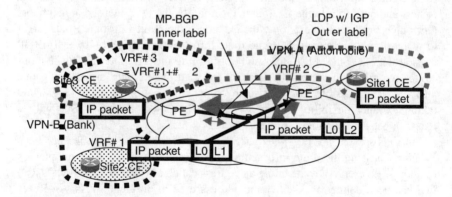

FIGURE 6.31 Communication between outer label and inner label.

REFERENCES

1. Davie, B. and Rekhter, Y., *MPLS Technology and Applications,* Academic Press, New York, 2000.
2. Awduche, D., Malcolm, J., Agogbua, J., O'Dell, M., and McManus, J., Requirements for Traffic Engineering over MPLS, RFC 2702, Sep. 1999; available on-line at http://www.ietf.org/rfc/rfc2702.txt.
3. Katz, D., Kompella, K., and yeung, D., Traffic Engineering (TE) extensions to OSPF version 2, RFC 3630, Sept. 2003.
4. Jamoussi, B., Andersson, L., Callon, R., Dantu, R., Wu, L., Doolan, P., Worster, T., Feldman, N., Fredette, A., Girish, M., Gray, E., Heinanen, J., Kilty, T., and Malis, A., Constraint-Based LSP Setup Using LDP, RFC 3212, Jan. 2002; available on-line at http://www.ietf.org/rfc/rfc3212.txt.
5. Awduche, D., Berger, L., Gan, D., Li, T., Srinivasan, V., and Swallow, G., RSVP-TE: Extensions to RSVP LSP Tunnels, RFC 3209, Dec. 2001; available on-line at http://www.ietf.org/rfc/rfc3209.txt.
6. Braden, R., Zhang, L., Bersonm, S., Herzog, S., and Jamin, S., Resource ReSerVation Protocol (RSVP), Ver. 1 Functional Specification, RFC 2205, Sep. 1997; available on-line at http://www.ietf.org/rfc/rfc2205.txt.
7. Durham, D. and Yavatkar, R., *Inside the Internet's Resource Reservation Protocol: Foundations for Quality of Service,* John Wiley & Sons, New York, 1999.
8. Pepelnjak, I. and Guichard, J., *MPLS and VPN Architectures,* Cisco Press, IN, USA., 2000.
9. Rekhter, Y. and Rosen, E., Carrying Label Information in BGP-4, RFC 3107, May 2001; available on-line at http://www.ietf.org/rfc/rfc3107.txt.

7 Structure of IP Router

The major functions of an IP router can be categorized into two function groups: a data-path function group and a control function group. Figure 7.1 shows the concept of the data-path function. The data-path function operates on all the IP packets passing through the router. This function includes a forwarding function, a switching function, a scheduling function at the output port, etc. When IP packets arrive at the IP-packet input block, they are transferred to the forwarding block. In the forwarding block, the output port for the IP packets is determined based on the destination address included in the header of the IP packets and the information in the IP routing table. The destination of the output port is attached to the IP packets, and the packets are transferred to the switch block, which distributes the IP packets to the desired output port. The output port has a buffer to accumulate the IP packets temporarily, as multiple IP packets from different input ports may arrive at the same time. The scheduling block in the output port determines in what order the accumulated IP packets should be output to the output port considering the fairness and the requirements of the class to which the IP packet belongs. The requirement for processing speed in the data-path function is strict, as this function must be performed for each incoming IP packet. Therefore, although the data-path function is mostly executed by software, it is implemented by hardware when high-speed processing is required.

The other router function group, the control function, includes such functions as system configuration, system management, and updating of IP routing table information by routing protocol. Because implementation of these functions is requested less frequently than the data-path function, which is executed for each incoming IP packet, the restriction of processing speed is not as strict, so it is implemented by software.

In this chapter, we describe the basic technologies of the data-path function, including switch architecture, packet scheduling, and the forwarding engine.

7.1 STRUCTURE OF ROUTER

There are various router structures available to fulfill different requirements. In this section, we describe the structures of router in three categories: low-end class, middle class, and high-end class. The low-end-class router is a primitive device with the simplest structure and the lowest cost. It is currently used as a local-area network (LAN) router in situations where high performance is not required. The middle-class router provides a higher level of performance than the low-end-class router and is used in metropolitan-area networks (MAN) or LANs that require relatively higher performance. The high-end-class router operates at the highest level of performance and is highest in cost. This is used mostly in wide-area networks (WAN).

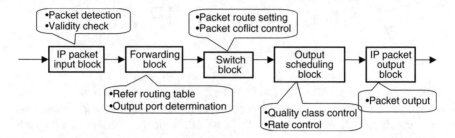

FIGURE 7.1 Data-path function in router.

7.1.1 Low-End-Class Router

Figure 7.2 shows the structure of a low-end router. This features of this primitive class of router include a forwarding function, a switching function, and an output scheduling function as well as a CPU (central processing unit), common bus, and common

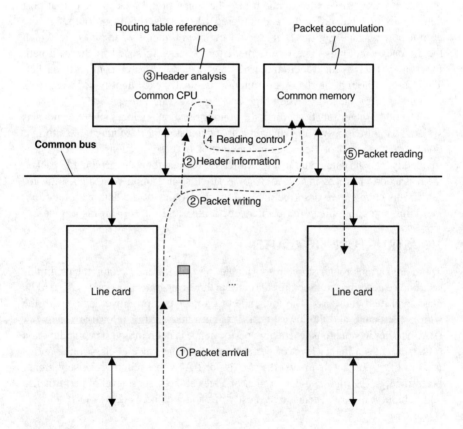

FIGURE 7.2 Structure of low-end router.

Structure of IP Router

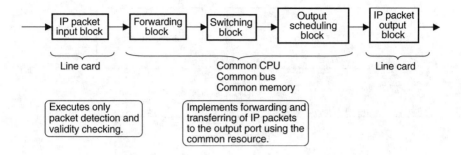

FIGURE 7.3 Arrangement of data-path functions in low-end router.

memory. Figure 7.3 shows how the data-path functions are arranged in the low-end router. The operation of the low-end router is described as follows using Figure 7.2.

1. IP packets arrive in the line card. The line card detects the arrival of packets and executes validity checking.
2. The line card transfers the packets to the common memory through the common bus. The common memory accumulates the packets. The header information is transferred to the common CPU.
3. The common CPU determines to which port the IP packets should be output by referring the IP routing table based on the header information. This is a forwarding function. To determine the output port, the CPU refers to the routing table stored in the common memory.
4. When the output port has been determined, the common CPU reads out the IP packets accumulated within the common memory.
5. The IP packets are read out from the common memory to the line card of the output port.

All of these operations are executed by software. Because implementation via software is costless compared with hardware, this structure is mostly utilized in low-end routers. Although CPU performance has improved significantly over the years, the ability to process and transfer the IP packets arriving in the router with a single common CPU is limited.

7.1.2 Middle-Class Router

Figure 7.4 shows the structure of a middle-class router. In the middle-class router, to overcome the performance limitations of low-end routers with a single common CPU, discrete CPUs and discrete memories to implement the forwarding functions are furnished. Figure 7.5 shows the arrangement of the data-path functions of a middle-class router. The operation of a middle-class router is described as follows using Figure 7.4.

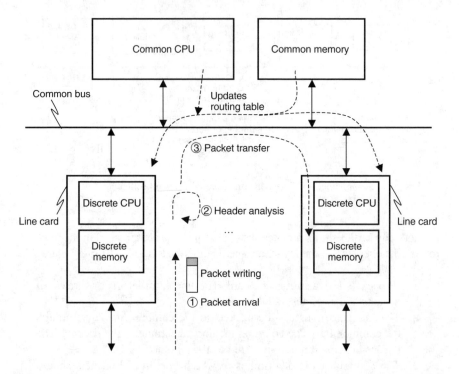

FIGURE 7.4 Structure of middle-class routers.

1. IP packets arrive in the line card, where they are written into the card's discrete memory. The line card detects the arrival of packets and executes validity checking.
2. The discrete CPU analyzes the header of the IP packets accumulated in the discrete memory of the line card and determines the output port for the IP packets. The output port is determined by referring to the routing table held within the discrete memory of the line card. The routing table in discrete memory is continually updated by the common CPU and common memory.

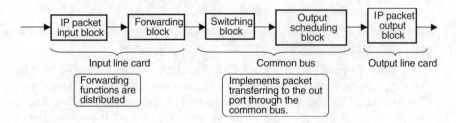

FIGURE 7.5 Arrangement of data-path functions in middle-class routers.

Structure of IP Router

3. Once the output port of the line card has been determined, the IP packets accumulated in discrete memory are read out and transferred to the line card's output port through the common bus. In this structure, because a common bus is used to transfer the IP packets, only one packet can be transferred at a time, even when multiple line cards are trying to transfer multiple packets at the same time.

The forwarding function to determine the output port is processed in parallel by a discrete CPU in each line card. The common CPU centrally manages the routing table using the common memory, and when the routing table is updated, the common CPU sends a copy of the table to the discrete memory of each line card. The cost of the middle-class router increases because each line card has its own discrete CPU and discrete memory. However, the router's performance is improved because the common CPU is no longer required to process the forwarding function. The load on the common CPU is thus greatly reduced compared with the common CPU in the low-end-class router.

One problem with middle-class routers is that when the line speed of the input/output ports becomes high or when the number of input/output ports is increased, the common bus becomes a bottleneck and limits the router's ability to process/transfer all of the incoming IP packets. Routers of this class do not execute quality-class controlling or rate controlling for packet output.

7.1.3 High-End-Class Router

Figure 7.6 shows the structure of a high-end router. This high-end-class router consists of a route control block, an input line card, an output line card, and a switch fabric.

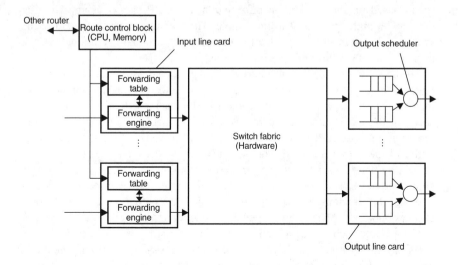

FIGURE 7.6 Structure of high-end routers.

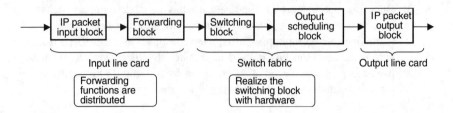

FIGURE 7.7 Arrangement of data-path functions in high-end routers.

The forwarding function, switching function, and output scheduling function are implemented in each input line card, switch fabric, and output line card, respectively. Figure 7.7 shows the arrangement of data-path functions of a high-end-class router.

In this high-end router, the IP packets processed by an input line card are transferred from each input line card to each output line card through an independent switch block. If the switch block has enough switching throughput, the bottleneck of transfer capacity can be removed. The data-path function is basically implemented by hardware. High-end-class routers are characterized by hardware implementation of a switch block. The data-path function is designed so that the IP packets with the minimum packet length can be processed at the same speed as the input line (i.e., at the wire speed). If the IP packets could not be processed at wire speed and IP packets successively arrived in the input line card, the IP packets would fall into arrears and the router could not achieve the expected performance. Therefore, it is important to implement the hardware so that each function block can process the IP packets at wire speed. In the forwarding block, the output port must be determined at a wire speed based on the destination information of the IP packet using a routing table and a high-speed forwarding engine. The technology implemented in the switching block must be capable of minimizing or eliminating conflict between packets entering from multiple input cards.

If the switching block is to improve the data transmission efficiency, it must have a method of controlling packet confliction, and the transmission speed of internal data in the switching block must be higher than the data transmission speed of the input/output line cards. When the IP packets are transferred from the input line card to the output line card, the IP packets are input to the output buffer of the output line card. The output buffer is necessary because there is a possibility that IP packets will be transferred from multiple input line cards to a single output line card. In this case, only one IP packet is transferred to the output line, and the remaining IP packets have to wait in the output buffer until the next output opportunity. The output scheduling function executes this output control.

The output scheduling function takes into consideration a service-quality class, a required rate, and fairness. The packets relating to output scheduling are divided by service-quality class (or by label-switched path, or by flow) and then, after they are stored in the service-quality-class buffer that corresponds to the output line block, they are transmitted to the output line. In the high-end router, packet scheduling is precisely executed so as to satisfy the required quality for the respective service-quality class. As a measure of service-quality class, there is an end-to-end delay. Audio and moving-picture applications require that this delay time be as short as possible. On the other hand, an application such as e-mail does not require such short delay times and thus can tolerate longer delay times. Service quality for e-mail is categorized into the best-effort class. Even if the required rates for different packets are the same, if they have different service-quality classes, they are processed according to a given priority assigned to their service-quality class. The best-effort class does not declare a required rate and effectively utilizes the residual bandwidth that is not used by service-quality classes that are higher than the best-effort class. At this time, the residual bandwidth must be allocated equally and fairly.

There are various packet-scheduling methods that can be used in the output-scheduling block. These include a perfect priority scheduling method that considers the priority based on the service-quality class, a WRR (weighted round-robin) scheduling method that considers the required rate, and a DRR (deficit round-robin) scheduling method. These and other scheduling methods are described in detail in Section 7.3.

The IP packets are processed in the input line card before being sent to the switching block. There are two methods of processing the packets in the input line card. One is a variable-length method that processes packets with variable length based on the IP packet, and the other is a cell method that divides the IP packet into several-fixed length packets called cells. (Note that the packet length need not be the same as the 534-byte cells used in Asynchronous Transfer Mode.) In the cell method, the IP packets are divided into cells before they are transferred from the input line card to the switching block. The switching block executes switching by cell unit and transfers the cell to the output line card. The output line card reassembles the multiple cells into an IP packet and transfers them to the output line. In the switching block of the high-end routers, where high speed and complex processing are required, the cell method is typically used because it facilitates packet synchronization, which prevents conflicts caused by forwarding packets/cells to the same output line. In Section 7.2, we describe the switching method of the switching block, focusing mostly on the cell method.

7.2 SWITCH ARCHITECTURE

7.2.1 CLASSIFICATION OF SWITCH ARCHITECTURE

Figure 7.8 shows the typical switch architectures. Looking at the switch architecture from the viewpoint of buffer arrangement, there are four types: an output-buffer-type switch, a common-buffer-type switch, a FIFO (first-in first-out) input-buffer-type switch, and a VOQ (virtual output queue) input-buffer-type switch.

In the output-buffer-type switch, each output line has an output buffer to avoid confliction of multiple cells going to the same output line from multiple

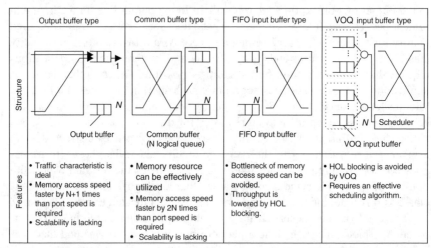

FIGURE 7.8 Typical switch architectures.

input lines. Consequently, this type of switch shows an ideal traffic characteristic in delay performance, etc. When the number of input lines is N, a maximum of N cells are simultaneously input to the output buffer, and it is required to output one cell. Therefore, the memory access speed must be $N + 1$ times faster than the speed of the input/output lines. If it is assumed that the cell length is L bits and the line speed is C bps, then the required memory access speed S_{access} must be faster than $(N + 1) \times C/L$. Therefore, N must be less than $(S_{access} \times L/C) - 1$, i.e., N is limited in inverse proportion to C. For example, if it is assumed that the access speed (S_{access}) of SRAM used in the output buffer is 200 MHz, C = 10 Gbps, and L = 64 bytes = 512 bits, then N must be equal to or less than 9.24, and the maximum number of input/output lines is 9. The problem is that it becomes difficult to obtain a switch of sufficiently large size because of the lack of scalability for expanding the switch size.

The common-buffer-type switch uses the same memory instead of N output buffers in the output-buffer-type switch as a logical queue to effectively utilize the memory resource. Because the same memory is shared, when there is no cell in a certain output port, another output port's cell can use the buffer area of that output port. Because of this memory sharing, the total amount of required memory can be greatly reduced compared with the output-buffer-type switch. However, the common-buffer-type switch has to input N cells in one cell time and has to output N cells in one cell time. Thus the memory access speed must be $2N$ times faster than input/output line speed. This requirement is stricter than for the output-buffer-type switch, so the lack of scalability in switch size again comes into play.

Structure of IP Router

There is a switch architecture — an input-buffer-type switch — that avoids the restriction of switch size caused by the bottleneck of memory access speed in the output-buffer-type or the common-buffer-type switches. This type of switch features a buffer in the input line side. Input-buffer-type switches can be divided into two types, depending on the structure of the input buffer: a FIFO (first-in first-out) input-buffer type and a VOQ (virtual output queue) input-buffer type. We describe the architecture of these input-buffer-type switches in the next section.

7.2.2 Input-Buffer-Type Switch

7.2.2.1 FIFO Input-Buffer-Type Switch

Figure 7.9 shows the structure of a FIFO input-buffer-type switch. Each input line has its own buffer. Between the input buffer and the output port, there is a lattice-type switch called a crossbar switch. The speed of the internal line of the crossbar switch is the same as the speed of the input/output lines. Thus, a maximum of one cell is input into the input buffer in one cell time, and a maximum of one cell is output from the input buffer in one cell time.

In the case of Figure 7.9, cells A and D have been input to the input buffer of input port 0, cells B and E have been input to the input buffer of input port 1, and cell C has been input to the input buffer of input port 2. The destinations of cells A, B, C, D, and E are output ports 1, 1, 0, 2, and 2, respectively. When looking at the top cell of the FIFO input buffer, the destinations of cells A and

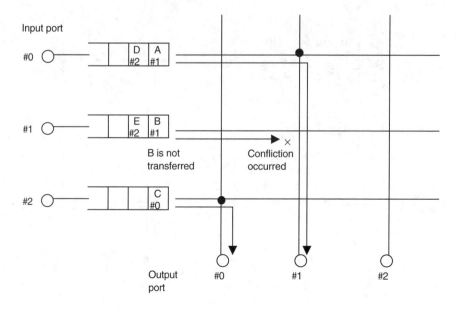

FIGURE 7.9 FIFO input-buffer-type switch.

B are port 1, but because there is a conflict, just one of these cells can be transferred. In this example, cell A in input port 0 is allowed to transfer to output port 1. Cell B is given an opportunity of outputting to output port 0 at the next cell time. The destination of cell C in input port 2 is output port 0, and there is no conflict. As a result, cell C is allowed to transfer to output port 0.

In input port 1, the destination of the second cell (cell E) is output port 2. However, even though output port 2 is available, cell E cannot be transferred because the top cell (cell B) has been defeated in competition. This is an example of HOL (head of line) blocking. In the case of Figure 7.9, only two of three output ports are utilized because of HOL blocking. This illustrates a limitation of the FIFO input-buffer-type switch, where throughput of the switch is limited by HOL blocking. When input traffic is addressed to each destination according to even distribution, the throughput of a FIFO input-buffer-type switch is 2 −2 = 58.6% [1].

7.2.2.2 VOQ Input-Buffer Type

The problem of throughput reduction by HOL blocking in the FIFO input-buffer-type switch can be solved by rearranging the input buffer so that stored cells at each input port can pass to their assigned output ports as soon as that port is available. This buffer is called a virtual output queue (VOQ).

Figure 7.10 shows the structure of this VOQ input-buffer-type switch. The example of input cells in Figure 7.10 is the same as that of Figure 7.9. However, because

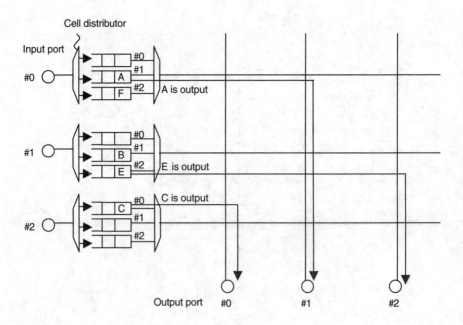

FIGURE 7.10 VOQ input-buffer-type switch.

Structure of IP Router

there are three output ports, three VOQs are allocated to each input port. For example, input port 0 has three VOQs that go to output ports 0, 1, and 2. The cells coming into input port 0 are stored to respective VOQs according to the destination information in the cell header through the cell distributor. Cell A is stored into a VOQ that addresses output port 1, while cell F is stored into a VOQ that addresses output port 2. Similarly, in input port 1, cell B is stored into a VOQ that addresses output port 1, while cell E is stored into a VOQ that addresses output port 2. In input port 2, cell C is stored into a VOQ that addresses output port 0.

To effectively utilize the input/output ports, scheduling is built to determine from which VOQ to which output port the cells should be transferred. In the case of Figure 7.10, as the result of scheduling, cell A is transferred from input port 0 to output port 1, cell E is transferred from input port 1 to output port 2 and cell C is transferred from input port 2 to output port 0.

In the FIFO input-buffer-type switch, because cell B was not transferred into input port 1, cell E was blocked even though output port 2 was available. But, in this VOQ input-buffer-type switch, because cells are stored to buffers coinciding with each output port, HOL blocking can be avoided. However, the performance of a VOQ input-buffer-type switch depends greatly on the scheduling algorithm adopted. Thus it is imperative that the scheduling algorithm be effective.

7.2.2.3 Maximum Size Matching

The problem of scheduling in a VOQ input-buffer-type switch can be considered by modeling it as shown in Figure 7.11. In the graph-G of Figure 7.11(a), the cells in input port 0 are stored in VOQs addressing output ports 1 and 2. In input port 1, cells are stored in VOQ addressing output ports 1 and 2. And in input port 2, a cell is stored in VOQ addressing output port 0. The arrows in the figure point from the input port where the cells are stored in VOQs to the corresponding output port. In scheduling in a VOQ input-buffer-type switch, given the fact that the internal speed of the crossbar switch is the same as the line speed of the input/output port, an aggregate of arrows is selected such that there is one arrow for each pair of input and output ports. This aggregate of arrows is called "matching-M." In the case of an $N \times N$ switch, a maximum of N arrows can be selected.

The maximum-size-matching algorithm is an algorithm that selects matching-M so that the number of arrows becomes maximum in graph-G [2]. However, the maximum-size-matching algorithm does not always maximize the throughput of the switch because it considers only the output request for the top cell in VOQ and not the status of the cells stored in VOQ. Also, in the maximum-size-matching algorithm, there is a problem of starvation where certain cells can never be output.

Figure 7.12 shows an example of starvation in maximum-size matching. In graph-G of Figure 7.12(a), (i,j) is a load of traffic going from input port i to

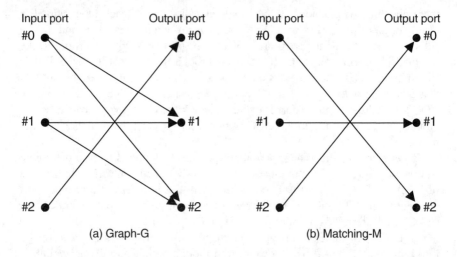

FIGURE 7.11 Scheduling in VOQ input-buffer-type switch.

output port j. Here, $(0,0) = (0,1) = (1,0) = 1.0$, and $(1,1) = 0.0$ are assumed. When an arrow going from input port 0 to output port 0 is selected, an arrow starting from input port 1 cannot be selected, and only one arrow can be selected as matching-M. When the arrow going from input port 0 to output port 1 is selected, the arrow going from input port 1 to output port 0 can also be selected, and the number of arrows that can be selected as matching-M thus becomes two. The maximum-size-matching algorithm is designed to select the latter matching-M. However, in this adopted matching-M, the traffic going from input port 0 to output port 0 cannot be transferred. Figure 7.12(b) shows the result of matching-M.

Another problem is the complexity of computing the maximum-size-matching algorithm. This is given as calculation order of $O(n^{5/2})$ where n is the switch size, which is too complex to implement in hardware.

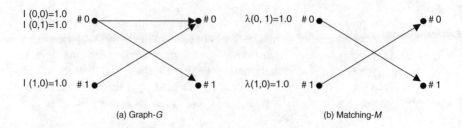

FIGURE 7.12 Starvation in maximum-size-matching algorithm.

7.2.2.4 Maximum Weighting Size Matching

The maximum-weighting size-matching algorithm can solve both problems in the maximum-size-matching algorithm: the difficulty of maximizing the switch throughput and the starvation problem.

In the maximum-size-matching algorithm, only the output request of the top cell of VOQ is considered, and the status of the cell stored in VOQ is not considered. In contrast, the maximum-weighting size-matching algorithm considers a weighting time after a cell becomes the top cell of the VOQ. In the maximum-weighting size-matching algorithm, assuming that the weighting time of the top cell in VOQ to store the cell going from input port i to output port j is $W_{i,j}$, it selects the matching-M that makes (i,j) $M \times W_{i,j}$ maximum. McKeown et al. [2] have demonstrated that if the process of cell arrival is independent and the input traffic does not cause an overload on the output port, then maximum-weighting size matching can achieve 100% of switch throughput. However, the complexity of computing of the maximum-weighting size-matching algorithm is $O(N^3 \log N)$, which, like the maximum-size-matching algorithm, is also too complex to implement in hardware.

7.2.2.5 Parallel Interactive Matching (PIM)

The maximum-size-matching and maximum-weighting size-matching algorithms were not realistic solutions because of the difficulty of implementing them in hardware. The complexity of computing the matching-M that makes an object function at maximum capacity was simply too high. A local maximum-size-matching algorithm has been proposed as a realistic implementation. Two representative local maximum-size-matching algorithms are PIM (parallel iterative matching) and "iSLIP". The special characteristic of the local maximum-size matching is that implementation is simple and easy because it searches the local maximum by simple iterative computation instead of searching for the maximum value of the objective function. We describe iSLIP in Section 7.2.2.6.

Figure 7.13 shows an example of PIM operation. PIM consists of three phases: request, grant, and accept phases [3].

Request phase: Each input port transmits a request signal to the output port of VOQ in which at least one cell is stored.
Grant phase: An output port that receives one or more requests from the input ports randomly selects one request and returns a grant signal back to the input port.
Accept phase: An input port that receives one or more grants from the output ports randomly selects one grant and transmits an accept signal to the output port.

In the example of Figure 7.13, as a matching-M, a pair of (0,1) and (2,0) is selected at the first iteration. Here, (i,j) represents input port i and output port j.

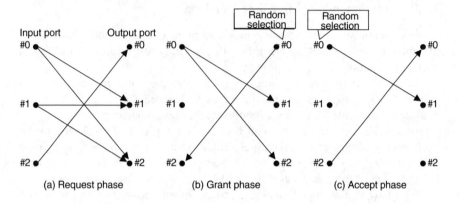

FIGURE 7.13 Example of PIM operation.

Next, in the second iteration, the object of selection is the input/output ports that were not included in matching-M at the time of the first iteration, and then (1,2) is selected and included in matching-M.

In using the local maximum-size-matching algorithm, a local maximum value is sought at each iteration, and the selected pair of input/output ports is accumulated until they converge. Although matching cannot be maximized, this algorithm is simpler than the maximum-size-matching or the local maximum-size-matching algorithms.

When the destination of input traffic is even, the throughput of PIM using only the first iteration approaches $1 - 1/e = 63\%$ when switch size N becomes large. As the number of iterations increases, the switch throughput also increases, and if iteration is executed N times, the throughput reaches 100%.

To obtain high throughput, many iterations are required. But, because PIM scheduling must be completed within one cell time, the number of iterations is limited, depending on the hardware performance. And in the grant and accept phases, because random selection is used, a random number must be generated. This algorithm for generating the random number is very complex. These are the problems relating to PIM.

7.2.2.6 *i*SLIP

*i*SLIP is one of the local maximum-matching algorithms and solves the two problems of PIM described in the last part of Section 7.2.2.5 [4, 5]. The difference from PIM is that it adopts a round-robin selection instead of random selection in the grant and accept phases. In round-robin selection, a request is searched by circulation starting from the round-robin pointer, as shown in Figure 7.14. In circulation, the request that was found first is selected. Then, to search the next cell time, the round-robin pointer is updated to the new point next to the selected request.

Structure of IP Router

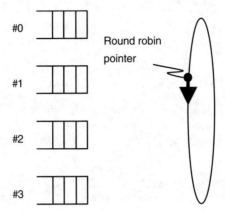

FIGURE 7.14 Round-robin scheduler.

Figure 7.15 and Figure 7.16 show examples of *i*SLIP operation. *i*SLIP also consists of a request phase, a grant phase, and an accept phase, just like PIM.

Request phase: Each input port transmits a request signal to the output port of VOQ in which at least one cell is stored.

Grant phase: An output port that receives one or more requests from input ports selects one request by round-robin selection and returns a grant signal back to the input port.

Accept phase: An input port that receives one or more grants from output ports selects one grant by round-robin selection and transmits an accept signal to the output port. The round-robin pointers are then updated for the output and input ports that exchanged grant and accept signals.

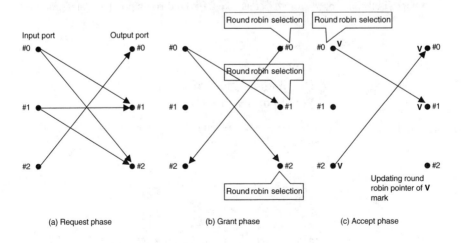

FIGURE 7.15 Example of *i*SLIP operation (first iteration).

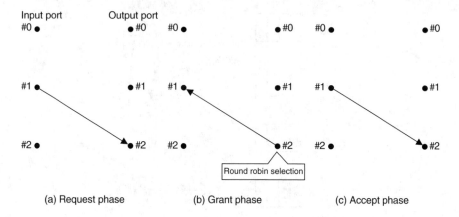

(a) Request phase   (b) Grant phase   (c) Accept phase

FIGURE 7.16 Example of *i*SLIP operation (second iteration).

Updating of the round-robin point is limited to only the first iteration because of the need to maintain the fairness of each input/output port. In Figure 7.15, (0,1) and (0,2) are selected as the matching-*M* by the first iteration. At this time, the round-robin pointers of input port 0, output port 1, input port 2, and output port 0 are updated, and other round-robin pointers are not updated. In Figure 7.16, (1,2) is selected by the second iteration from among the input/output ports that were not selected as matched ports at the first iteration.

The special feature of *i*SLIP is that it can achieve 100% throughput via a single iteration for input traffic of which destinations are even. The principle of operation of *i*SLIP is shown in Table 7.1. We assume that the input load is 1.0. In the cases where input-traffic destinations are even, because cells are stored evenly to all VOQs, after enough time has passed, VOQ comes to such a state that a sufficient number of cells have been stored. Let us take the round-robin pointers corresponding

**TABLE 7.1
Distribution Effect of Round-Robin Pointer in *i*SLIP**

Pointer	\multicolumn{6}{c}{Time (t)}					
	0	1	2	3	4	5
P_{i0}	0	1	2	0	1	2
P_{i1}	0	0	1	2	0	1
P_{i2}	0	0	0	1	2	0
P_{o0}	0	1	2	0	1	2
P_{o1}	0	0	1	2	0	1
P_{o2}	0	0	0	1	2	0

Structure of IP Router

to the respective input/output port as P_{i0}, P_{i1}, P_{i2}, P_{o0}, P_{o1}, and P_{o2}. Here, the pointer subscripts designate a selected input/output port. In Table 7.1, for example, the number of pointer P_{i0} of input port = 0 at $t = 0$ indicates that the output port is 0. At time $t = 0$, the values of all pointers are set to 0. When $t = 0$, (0,0) is selected as the matching-M from the pointer values, and the corresponding pointer value is updated. When $t = 1$, (0,1) and (1,0) are selected, and the pointer values are updated. When $t = 2$, (0,2), (1,1), and (2,0) are selected, and the pointer values are updated. In this way, as the pointer values corresponding to the selected ports are updated and the pointer values corresponding to the unselected ports are not updated, the pointer values of respective ports are shifted accordingly. Thus, the pointer values at $t = 2$ are all different, and matching of three input/output pairs is achieved. Thereafter, the pointer values are continuously updated, keeping the matched state between three input/output pairs. As a result, throughput reaches 100%. Because the pointer value is shifted automatically, this is called the distribution effect of a round-robin pointer. In the *i*SLIP algorithm, it is possible to achieve 100% throughput by a single iteration because of the distribution effect of the round-robin pointer.

Figure 7.17 shows a comparison of switch throughput for various input-buffer-type switches where the input traffic of the destinations is even. The results of PIM and *i*SLIP correspond to the case of single iteration. Because the VOQ input-buffer-type switch of *i*SLIP can achieve high throughput using simple hardware, it is widely implemented in commercial routers [4].

7.2.2.7 Application of *i*SLIP to Three-Stage Cross Network Switch

In the *i*SLIP algorithm, a single-stage crossbar switch has been utilized. However, in the crossbar switch, the number of switching elements at the cross points in the crossbar increases in proportion to square of the number of input/output ports, and the number of input/output ports that can be implemented is limited by the

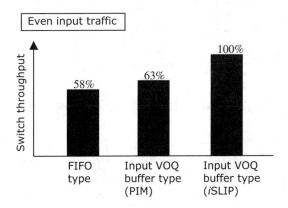

FIGURE 7.17 Comparison of switch throughput.

number of pins when the crossbar switch is integrated as an LSI (Large-Scale Integration) switch. Thus, the crossbar switch cannot be implemented by hardware when the number of input/output ports is increased.

To resolve the problem of limitation in the crossbar switch, Clos [6] has proposed a three-stage cross-network switch that connects switches with the network and deals with these connected switch group as a single switch. In an IP router, cells (packets) that were input into the first-stage switch select the appropriate second-stage switch and are directed to the third-stage switch, which has the destination output lines. To obtain high throughput, cell conflict must be avoided. In this case, it becomes a problem of which of the second-stage switches the cells or packets should go through. This confliction-control method in a three-stage cross-network switch is called a dispatching system.

The concurrent round-robin dispatching (CRRD) system applies the round-robin arbiter adopted in the *i*SLIP algorithm and its pointer updating method to a three-stage cross-network switch, [7, 8]. Figure 7.18 shows a structure of this CRRD system. In the first-stage switch, a VOQ corresponding to the output port is set up. Cells that are input into the first-stage switch from the input port are stored into VOQ according to the destination of the output port of the third stage. In CRRD, there are three round-robin arbiters. The first round-robin arbiter selects the output line of the first-stage switch or the second-stage switch. The second round-robin arbiter selects VOQ. The third round-robin arbiter selects the first-stage switch. When determining the matching-*M*-considering selection of the second-stage switch, it goes through the request phase, the grant phase, and the accept phase just like the *i*SLIP. Updating the round-robin pointer is executed only on the round-robin arbiter that has been related to the matching process. Thus, conflict can be avoided because of the distribution effect of the round-robin pointer in which the pointer

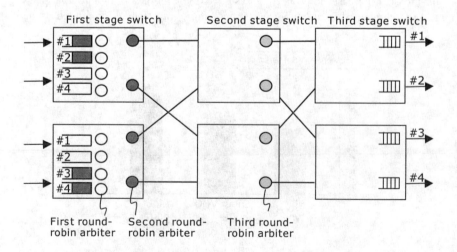

FIGURE 7.18 Structure of CRRD system.

TABLE 7.2
Types of Scheduling in Input-Buffer-Type Switches

Switch Type	Scheduling Type	
	Random	Round Robin
Crossbar	PIM	*i*SLIP
Three-stage cross network	RD	CRRD

value is automatically shifted. It has been shown that, in a CRRD system, it is possible to achieve 100% of throughput where the input traffic of the destinations is even.

On the other hand, a random dispatching (RD) method that adopts the random-selection method instead of the round-robin selection method has also been proposed [9]. However, in this RD method, the requests conflict owing to random selection, as in the PIM algorithm, and it can achieve only 63% of switch throughput unless the switch is internally made faster.

Table 7.2 shows a classification of scheduling types in input-buffer-type switches. For the crossbar and three-stage cross-network switches, there are PIM and RD in random selection, and *i*SLIP and CRRD in round-robin selection, respectively.

7.3 PACKET SCHEDULING

The output line card of routers (Figure 7.6) transfers the packets/cells that were transferred from the switch block to the output lines. At this time, the packets are classified into service classes (label-switched path or flow) and sent out to output lines according to the packet scheduling after being stored to the buffer of the corresponding service class in the output line card. In this section we describe the representative packet-scheduling methods.

There are several types of packet scheduling.

- FIFO (first-in first-out) queuing
- Complete priority scheduling
- Generalized processor sharing
- Packetized general processing sharing
- Weighted round-robin (WRR) scheduling
- Deficit round-robin (DRR) scheduling

7.3.1 FIFO (FIRST-IN FIRST-OUT) QUEUING

FIFO is the simplest packet-scheduling method. Packets of all the service classes are input into a single buffer and are read out according to the order of input. A FIFO queuing method executes a shaping function that stores the packets temporarily in the FIFO so as to be able to read out at the output line speed when the writing speed of the input packets is faster than the readout speed. Because this method is easy to implement, it is mostly implemented in the output block to high-speed lines in the intermediate routers in the backbone network. However, it is not possible to provide various QoS (quality of service) corresponding to each service class with only this FIFO queuing method.

7.3.2 Complete Priority Scheduling

Complete priority scheduling has buffers corresponding to each priority class, and packets are stored into buffers corresponding to the respective priority class. Figure 7.19 shows an example of priority class. Priority class 1 has the highest priority, and priority class 3 has the lowest priority. The priority class 1 packets can be read out regardless of whether or not the packets are stored in the priority class buffers. The priority class 2 packets can be read out only when the priority class 1 packets do not exist, and regardless of whether the priority class 3 packets are stored or not. The priority class 3 packets can be read out only when both priority class 1 and 2 packets do not exist.

Because the higher-priority class packets are read out prior to the lower-priority class packets, it is possible to achieve a low-latency service. For example, a service class for voice communication that requires a high latency quality can be allocated to the highest priority class.

7.3.3 Generalized Processor Sharing

The generalized processor sharing (GPS) method is a scheduling method that intends to share the bandwidth ideally. A buffer is provided to each flow, and the packets are stored into the buffer corresponding to the flow. We express the presence of packets in a buffer as "flow is active." Conversely, we express the absence of packets in a buffer as "flow is inactive." In this generalized processor-sharing method, bandwidth is shared among the active buffers depending on a weight.

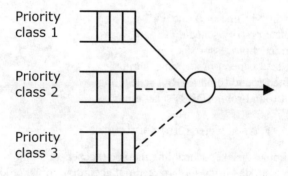

FIGURE 7.19 Complete priority scheduling.

Structure of IP Router

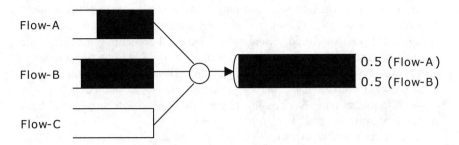

FIGURE 7.20 Generalized processor sharing (GPS).

Consider the case in which the weight is the same for all the flows. In the case of Figure 7.20, buffers are provided to flows A, B, and C; flows A and B are active, and flow C is inactive. When an input bandwidth of the sum of flows A and B exceeds the bandwidth of the output line and both of the input bandwidths exceed 50% of output line bandwidth, each flow shares 50% of bandwidth equally.

In the generalized processor-sharing method, flow is treated just like a water stream. Because the scheduler reads out data not packet by packet, but by the infinitely small amount by flow A, flow B, flow A, etc., it can provide an equal sharing of bandwidth. However, from a practical standpoint, it is impossible to deal with flow as a fluid and to read out the infinitely small amount of data step by step. The generalized processor-sharing method is just an idealized scheduling method and is not an actual solution.

7.3.4 PACKETIZED GENERALIZED PROCESSOR SHARING

The packetized generalized processor sharing (PGPS) method is also called the weighted fair queuing (WFQ) method [10], and the names PGPS and WFQ are used interchangeably. In PGPS, scheduling is executed by selecting the packets to be read out packet by packet, as shown in Figure 7.21, so as to approach as closely as possible the idealized generalized processor sharing described in Section 7.3.3.

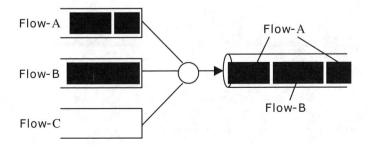

FIGURE 7.21 Packetized generalized processor sharing (PGPS).

In selecting the packets to be read out, PGPS uses a concept of "finish time." A finish time is assigned to each packet corresponding to the time when reading of the packet was completed, assuming that the ideal scheduling method or the generalized processor-sharing method was utilized. In PGPS, the finish time for each packet is calculated, and the packet with the earliest finish time — determined by comparing it with the finish time of the packet at the top position of buffer — is read out. In this manner, it is possible to simulate approximately the generalized processor-sharing method.

One of the advantages of the packetized generalized sharing method is the ability to determine the upper limit of delay time within the network. Because the upper limit of delay time in the case of using the packetized generalized sharing method is given to each node against the amount of traffic flow permitted for the network, it is possible to determine, based on this, the upper limit of delay time within the network. Traffic flow is monitored by using a device called a "token backet" at the entrance of the network. The upper limit of delay time within the network can be used to control the flow of reception.

A problem of packetized generalized sharing is that calculation of the finish time is very complex. A finish time must be calculated every time a packet is input to the buffer, and if the output line is high speed and the number of buffers is large, implementation of packetized generalized sharing becomes difficult. Therefore, this packetized generalized sharing method is mostly used when the speed of the line is slow and the number of buffers is small.

7.3.5 WEIGHTED ROUND-ROBIN (WRR) SCHEDULING

As one of the solutions to resolve the complex implementation problem of packetized generalized sharing, there is a WRR method. For active flow, this method selects buffers n times in proportion to a given weight and reads out the packets [11].

Here we describe the WRR method using Figure 7.22. The flow of each buffer is active. The round-robin method of Figure 7.22(a) selects the buffers sequentially (for equally weighted buffers) and reads out packets from the selected buffers. Figure 7.22(b) and Figure 7.22(c) illustrate two implementation methods

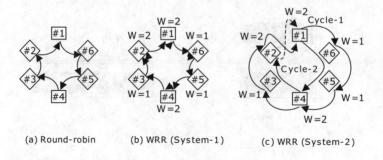

FIGURE 7.22 Weighted round-robin (WRR).

Structure of IP Router

of WRR scheduling. Buffers 1, 2, and 3 are given heavier weight by two times than the other buffers. System 1 of Figure 7.22(b) reads out the packets two times from buffers 1, 2, and 3, and one time from the other buffers while the round-robin arbitration goes around one cycle. On the other hand, in system 2 of Figure 7.22(c), the round-robin arbitration consists of two cycles. In cycle 1, packets are read out from all the buffers, but in cycle 2, a packet is read out only from the buffers that have a weight of two times greater. The number of cycles increases according to the value of weight.

When comparing systems 1 and 2, system 1 is easier to implement, but the time required for round-robin arbitration to go around one cycle is longer than for system 2, which consists of multiple cycles. Therefore, the packet delay time of system 1 becomes longer than that of system 2.

In the WRR method, packets are read out by the number of times corresponding to their weight regardless of the packet length. Therefore, the amount of data is not proportional to weight. When an average packet length of a certain buffer is longer than the average packet length of the other buffers, that buffer transfers more data than the others. If one wishes to consider the weight for the amount of data, it is necessary to know the average packet length beforehand and set up the value that is proportional to "weight for amount of data/average packet length" as a weight factor of WRR.

7.3.6 Weighted Deficit Round-Robin (WDRR) Scheduling

In the WRR method, it is necessary to know the average packet length beforehand if we want to output an amount of data that is proportional to its weight. However, it is difficult to forecast the average packet length for real traffic. In this WDRR (weighted-deficit round-robin) method, it is not necessary to know the average packet length beforehand to output the amount of data corresponding to the weight that was set up [12]. Because the WDRR method is based on the round-robin arbitration (just like the WRR method), implementation is easier than the packetized generalized processor sharing method.

Here we describe the WDRR method using Figure 7.23. In the WDRR method, each buffer has a deficit counter. The initial value of the deficit counter has been set

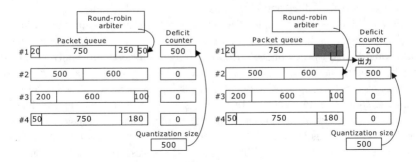

FIGURE 7.23 Weighted-deficit round-robin (WDRR).

to 0. A quantization size in inverse proportion to respective weight has been given to each buffer. In the case of Figure 7.23, it is assumed for simplicity that weight is uniform, that all the quantization sizes are 500, and that the quantization size is added to the deficit counter value in turn of buffers selected in a round-robin method.

First, in buffer 1, a quantization size of 500 is added to the deficit counter value of 0. As a result, the deficit counter value for buffer 1 becomes 500. Buffer 1 can transfer packets if the packet length is smaller than the deficit counter value. In this example, because the length of the top packet is 50, this top packet can be transferred. When the packet is transferred, the deficit counter value is subtracted by packet length transferred, that is, it becomes 450 (= 500 − 50). And because the length of the second packet is 250 (<450), the second packet can also be transferred, and the deficit counter value becomes 200 (= 450 − 250). The length of the third packet is 750 (>200), so the third packet cannot be transferred. Then, the round-robin arbiter moves to buffer 2, and the quantization size 500 is added to the deficit counter value of 0 to make the deficit counter value 500. Because the length of the top packet is 600 (>500), the top packet cannot be transferred. Hereafter, the same procedures are repeated, and when there are no longer any packets in the buffer, the deficit counter value is reset to 0.

In the WDRR method, the length of the packet transferred is subtracted from a deficit counter value. Because the deficit counter reflects the weight of the packet length, it is not necessary to know the average packet length in advance, which was a problem in the WRR method. Because it is based on a round-robin arbiter and is easy to implement, WWDR is widely adopted in today's routers.

7.4 FORWARDING ENGINE

7.4.1 ROUTE LOOKUP

The packets arriving at the router from the input line are processed in the forwarding block. The forwarding block determines to which output port the IP packets should be sent based on the destination address in the header of the IP packet and transfers the packets to the switch block. The operation to determine the output port from the routing table is called a route lookup.

Table.7.3 shows an example of a routing table, which contains such columns as prefix/mask length, next hop, and output port. A routing table can be created automatically by a protocol such as OSPF (Open Shortest Path First), etc., or it can be created manually. For example, assume that packets with a destination IP address of 176.10.15.28 were input. The mask length indicates to which bit from the most significant bit (leftmost bit) the bits are important, and the bits that are located to the right of that bit are ignored. Route lookup searches the table of prefixes and chooses the one that coincides with the IP address over the longest number of bits. In the case of Table.7.3, the IP address 176.10.15.28 coincides partly with 176/8 and 176.10.25/24, but it also coincides in a longer number of

TABLE 7.3
Example of Routing Table

Prefix/Mask Length	Next Hop	Output Port
129.66/16	129.30.166.55	3
129.66.103/24	129.30.167.13	4
176/8	129.30.168.20	6
176.10/16	129.30.169.21	5
176.10.25/24	129.30.170.169	1

bits with 176.10/16. Thus, it determines that the most-matched prefix over the longest bits with the IP address is 176.10/16. In this case, the IP address of the next hop is 129.30.169.21 and output port 5, so the IP packets are transferred to output port 5. Searching the prefix that coincides over the longest bits with the IP address is called a "longest match."

In route lookup, if there are too many prefix entries registered to the routing table, it takes a long time to determine the output port. When the speed of the input line is high and the number of entries is high, the route-lookup routine can become a bottleneck. To avoid a bottleneck, the system must determine the output port during the time the IP packet is transferred. For example, assume that IP packets of 40-byte length are continuously input to a 10 Gbps input line. In this case, the required time for route lookup for one packet is $40 \times 8/10^9 = 32$ nsec. The route-lookup routine must be completed within this allotted time.

7.4.2 Design of Route Lookup

A forwarding engine executes route lookup. As described in Section 7.1, in high-end routers, a forwarding engine is furnished to each input line. When the forwarding engine must have a high-speed capability, a network processor is dedicated as a forwarding engine. The network processor, unlike a conventional general purpose processor, has been developed for special communications applications. Although the functions of the network processor can be changed by software, it is not as flexible as a general purpose processor. In general, because the network processor is limited in its application area, its cost is higher than general purpose processors, and it is used mostly for applications that require high-speed performance.

In implementing the route-lookup system, there are two methods. One is to use RAM (random access memory), and another is to use CAM (content addressable memory).

In the method that uses RAM, the information in the routing table is modified to a data structure that facilitates the table lookup, and this data structure is held in RAM. To obtain a prefix with the longest match from a single IP address,

184   GMPLS Technologies: Broadband Backbone Networks and Systems

multiple accessing and searching of RAM is necessary. In implementing the route-lookup system using RAM, the following items are taken into consideration.

Number of times of searching
Size of data structure used for searching the route lookup
Ease of updating the data structure

High-speed route lookup is enhanced by reducing the number of times of searching as much as possible. The number of times of searching corresponds to the number of times RAM is accessed, and the memory access time greatly affects the route-lookup time. To avoid a bottleneck, the number of times that memory is accessed must be reduced. Data-structure considerations are also important. As the number of prefix entries increases, the data structure becomes larger, causing a shortage in the amount of free memory. To reduce the implementation cost, the size of the data structure must be reduced. Another important consideration is the ease of updating the data structure. As the network size increases, the frequency of updating the network topology also increases. If topology is updated, the routing table must also be updated. Thus, the data structure used for searching the route lookup must be updated frequently. Corresponding to a change of network topology, it is desirable if the data structure can be updated easily.

Relating to route lookup using RAM, we describe the representative algorithms of route lookup in Sections 7.4.3 through 7.4.5, focusing attention on the three items identified here.

The method using CAM obtains the longest match with a single access to memory using hardware logic. Because the price of CAM is higher than ordinary RAM, if the number of prefix entries increases, the effect of cost increase using CAM becomes great. However, the merit of reducing the route-lookup time is very significant. We describe the route-lookup system using CAM in Section 7.4.6.

7.4.3 Trie Structure

Here we describe the route-lookup system based on the Trie structure [13] as shown in Figure 7.24. In this case, the prefix entries of 00*, 0001*, 001*, 0101*, 101*, 10100*, and 111* have been registered.

When a destination IP address is input, it is searched to find which prefix it belongs to. To make searching easier, a Trie structure is constructed. A Trie structure has a 0-pointer and a 1-pointer at the branch point, which is also called a node. Searching the prefix that creates the longest match is done in such a way that, starting from top bit of the destination IP address, each bit is judged as 0 or 1 at each node in turn and moves to next node (child node) to be judged as 0 or 1, and so on. The hth node judges 0 or 1 of the hth bit of the destination IP address.

Next, we describe this operation using an example. Let us assume that 101011 (for simplicity, six bits assumed) was input as the destination IP address. First, because the first bit is 1, it goes to branch point 1 at the first node. Because the second bit is 0, it moves to branch point 0. Similarly, the third bit and the fourth bit are judged. At the node that judges the fourth bit, prefix 101* (e in Figure 7.24)

Structure of IP Router

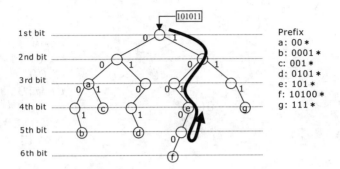

FIGURE 7.24 Trie structure.

has been registered. Then, it proceeds to the fifth bit. The fifth bit is 1, but there is no branch point corresponding to this value. Therefore, it returns to the fourth bit's node, and determines the prefix 101* as the longest match.

As a special feature of Trie structure, the data structure is easily updated. For example, if we want to add 1* as a new prefix, we can do it just by adding the prefix to the relevant location. If we want to delete a prefix, we can do it simply by deleting the prefix of the relevant node. And if the node with a deleted prefix does not have a child node, the pointer relating to it is also deleted. Then, going up to the upper-level node, it judges whether the upper-level node should be deleted or not. This process is repeated until the longest match can be obtained.

A problem of Trie structure is that the number of times a search should be executed corresponds to the number of bits of the destination IP address. The number of bits of IP addresses corresponding to IP v.4 and IP v.6 are 32 bits and 128 bits, respectively. Therefore, in route lookup by Trie structure, a maximum 128 searches are required for IP v.6.

7.4.4 Patricia Tree

To reduce the number of times of searching the Trie structure, a Patricia tree has been proposed [14]. Figure 7.25 shows an example of a Patricia tree. This example of prefix is the same as Figure 7.24. In Trie structure, a hierarchical structure was created corresponding to the number of bits of the IP address. The number of hierarchical levels is reduced in a Patricia tree. For example, in Figure 7.24, there is a node that has no prefix between prefix 00* (indicated by a in Figure 7.25) and 0001* (indicated by b in Figure 7.25). Because the pointer of this node is just one (0-pointer), it can be omitted. So, in a Patricia tree, the node that has no prefix and has just one pointer is deleted by placing the prefix 0001* (indicated by b in Figure 7.25) just below the prefix 00* (indicated by a in Figure 7.25). The node with prefix 00* (indicated by a in Figure 7.25) has a 01 pointer and a 0 pointer. In this way, it is possible to reduce the number of times of searching by deleting unnecessary nodes as much as possible.

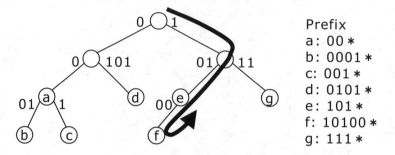

FIGURE 7.25 Patricia tree.

Next, we describe the operation of a Patricia tree using an example of Figure 7.25. Here we assume that the destination IP address 101011 was input similarly to the example in Section 7.4.3. First, taking the first bit as the reference at the first node, because the first bit is 1, it advances to branch point 1. Then, taking the second and third bits as the reference, because they are 01, it advances to branch point 01. In the node to which it reached, there is a prefix 101* (indicated by a in this figure). And then, taking the fourth and fifth bits as the reference, although they are 01, there is no branch point of 01, so it returns to the node that has a prefix and selects the prefix 101* as the longest match.

In this way, in a Patricia tree, the number of hierarchical layers is reduced, and by judging the multiple bits at the same time, the average number of search times is reduced. Also, the amount of data held in the data structure was reduced in comparison with a Trie structure. As for updating the data structure, it is just a little bit more complex than for the Trie structure. However, in a Patricia tree, there is a case that the number of hierarchical layers cannot be reduced, depending on the prefix of the entry. The worst-case value of the number of search times in a Patricia tree is the same as the one in the Trie structure, that is, the number of bits of the destination IP address.

7.4.5 Binary Search Method

To reduce the search time during route lookup in a Trie structure or a Patricia tree, a binary search method using a binary search tree has been developed [15]. In the algorithm of the binary search method, two IP addresses, maximum and minimum, are first created for one prefix. IP addresses created from all the prefixes are sorted in ascending order (or in descending order). Then, when the destination IP addresses are judged to be between the entries to which they belong, the longest match is determined. From an experiment using an actual routing table with 38,000 entries, it has been reported that the search time using the binary search method was reduced to 20% of the search time in a Patricia tree [15].

Here we describe the route-lookup operation by binary search method using Figure 7.26 as an example. Again, for simplicity, we consider a case where the address

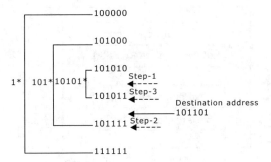

FIGURE 7.26 Binary search method.

is six bits. Three prefixes are given: 1*, 101*, and 10101*. Because these are prefixes, even if we sort these and apply a binary search method, the longest match cannot be obtained. So, to apply the binary search method to route lookup, the following idea is implemented. That is, we consider that 1* indicates the range from the minimum value of 10000 to the maximum value of 11111. Similarly, we consider that 101* indicates the range from the minimum value of 10100 to the maximum value of 10111, and that 10101* indicates the range from the minimum value of 101010 to the maximum value of 101011. Therefore, six entries of IP address, 10000, 11111, 10100, 10111, 101010, and 101011 were created from three prefixes. Next, if we sort these in ascending order, these are arranged as shown in Figure 7.26.

Then, by using a popular binary search method, the destination IP address 101001 is examined to see between which entries it belongs. In step 1, six IP addresses are divided into two groups, and the IP address is examined to see which group it belongs to. As a result, it is judged that the IP address belongs to the lower group. In step 2, the lower group is further divided into two groups, and the IP address is examined again to find the group to which it belongs. As a result, it is judged that the IP address belongs to the upper group. In step 3, the same operation is repeated. Finally, it is determined that the IP address belongs between 101011 and 101111. If we focus attention on the range of the prefixes, the belonging prefix in the smallest range becomes 101*. This prefix 101* is equal to the result of a longest match.

In this way, it is possible to apply the binary search method to route lookup by taking the range of prefixes into consideration. The search time in Trie structure and Patricia tree methods was increased in proportion to the number of bits of the IP address. In the binary search method, however, the reach time depends on the number of entries, and not on the number of bits of the IP address. If the number of entries is N, the order of the search time is given by $O(\log N)$, and the search time is reduced. But, it should be noted that, because the entries must be sorted in ascending order during the binary search method, it may take time to update the data structure if entries are frequently created or deleted.

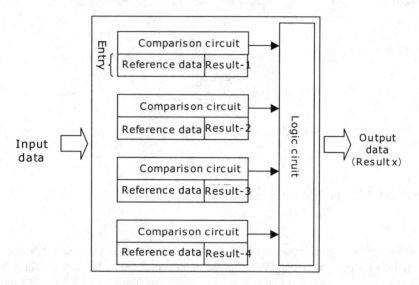

FIGURE 7.27 Typical structure of route-lookup system using CAM.

7.4.6 Route Lookup with CAM

Figure 7.27 shows a typical structure of a route-lookup system using CAM. In CAM, each entry has a reference data field and a result of comparison. When data is input, it compares each bit of all the reference data and the input data at the same time, and if there are any matched entries, it outputs the result as the output data. By accessing CAM just one time, we can get the result. However, in conventional CAM, because the length of the reference data to be compared is fixed, it is difficult to apply it as is for longest matching of a route-lookup system where the entry is a prefix.

Therefore, when the prefix is used as the reference data, as shown in Figure 7.28, only the bits corresponding to the mask length are compared. Because comparison is done only on the bits corresponding to the mask length, the bits other than the mask length can be ignored. Such CAM with a three-value logic is called ternary CAM (TCAM). The entries have been sorted in ascending order of mask length in advance. As input data, the IP address is input. Each comparator simultaneously examines the IP address to see if it coincides with the bits of mask length. Because TCAM is using the mask length, there is a possibility of multiple matching of entries. In this case, the entry with the longest mask length is selected from the matched multiple entries by priority encoder. As a result, output port 3 is selected in this example. The priority encoder in Figure 7.28 has been designed to output the entry with the longest mask length from the matched entries.

In this way, by using TCAM in route lookup, it is possible to obtain the longest match with hardware logic with just a single access to memory. However,

Structure of IP Router

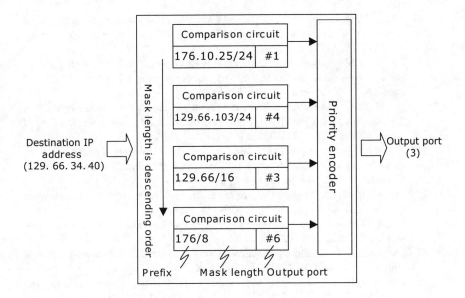

FIGURE 7.28 Ternary CAM (TCAM).

because the price of CAM is higher than ordinary RAM, if the number of prefix entries increases, the effect of a cost increase by CAM becomes great. And because the mask lengths must be sorted in ascending order, it may take a long time to update the data structure if entries are frequently created or deleted. In a route-lookup system using a TCAM, lowering the cost of memory and increasing the speed of updating the entries are current issues that need to be resolved [16].

REFERENCES

1. Karol, M., Hluchyj, M., and Morgan, S., Input versus output queueing on a space division switch, *IEEE Trans. Commun.*, 35, 1347–1356, 1988.
2. McKeown, N., Mekkittikul, A., Anantharam, V., and Walrand, J., Achieving 100% throughput in an input-queued switch, *IEEE Trans. Commun.*, 47, 1260–1267, 1999.
3. Anderson, T., Owicki, S., Saxe, J., and Thacker, C., High speed switch scheduling for local area networks, *ACM Trans. Comput. Syst.*, 11, 319–352, 1993.
4. McKeown, N., The *i*SLIP scheduling algorithm for input-queues switches, *IEEE/ACM Trans. Networking*, 7, 188–200, 1999.
5. McKeown, N., A fast switched backplane for a gigabit switched router, *Bus. Commun. Rev.*, 27, 1997.
6. Clos, C., A study of non-blocking switching networks, *Bell Sys. Tech. J.*, 406–424, 1953.
7. Oki, E., Jing, Z., Rojas-Cessa, R., and Chao, H.J., Concurrent round-robin dispatching scheme in a Clos-network switch, *IEEE ICC*, 107–111, 2001.

8. Oki, E., Jing, Z., Rojas-Cessa, R., and Chao, H.J., Concurrent round-robin-based dispatching schemes for Clos-network switches, *IEEE/ACM Trans. Networking*, 10, 830–844, 2002.
9. Chiussi, F.M., Kneuer, J.G., and Kumar, V.P., Low-cost scalable switching solutions for broadband networking: the Atlanta architecture and chipset, *IEEE Commun. Mag.*, Vol. 35, Issue 12, 44–53, 1997.
10. Parekh, A.K. and Gallager, R.G., A generalized processor sharing approach to flow control in integrated services networks: the single-node case, *IEEE/ACM Trans. Networking*, 1, 344–357, 1993.
11. Katevenis, M., Sidiropoulos, S., and Courcoubetis, C., Weighted round-robin cell multiplexing in a general-purpose ATM switch chip, *IEEE J. Selected Areas Commun.*, 9, 1265–1279, 1991.
12. Shreedhar, M. and Varghese, G., Efficient fair queuing using deficit round-robin, *IEEE/ACM Trans. Networking*, 4, 375–385, 1996.
13. Knuth, D., Sorting and Searching, Vol. 3 of *IT Fundamental Algorithms,* Addison-Wesley, Reading, MA, 1973.
14. Morrison, D.R., ìPatricia ó practical algorithm to retrieve information coded in alphanumeric, *J. ACM,* 17, 1093–1102, 1968.
15. Lampson, B., Srinivasan, V., and Varghese, G., IP lookups using multiway and multicolumn search, *IEEE/ACM Trans. Networking*, 7, 324–334, 1999.
16. Liu, H., Routing table compaction in ternary CAM, *IEEE Micro.*, 22, 58–64, 2002.

8 GMPLS (Generalized Multiprotocol Label Switching)

GMPLS is a protocol that is applied to the TDM (time-domain multiplexing) layer, the wavelength-path layer, and the fiber layer by generalizing the label concept of MPLS that has been applied to the packet layer for transferring IP packets. GMPLS makes it possible to execute distributed control — a feature of MPLS — thereby simplifying the operation. It is also possible to totally engineer the traffic based on the traffic information or topology information of each layer and to improve the utilization efficiency of the network. In this chapter, we describe the history of the development from MPLS to MPλS/GMPLS, provide an outline of GMPLS and its protocol, and present an application example of GMPLS.

8.1 FROM MPLS TO MPλS/GMPLS

Recently, a WDM (wavelength-division multiplexing) technology that transmits high-volume information using multiple wavelengths through a single fiber transmission line has been greatly advanced. In the initial fiber optics transmission technology, information was transmitted by using just one wavelength on a single fiber transmission line between two nodes. But if wavelength multiplexing technology is adopted, the transmission capacity increases in proportion to the number of wavelengths accommodated within the same single fiber, making it well suited for high-volume transmission of information.

An IP/MPLS network is built on the SDH/SONET (synchronous digital hierarchy/synchronous optical network) path network, and most SDH/SONET path networks are constructed on an optical-fiber network. To each optical fiber, a wavelength-multiplexing technology is applied. Usually, multiple SDH/SONET paths are allocated to one wavelength band. Figure 8.1 shows the layer structure of this network. So far, as shown in Figure 8.1(a), the network has been structured by a fiber layer, a TDM layer, and a packet layer (from bottom to top). When traffic demand between nodes becomes larger and the number of wavelengths to be multiplexed increases, network utilization efficiency can be improved by using a wavelength with a bandwidth greater than that of SDH/SONET as a wavelength path by assigning the nodes that execute switching on each wavelength. Therefore, as shown in Figure 8.1(b), it is possible to save

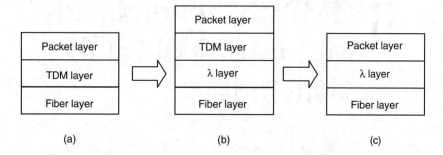

FIGURE 8.1 Network layer structure.

the total cost of the network by using a wavelength-path layer that is inserted between the TDM layer and the fiber layer to deal with the SDH/SONET path. This wavelength-path layer is also called a λ-layer, because λ is generally used as a symbol to express wavelength. The node that executes switching by wavelength unit or fiber unit and the node that executes switching the TDM path are referred to as OXC (optical cross-connect) and DXC (digital cross-connect), respectively. When a large traffic unit is treated, the OXC gets a cost advantage over the DXC. The higher the layer is, the finer the unit of path switching becomes. When IP traffic volume increases and the IP packet transmission becomes dominant, network cost can be saved by taking a simple layer structure in which the packet layer is put directly above the λ-layer by eliminating the TDM layer, as shown in Figure 8.1(c). The image in Figure 8.2 projects the dates of transition to the network layer structures illustrated in Figure 8.1. The speed of penetration of the layer structures shown in Figure 8.1(b) and Figure 8.1(c) will depend on the speed of progress in wavelength-path network construction technology and on the growth in the volume of IP traffic.

Addressing the progress of the λ-layer in an optical network, an MPλS (Multiprotocol Lambda Switching) has been proposed that applies the distributed control technology of MPLS in the packet layer to manage the λ-layer network [1]. Here, λ means wavelength. In MPLS, it is possible to set up a LSP (label-switched path) by exchanging the link information between nodes with a routing protocol and by using a signaling protocol [2, 3]. In MPLS, an LSP is created by attaching to an

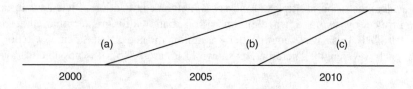

FIGURE 8.2 Image of transition of network layer structure.

IP packet a label that is uniquely defined for each link between two nodes, and the IP packets are transferred along the LSP by swapping the label in a LSR (label-switching router). In MPλS, λ or wavelength inside the fiber is treated as a label just like in MPLS, and it is possible to build the wavelength path by connecting the input-side wavelength and the output-side wavelength in OXC. In MPλS, similar to MPLS, distributed control is also possible by exchanging the link information between nodes with a routing protocol and by setting up the wavelength path using a signaling protocol. That is, we can say that MPλS is the protocol that applies the concept of a label used in MPLS to the λ-layer. Further, it is through a generalized MPLS (GMPLS) that the more generalized concept of label was also applied to the TDM layer and the fiber layer.

8.2 GENERAL DESCRIPTION OF GMPLS

MPLS architecture has been defined to support data transmission based on the concept of a label [4]. In RFC 3031, an LSR is defined as a node that has a data-transmission plane that can identify the boundary of an IP packet or cell (IP packet with a label attached) and execute the data-transmission task according to the content of the IP packet header or cell header. In GMPLS, an LSR includes not only the node that executes the data-transmission task according to the content of the IP packet header or cell header, but also the device that executes data transmission according to the information of time slot, wavelength, and physical port of the optical-fiber network.

The LSR interface in GMPLS is categorized into four types depending on its switching capability: PSC (packet-switch capable), TDM (time-division-multiplex capable), LSC (lambda-switch capable), and FSC (fiber-switch capable). Figure 8.3 shows the concept of label for the four-layer network structure identified in Figure 8.1(b).

PSC: The PSC interface can identify the boundary of an IP packet or cell and executes the data-transmission task according to the contents of the IP packet header or cell header. In the packet layer of Figure 8.3(a), a label that is uniquely defined for each link is attached to the IP packet to form the LSP. The link in Figure 8.3(a) indicates the link that was defined between two LSRs to transfer the IP packet. In the case where the IP packet is transferred via SDH/SONET, this link is called an SDH/SONET path, and in the case where the IP packet is transferred via Ethernet, this link is called an Ethernet path.

TDM: The TDM interface is repeated periodically and executes the data-transmission task according to a time slot. In the TDM layer of Figure 8.3(b), the label corresponds to a time slot. As an example of a TDM interface, there is a DXC interface in which the TDM path or SDH/SONET path is formed by connecting the time slot assigned to the input side and the one assigned to the output side. The link can correspond to the wavelength path or simply to the fiber.

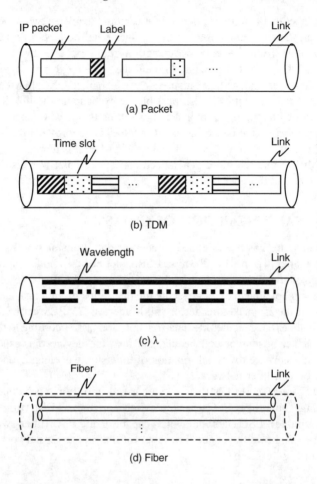

FIGURE 8.3 Concept of label.

LSC: The LSC interface executes the data-transmission task according to the wavelength inside the optical fiber through which data is transmitted. In the λ-layer of Figure 8.3(c), the label corresponds to the wavelength. As an example of TDM interface, there is an OXC interface in which the λ-path is formed by connecting the wavelength assigned to the input side and the one assigned to the output side. The OXC interface with LSC executes switching by wavelength unit.

FSC: The FSC interface executes the data-transmission task according to the position of the actual physical port of the optical fiber through which data is transmitted. In the fiber layer of Figure 8.3(d), the label corresponds to the optical fiber itself. As an example of FSC interface, there is an OXC interface in which the fiber path is formed by connecting the input-side fiber and the output-side fiber to each other.

GMPLS (Generalized Multiprotocol Label Switching)

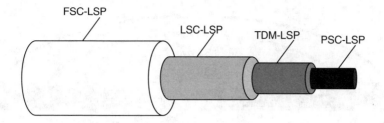

FIGURE 8.4 Hierarchization of LSP.

The OXC interface with FSC executes switching by fiber unit. The link in this case means a physical aggregate of optical fiber, such as conduit, for example.

It is possible to use these switching-capability interfaces by hierarchizing them. In this case, the hierarchy is formed by FSC, LSC, TDM, and PSC in turn from the bottom to the top level. In GMPLS, the path corresponding to respective switching capability is also called an LSP. Figure 8.4 shows a hierarchical structure of LSP. In the case of the layer structure as shown in Figure 8.1(b), PSC-LSP belongs to TDM-LSP and the link of PSC-LSP becomes TDM-LSP. TDM-LSP belongs to LSC-LSP, and the link of TDM-LSP becomes LSC-LSP. LSC-LSP belongs to FSC-LSP, and the link of LSC-LSP becomes FSC-LSP. In the case of the layer structure as shown in Figure 8.1(c), the TDM layer is eliminated, and PSC-LSP belongs to LSC-LSP, with the link of PSC-LSP becoming LSC-LSP. The relationship between LSC-LSP and FSC-LSP is the same as the case of Figure 8.1(b). As one moves down to the lower layer, the bandwidth of the LSP becomes greater.

The term "region" means the range (aggregate) of network that operates with a certain switching capability. However, in this chapter, the term "layer" is used instead of "region" to be consistent with usage elsewhere in this book.

Figure 8.5 shows the relationship between the switching capability and the layer. In between the PSC interfaces, PSC-LSP is formed. The area between the PSC interfaces is called a PSC layer, and the region is the domain that the relevant layer can reach. The interface with TDM forms a TDM-LSP. The area between the TDM interfaces is called a TDM layer. The LSC layer and FSC region are also defined in the same way.

GMPLS provides several advantages. Figure 8.6 shows the current IP/MPLS network. In the packet layer, the network is distributedly controlled using a routing protocol or signaling protocol. In the TDM layer and the λ-layer, the network is centrally controlled by setting up the route or path. In this environment, network operators had to learn how to operate each network corresponding to the respective layer, because different network control methods were used in the different layers.

However, by using GMPLS, it becomes possible to control the network distributedly by expanding MPLS to the TDM layer and the λ-layer, as shown

196 GMPLS Technologies: Broadband Backbone Networks and Systems

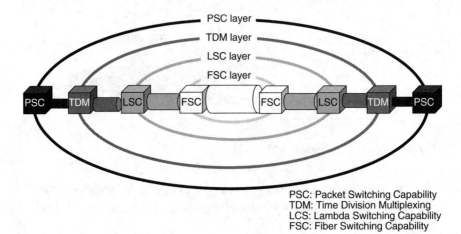

PSC: Packet Switching Capability
TDM: Time Division Multiplexing
LCS: Lambda Switching Capability
FSC: Fiber Switching Capability

FIGURE 8.5 Relationship between the switching capability and the region.

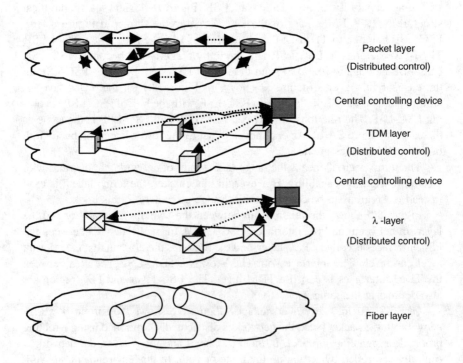

FIGURE 8.6 Current IP/MPLS network.

GMPLS (Generalized Multiprotocol Label Switching)

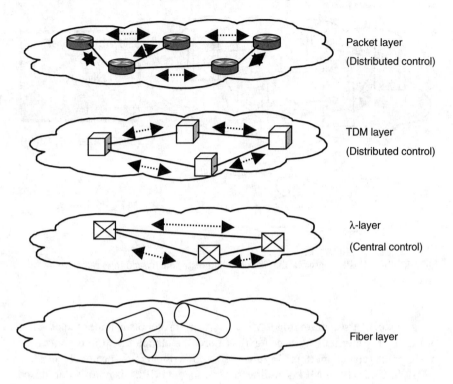

FIGURE 8.7 GMPLS network of which each layer is controlled distributedly.

in Figure 8.7. In this environment, the functions that were executed by a central control unit are now allocated to each node and controlled distributedly. Thus it becomes possible to flexibly address addition or deletion of nodes or links, resulting in improved scalability against the capacity of the network. Moreover, because each layer is operated based on the same GMPLS, human resources can be used more effectively because, after learning and mastering the protocol of GMPLS, the network operator will be able to operate all of the layers.

Furthermore, in a GMPLS network, it is possible to control the network distributedly and integratedly with multiple layers, as shown in Figure 8.8. The integrated traffic engineering for multiple layers is called "multilayer traffic engineering." Figure 8.9 shows the concept of this multilayer traffic engineering. In this example of Figure 8.9, the network is composed of the packet layer, the λ-layer, and the fiber layer. The topology of the λ-layer changes corresponding to the change in traffic demand in the packet layer, and the route that PSC-LSP selects is converted according to the change of topology in the λ-layer. Because multiple layers can collaboratively adapt the topology or route of each layer to accommodate changes in traffic demand, multilayer traffic engineering increases the efficient use of network resources.

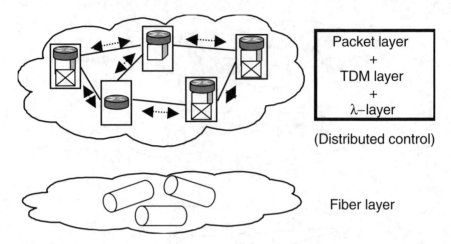

FIGURE 8.8 GMPLS network that is controlled distributedly and integratedly with multiple layers.

The work of standardizing GMPLS network protocols is under the purview of the IETF (Internet Engineering Task Force), a standardization organization on Internet-related matters. Figure 8.10 shows the major protocols used in GMPLS. Under the GMPLS architecture defined, GMPLS is mainly composed of the OSPF (Open Shortest Path First) extension of the routing protocol, the RSVP-TE extension of the signaling protocol, and the Link Management Protocol (LMP).

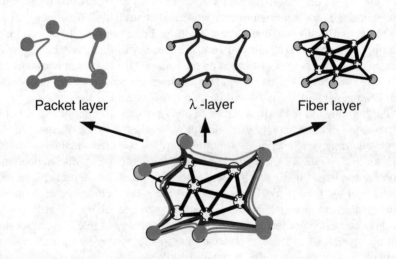

FIGURE 8.9 Multilayer traffic engineering.

GMPLS (Generalized Multiprotocol Label Switching)

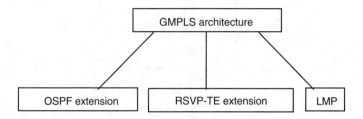

FIGURE 8.10 Major protocols of GMPLS.

8.3 SEPARATION OF DATA PLANE FROM CONTROL PLANE

One of the features of a GMPLS network is that the data plane is separated from the control plane. In an IP/MPLS network, it is assumed that all of the nodes supporting the IP/MPLS protocol have an interface that can identify and process the packet. In this way, the control packet of the routing protocol or signaling protocol can be transferred through the same physical media as the data packet. However, in a GMPLS network, all of the interfaces need not be able to identify and process the packet. Indeed, there are interfaces that cannot identify and process the packet such as TDM, LSC, or FSC interfaces. Therefore, transfer of control packets is executed logically using a different interface than that used for data transmission. Thus, in a GMPLS network, the data plane and the control plane are logically separated from each other. The interface of the node in the control plane identifies and processes the packet.

The control packets of the routing protocol, the signaling protocol, and the LMP are described in the next section. In a GMPLS network, these packets are transferred through the control plane, which is logically separated from the data plane.

8.4 ROUTING PROTOCOL

8.4.1 OSPF Extension

The current routing protocols typically used in an IP/MPLS network are OSPF (Open Shortest Path First) or IS-IS (Intermediate System to Intermediate System). In a GMPLS network, the OSPF that has been utilized in IP network is extended [5]. In the OSPF extension, such concepts as a traffic-engineering (TE) link, hierarchization of the LSP, unnumbered links, link bundling and LSA (link-state advertisement) were introduced [6].

In a GMPLS network, as illustrated by the hierarchization in Figure 8.4, a lower-layer LSP can become a link of an upper-layer LSP. For example, when an LSP is set on a certain TDM path, the TDM path behaves like a fixed link that has been there permanently for a long time. When the lower-layer LSP is

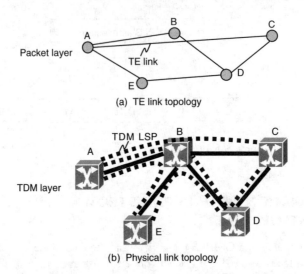

FIGURE 8.11 Concept of traffic engineering.

set, the originating node of the LSP, when viewed from the upper layer, is advertised within the network as an upper-layer link. This LSP is called a TE link. Figure 8.11 shows the concept of this TE link. The broken line seen in Figure 8.11(b) is a TDM path. There is a direct path between A-C, but not between B-C. In this case, the TE link takes a topology such as that shown in Figure 8.11(a). In the TDM layer, the LSP (TDM-LSP) acts as a TE link between packet and layer. When PSC-LSP is set up, the route is selected according to the topology that is constructed by TE links. Although the TE is an abstract link, in the case where it is used for traffic engineering, such as route selection when setting up the LSP, it goes on by just referring to the topology constructed by TE links without considering the physical structure. In general, in the topology of a GMPLS network, a physical link, such as an optical fiber, is also called a TE link; there is no distinction between physical and abstract links.

Next, we describe the unnumbered link. The interface of a link in an MPLS network is generally assigned an IP address. According to this IP address, it is possible to identify the link inside the network. However, in a GMPLS network, because it is possible to accommodate more than 100 wavelengths per one optical fiber, the number of required IP addresses becomes huge if an IP address is assigned to each interface of these wavelengths. Furthermore, because the LSP of each layer is advertised to the upper layer as a TE link, the supply of IP addresses may be depleted if an IP address is assigned to each TE link. Therefore, in GMPLS, to identify the link (hereinafter, a TE link is referred to simply as a "link"), a link identifier (link ID) that

GMPLS (Generalized Multiprotocol Label Switching)

is assigned to the interface of the link has been introduced. Although an IP address must still be assigned globally, this link ID is good if it is unique just within the router. It is possible to identify the link inside the network from a combination of the router ID and the link ID. A link that is expressed by a combination of the router ID and the link ID is called an "unnumbered link," meaning that an IP address is not assigned to each interface of the link. So, in GMPLS, if the number of wavelengths or TE links increases, there is no problem of a shortage of IP addresses.

Next, we describe link bundling, the purpose of which is to abstract multiple links with the same nature by integrating them into one TE link [7]. The conditions of the same-nature link are: (1) the links are set up between the same nodes; (2) the links are the same type of link (point-to-point/point-to-multipoints); (3) the links are in the same TE metric; and (4) the links are in the same resource class. The purpose of link bundling is to improve the scalability of routing by reducing the amount of advertising of link states by routing protocol. The bundled link is composed of individual resources. The methods to manage each resource are described in Section 8.6. Although the amount of advertising can be reduced by integrating and abstracting the multiple links to a single TE link with link bundling, it may happen that individual resource information is missed. For example, the maximum vacant capacity of a bundled link is set to the maximum value of vacant capacity for an individual resource. There is a trade-off between the effect of reducing the amount of advertising and the holding capability of resource information.

8.4.2 TE Link Advertisement

In an IP/MPLS network, the link states between routers are advertised using a router LSA of LSA type-1 [8]. To advertise the link state of a TE link into the GMPLS network, an "opaque" LSA is utilized, as shown in Figure 8.12 [9, 10]. "Opaque" is derived from its meaning of "uncertain." In GMPLS that is under discussion in IETF, type-10 is utilized because it is dealing with a routing protocol within the area. Opaque LSA is advertised according to the TLV (type, length, value) format storing opaque information. There are two types of TLV format. One is a router TLV that expresses the router information, and the other is a link TLV that expresses the link information. Link TLV has sub-TLV under itself. In the OSPF extension of GMPLS, the sub-TLV of link TLV has been defined as shown in Figure 8.13. From Type-1 to Type-9, nine types of sub-TLV have been defined as extensions for MPLS traffic engineering [10]. Adding to this, there are the following additions as extension for GMPLS [6].

- *Sub-TLV = 11 (Link local/remote identifier)*: Length is 8 octets. Link local identifier and link remote identifier are assigned with 4 octets per field to each. "Local" means the link's own node side, and "remote" means the counterpart node side of the link. The Link local/remote identifier is used in the case of an unnumbered link. If the remote TE link identifier is unknown, it is set to 0.

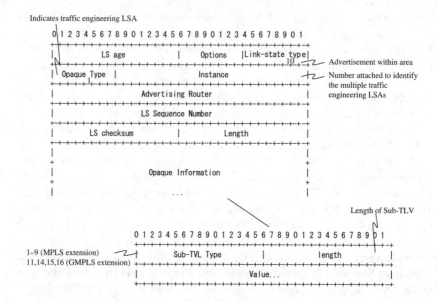

FIGURE 8.12 Opaque LSA format (RFC 2370).

- *Sub-TLV = 14 (Link protection type)*: Length is 4 octets. The link protection type indicates the reliability of the link. The first 1 octet has been defined as the following link-protection types.
  0×01 (Extra traffic type): This is a link for protecting the other links. Traffic of best-effort type flows on this link. When a fault occurs on the link to be protected, LSP data on the other link that was protected

Sub-TLV Type	Length	Name	
1	1	Link type	
2	4	Link ID	
3	4	Local interface IP address	
4	4	Remote interface IP address	
5	4	Traffic engineering metric	
6	4	Maximum bandwidth	
7	4	Maximum reservable bandwidth	
8	32	Unreserved bandwidth	
9	4	Administrative group	
11	8	Link Local/Remote Identifiers	← Added for GMPLS
14	4	Link Protection Type	← Added for GMPLS
15	variable	Interface Switching Capability Descriptor	← Added for GMPLS
16	variable	Shared Risk Link Group	← Added for GMPLS

FIGURE 8.13 Sub-TLV of opaque LSA in GMPLS OSPF.

GMPLS (Generalized Multiprotocol Label Switching)

is directed to flow to this link. Therefore, the LSP data that had been flowing on this link is lost.

0×02 (Unprotected): This link is not protected by other links. When a fault occurs, LSP data on this link is lost.

0×08 Shared type: There is one or more extra-traffic-type links that protect this link. The route of a shared-type link and the route of an extra-traffic-type link are independent of each other. The extra-traffic-type link is shared by one or more shared-type links.

0×08 1:1 type: There is one extra-traffic-type link that protects a 1:1 type. The route of a 1:1 type and the route of an extra-traffic type are independent of each other.

0×10 1+1 type: There is one dedicated-independent-route link that protects a 1+1 type. But, the link that protects this 1+1 type cannot be used to select the LSP route because it is not advertised as a link state.

0×20 Enhanced type: This is more reliable than the 1+1 type. For example, there are two or more dedicated and independent routes that protect the 1+1 type.

- *Sub-TLV = 15 (Interface switching capability identifier)*: Length is variable. There is a 1-octet field that indicates the switching capability, a 1-octet field that indicates the encoding type, and a 1-octet field that indicates the maximum LSP bandwidth for each priority. The maximum number of priorities supported is eight. The link is connected to the node via interface. As shown in Figure 8.11, in a GMPLS network, each interface has a different switching capability. For example, while an interface of a certain link may not identify the packet, it is possible to execute switching by a channel unit inside the SDH payload. The interfaces of both ends of the link need not have the same switching capability. There are several types of switching capability: PSC, TDM, LSC, and FCS. The types of encoding include packet, Ethernet, digital wrapper, λ, fiber, etc. The encoding type indicates which encoding type the interface can support. The relationship between switching capability and link is shown in the following list. Here, X and Y in (X, Y) indicate the switching capability at both ends of the interface.

  (PSC, PSC): Link between IP routers

  (TDM, TDM): Link between DXC and DXC

  (LSC, LSC): Link between OXCs

  (PSC, TDM): Link between IP router and TDM

  (PSC, LSC): Link between IP router and OXC

  (TDM, LSC): Link between TDM and LSC

  (PSC, PSC+LSC): Node that has the function of an IP router and both the functions of an IP router and OXC (Here, the node that has both the functions of an IP router and OXC has two switching capabilities for one interface. This functions like an IP router, and it is possible to set up PSC-LSC; it also functions like the OXC, and it is possible to set up LSC-LSP.)

- *Sub-TLV = 16 (Risk-shared link group)*: Length is variable. Risk-shared link group is an aggregate of links that are affected by a certain fault. For example, there is such a case that multiple wavelengths belong to a single fiber, and multiple LSC-LSPs are set as a link using a wavelength of the same fiber. Links between originating and destination nodes of these LSC-LSPs may differ from each other. When a failure occurs in a fiber link, these links (LSC-LSP) are affected simultaneously. If the upper layer is TDM and each of these is treated as an independent link looking from the TDM layer, the reliability of the TDM layer is not secured. Therefore, in a GMPLS network, each link can select an independent route taking a shared risk into consideration by indicating whether it belongs to the risk-shared-link group or not. One risk-shared-link group is assigned an independent value within the network and is expressed by 4 octets. The link can belong to multiple risk-share-link groups, and it is possible to include all of the risk-shared-link groups to which the link belongs to the field of the risk-shared-link group. Examples of selecting an independent route utilizing the risk-shared-link group are presented in the literature [11–13]. Traffic engineering utilizing the risk-shared-link group is described in Section 8.6.3.

8.5 SIGNALING PROTOCOL

8.5.1 RSVP-TE Extension of RSVP-TE and GMPLS

The signaling protocol is a protocol that sets up the LSP and manages the setup status of the LSP. Efforts are currently under way to standardize the protocol that extends the RSVP-TE, which is a signaling protocol of MPLS, to GMPLS [14–16]. In this section, we describe the RSVP-TE extension as a signaling protocol of GMPLS.

First, we briefly describe the RSVP-TE of MPLS again. As a typical message used in RSVP-TE, there is a Path message and an RSVP message. As shown in Figure 8.14, when an LSP is set up, the originating node transmits a Path message. This Path message reaches a destination node via the nodes on the route of the LSP. In this Path message, a label must be attached to each link for the nodes on the route. When the destination node receives the Path message, it transmits the RESV message back toward the nodes on the route in the reverse direction of the Path message. At that time, the destination side sets up the labels for the links that are connected on the node's upstream side. In addition to this, bandwidth may be reserved. Also the intermediate nodes, similar to the destination node, set up the labels for the links that are connected on the upstream side of the node itself when it receives the RESV message. When the intermediate node transmits the RESV message to the upstream node, it is required to set up the table and switching that homologize the label to the labels that were set for both the upstream-side and the downstream-side links. When the RESV message reaches the originating node, the setup of the LSP is completed by

GMPLS (Generalized Multiprotocol Label Switching)

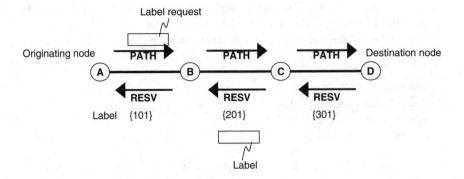

FIGURE 8.14 Path message and RESV message.

setting up the table and switching that homologize the label to that of the originating node.

In RSVP-TE, to manage the setup status of the LSP and to maintain the state of the LSP, the node on the route after the LSP was set up transmits the Path message or the RESV message periodically. This Path message or RESV message is also called a "refresh message." The state of the LSP of each node is verified by this refresh message, and the LSP is maintained. If a certain node does not receive the refresh message for any reason, the relevant node judges that an error has occurred and deletes the state of the LSP within the node itself, and at the same time, it transmits a Path ERROR message and a Path TEAR message toward both the upstream and downstream sides. A node that receives the Path message deletes the state of the LSP. When the originating node receives the error message, it transmits the Path TEAR message toward the downstream side to disconnect the LSP.

RSVP-TE manages the state of the LSP according to the refresh message at each node and disconnects the LSP depending on the network condition. Such a method for managing the LSP is called "management by soft state."

RSVP-TE signaling of an MPLS network is extended for use in a GMPLS network. Setting up the LSP means to switch the packets according to the label-conversion table, in which correspondence between the label of the input link and the label of the output link of the node on the route of the LSP has been set, by assigning the label of the link through which the LSP passes. In MPLS, label assignment executes just to set up the LSP route, but not to assign the bandwidth or network resource.* In GMPLS, as shown in Figure 8.3 in Section 8.2, the label corresponds to the time slot in the TDM layer, to the wavelength in the λ-layer, and to the fiber in the fiber layer. Therefore, assigning the label in a GMPLS network means to assign the bandwidth or network resource in layers other than

---

* Because the number of labels that can be assigned per link is limited, we can consider the label to be a part of the network resources. The number of bits to express the label is 20, so it is possible to set up 220 labels. Thus we can consider that there is no limit to the number of labels as long as no LSPs containing more than 220 labels are set up per link.

the packet layer. This point that label assignment means to assign the bandwidth or network resource is a characteristic of the extended RSVP-TE for GMPLS.

Examples of the utilization of this characteristic of RSVP-TE extension include:

Label request
Dual direction path signaling
Label setting
Architectural signaling

We describe these characteristics of the GMPLS RSVP-TE in Sections 8.5.2 to 8.5.5.

8.5.2 GENERAL LABEL REQUEST

As described in Section 8.2, in a GMPLS network, an LSP is defined for each layer. In an MPLS network, the path is intended only for a PSC-LSP in which the switching capability of the interface through which the LSP passes is PSC. In GMPLS RSVP-TE, it sets up a path to TDM-, LSC-, and FSC-LSP besides PSC-LSP and manages their setting states. When requesting the label with Path message, Path message has a label-request object. The label request in GMPLS requires a general label for GMPLS. As an extension of label request in GMPLS, an LSP encoding type, switching type, and G-PID (generalized payload ID) have been added. Figure 8.15 illustrates a format of the label-request object.

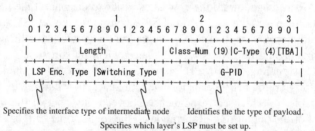

```
 0                   1                   2                   3
 0 1 2 3 4 5 6 7 8 9 0 1 2 3 4 5 6 7 8 9 0 1 2 3 4 5 6 7 8 9 0 1
+-+-+-+-+-+-+-+-+-+-+-+-+-+-+-+-+-+-+-+-+-+-+-+-+-+-+-+-+-+-+-+-+
|            Length             | Class-Num (19) |C-Type (4) [TBA]|
+-+-+-+-+-+-+-+-+-+-+-+-+-+-+-+-+-+-+-+-+-+-+-+-+-+-+-+-+-+-+-+-+
| LSP Enc. Type |Switching Type |              G-PID             |
+-+-+-+-+-+-+-+-+-+-+-+-+-+-+-+-+-+-+-+-+-+-+-+-+-+-+-+-+-+-+-+-+
```

Specifies the interface type of intermediate node
Specifies which layer's LSP must be set up.
Identifies the the type of payload.

LSP Enc.	Type	Value	Type
1	Packet	1	Packet-Switch Capable-1 (PSC-1)
2	Ethernet V2/DIX	2	Packet-Switch Capable-2 (PSC-2)
3	ANSI PDH	3	Packet-Switch Capable-3 (PSC-3)
4	ETSI PDH	4	Packet-Switch Capable-4 (PSC-4)
5	SDH	51	Layer-2 Switch Capable (L2SC)
6	SONET	100	Time-Division-Multiplex Capable (TDM)
7	Digital Wrapper	150	Lambda-Switch Capable (LSC)
8	Lambda (photonic)	200	Fiber-Switch Capable (FSC)
9	Fiber		
10	Ethernet 802.3		

FIGURE 8.15 Format of a label-request object.

GMPLS (Generalized Multiprotocol Label Switching)

LSP encoding type: This is represented by an 8-bit field. It indicates which interface of an intermediate node on the LSP route must support which encoding technology. As typical examples of this encoding type, there are packet, Ethernet, SDH, digital wrapper, λ, and fiber. For example, in the case where the encoding type is SDH, the interface of an intermediate node on the LSP route can identify and process the SDH frame. In the case where the encoding type is λ at an intermediate node, optical-electrical-optical conversion is not executed, and the transmission rate is not identified.

Switching type: This is represented by an 8-bit field. It indicates which layer's LSP must be set up, i.e., it specifies which layer's label must be set up. The switching types includes PSC, TDM, LSC, and FCS. For example, in the case where the switching type is TDM, it requires a time slot as the label, and in the case where the switching type is LSC, it requires a wavelength in fiber as the label.

G-PID: This is represented by a 16-bit field. It is an identifier of the payload that the LSP transmits. That is, it indicates the technology by which the originating/destination node can process the payload. G-PID has been defined in detail as shown in Figure 8.16. For example, when G-PID = 28, the technology used is POS (packet over SONET), no scrambling, and 16-bit CRC. And when G-PID = 31, the technology used is POS with scrambling and 32-bit CRC. Unless the interfaces at both ends of the LSP support the same G-PID, the content of the payload cannot be decoded, and communication through the LSP cannot be executed.

Because the switching capability and the encoding type of the interface-switching-capability identifier that was described in Section 8.4.2 have been advertised by GMPLS routing protocol, the originating node knows the LSP encoding types and switching types that the nodes within a GMPLS network are supporting.

Here we describe the method of setting up the LSP using Figure 8.14 as an example. The method of setting up an LSP described here is common to any layer's LSP. Let us consider the case that the LSP must be set up from node A, via node B and node C, to node D. In this case, the originating node (node A) transmits a Path message including the label request to node B. Then, node B checks whether its interface can support the LSP encoding type and switching type or not, and if it is okay, it transmits the Path message to node C. Node B creates the status of the relevant LSP and stores it before sending the Path message to node C. Node C operates just like node B, and it transmits the Path message to node D if there is no problem. The destination node (node D) checks whether its interface can support the LSP encoding type and switching type, as well as the G-PID, and creates the status of the LSP and stores it if there is no problem. Then, node D gives a label value 301 to the link between node C and node D. Node D sets this label value 301 to the input-side link of the label-conversion table and transmits the RESV message including the label value 301 to node C. When node C receives the RESV message, it sets the label value 301 to the

Value	Type	Technology
0	Unknown	All
1	Reserved	
2	Reserved	
3	Reserved	
4	Reserved	
5	Asynchronous mapping of E4	SDH
6	Asynchronous mapping of DS3/T3	SDH
7	Asynchronous mapping of E3	SDH
8	Bit synchronous mapping of E3	SDH
9	Byte synchronous mapping of E3	SDH
10	Asynchronous mapping of DS2/T2	SDH
11	Bit synchronous mapping of DS2/T2	SDH
12	Reserved	
13	Asynchronous mapping of E1	SDH
14	Byte synchronous mapping of E1	SDH
15	Byte synchronous mapping of 31 * DS0	SDH
16	Asynchronous mapping of DS1/T1	SDH
17	Bit synchronous mapping of DS1/T1	SDH
18	Byte synchronous mapping of DS1/T1	SDH
19	VC-11 in VC-12	SDH
20	Reserved	
21	Reserved	
22	DS1 SF Asynchronous	SONET
23	DS1 ESF Asynchronous	SONET
24	DS3 M23 Asynchronous	SONET
25	DS3 C-Bit Parity Asynchronous	SONET
26	VT/LOVC	SDH
27	STS SPE/HOVC	SDH
28	POS - No Scrambling, 16 bit CRC	SDH
29	POS - No Scrambling, 32 bit CRC	SDH
30	POS - Scrambling, 16 bit CRC	SDH
31	POS - Scrambling, 32 bit CRC	SDH
32	ATM mapping	SDH
33	Ethernet	SDH, Lambda, Fiber
34	SONET/SDH	Lambda, Fiber
35	Reserved (SONET deprecated)	Lambda, Fiber
36	Digital Wrapper	Lambda, Fiber
37	Lambda	Fiber

FIGURE 8.16 Types of G-PID.

GMPLS (Generalized Multiprotocol Label Switching)

output-side link of the label conversion table. Node C then gives the label value 201 to the link between node B and node C. Node C sets this label 201 to the input-side link of the label conversion table and, after it sets itself up so that its switching operation is in accord with the label-conversion table, it transmits the RESV message including the label value 201 to node B. Similarly, the RESV message is transferred to node B and node A. When the RESV message arrives at the originating node A, the LSP setup is completed by setting up the table and switching that homologize the label to the one in the originating node.

In GMPLS's label request, it is key that it checks whether or not the interface of the nodes on the LSP route support the LSP encoding type, the switching type, and G-PID specified in the Path message.

8.5.3 Bidirectional Path Signaling

An LSP in an MPLS network is a one-way path. However, when communication is extended to the TDM layer, λ-layer, and fiber layer in a GMPLS network, because the SDH/SONET path, the wavelength path, and the fiber path are considered to be bidirectional paths in principle, the signaling must be extended to accommodate a bidirectional path. In extending one-way signaling to bidirectional signaling, the most obvious solution would simply be to apply different one-way signaling in each direction. However, this approach has not been adopted for a variety of practical reasons, including the length of the setup time, the doubling of the amount of signaling messages, and so on. In a GMPLS network, a bidirectional path is set up by having the signaling go and return between the originating node and the destination node using the Path message and the RESV message, much like in one-way signaling, by using an upstream label.

Here we describe setting up bidirectional path signaling by using Figure 8.17 as an example. In bidirectional path signaling, the node that transmits the Path message is called an "initiator," and the node that transmits the RESV message is called a "terminator." The LSP that transfers data from the originating node to the destination node is called a "downstream path," and inversely, the LSP that

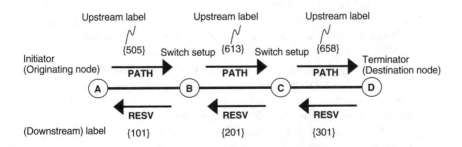

FIGURE 8.17 Upstream label.

transfers data from the destination node to the originating node is called an "upstream path." In setting up the one-way path, the state of the LSP is made with a Path message and the label setup is executed when an RESV message is transmitted. Label setup for the downstream path in a bidirectional path is executed when an RESV message is transmitted in the same way as in a one-way path. Line setup for the upstream path is executed when the Path message is transferred. This is characteristic in label setup in the case of a bidirectional path.

Before node A as an initiator transmits the Path message to node B, 505 is set up as the label value of the upstream path of the link between node A and node B, and setup of switching is executed for the upstream path. Node A puts the upstream label value 505 on the Path message and transmits it to node B. Node B sets up the label value 505 of the output link 2 and, at the same time, sets up the upstream label value 613 of the input link between node B and node C to the label-conversion table, and executes setup of switching for the upstream path according to the label-conversion table. Similarly, the Path message is transferred to node C and node D. When setup of the label-conversion table for the upstream path and setup of the switching are completed at node D, node D can transmit the upstream path to node A. On the other hand, as for the method of setting up the downstream path, it is the same as the method of setting up the one-way path, and the label of the downstream path is included in the RESV message.

Using this signaling procedure, it becomes possible to set up a bidirectional path in just one round-trip of the Path message and the RESV message.

8.5.4 Label Setting

Label setting has been introduced to apply to signaling in the λ-layer. In this case, the label corresponds to the wavelength. As described in Section 8.5.2, usually the label value of the input-side link is determined by the downstream-side node when the RESV message is transmitted and notified to the upstream-side node. However, the fact that the label corresponds to wavelength in the λ-layer introduces some potential problems:

- The case in which a transmission device of the upstream-side node does not support the wavelength corresponding to the label value that the downstream-side node has determined. Variable-wavelength lasers are expensive, and the output wavelengths are limited.
- The case in which there is a restriction in the wavelength-conversion function at an intermediate node. In this case, it is not possible to set up the switching that can homologize the label of the input-side link to the label of output-side link. That is, a wavelength-conversion switch that has a relationship with certain restrictions is required. In its extreme case, there is no wavelength-conversion function.

Although label setting of the packet layer means to set up the route of the LSP, in the λ-layer, it also includes the meaning of setting up the network resource

GMPLS (Generalized Multiprotocol Label Switching) 211

beside the LSP route, and a restriction in the transmission wavelength of the transmitter and in the wavelength-conversion function is generated. In such a case, the upstream-side node is required to impose a restriction of the label value to the downstream node when the upstream node transfers the Path message. The label set is defined as a group of restricted labels that the upstream node permitted.

Here, we describe the utilization of the label set using Figure 8.18 as an example. It is assumed that each node has no wavelength-conversion functions. With node A as an originating node, the wavelengths that the transmitter can supporting are red, yellow, green, and blue, and the label values corresponding to each wavelength are 101, 115, 120, and 150, respectively. In each node, correspondence between the link that is connected to a relevant node, and the label value has been set uniquely for each link. These correspondences are also the same in the opposed node for the relevant link. For example, in node B, the label values corresponding to the wavelengths red, yellow, green, and blue are 101, 115, 120, and 150, respectively. Node A transmits the Path message, including the label set (101, 115, 120, 150) as the restriction of wavelengths that the transmitter supports, to node B. In node B, because it has no wavelength-conversion function, it must use the label corresponding to the label set also for the link between node B and node C. The correspondences between the wavelength and the label are shown in Figure 8.18. Because the label is uniquely assigned to each link, the labels for the link between node A and node B and the one between node B and node C may indicate either the same value or different values. In the link between node B and node C, the label 215

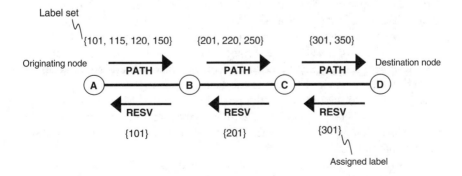

FIGURE 8.18 Label set.

corresponding to the yellow wavelength is not available in this example. Because node B has no wavelength-conversion function and can use only red, green, or blue, node B transmits the label set (201, 220, 250) to node C. In the link between node C and node D, the green wavelength is not available. Therefore, in node C, the label set is further restricted, and node C transmits the label set (301, 320) to node D. When node D receives the label set (301, 320), it can select one label from the received label set. In this example of Figure 8.18, node D selects the label 301 and transmits the RESV message including this label value to node C. Then node C selects 201 from (201, 220, 250) to create the LSP of red wavelength based on the correspondence table. Similarly, node B selects the label 101 and notifies it to node A by RESV message. As a result, an LSP that uses the red wavelength is created between node A and node D.

If there is no restriction of the label set, node D will select the label 315 that is available in the link between node C and node D, but because node C has no wavelength-conversion function, it is not possible to create the LSP. In this way, in cases where there is a restriction of wavelength in the node on the LSP route, the LSP can be set up effectively by using the label set.

8.5.5 Architectural Signaling

In a GMPLS network, when a concept of hierarchization of the LSP is introduced and a signaling that operates distributedly is used, it becomes possible to set up the LSP at a lower layer by triggering an LSP setup request at a higher layer. This is called "hierarchical signaling."

Figure 8.19 shows an example of hierarchical signaling. In this example, the network consists of node 1, node 2, node 4, and node 5 that have a PSC interface and node 3 that has an LSC interface, and there is a packet layer on the λ-layer. Here we try to set up the PSC-LSP from node 1 to node 5, assuming that LSC-LSP is not yet established between node 2 and node 4.

First, node 1 transmits a Path message (PSC) to node 2. The PSC in parentheses in Figure 8.19 indicates the switching type of label-request object that was described in Section 8.5.2. Because an LSC-LSP has not yet been established between node 2 and node 4, when node 2 receives the Path message (PSC), it transmits the Path message (LSC) to node 3 to establish the LSC-LSP. The Path message (LSC) is transferred up to node 4, and the RESV message (LSC) arrives at node 2 through node 3, and LSC-LSP is set up. The LSC-LSP between node 2 and node 4 becomes a link when it is seen from the packet layer, and node 2 transmits the Path message (PSC) to node 4. After this, according to the ordinary setup procedure of PSC-LSP, the Path message (PSC) arrives at node 5, and when the RESV message (PSC) is transferred in reverse direction on the Path message path and arrives at node 1, setup of PSC-LSP is completed.

As seen in this example, by using a hierarchical signaling, it becomes possible to set up automatically a lower-layer LSP by using a trigger from a higher layer to initiate LSP setup.

GMPLS (Generalized Multiprotocol Label Switching)

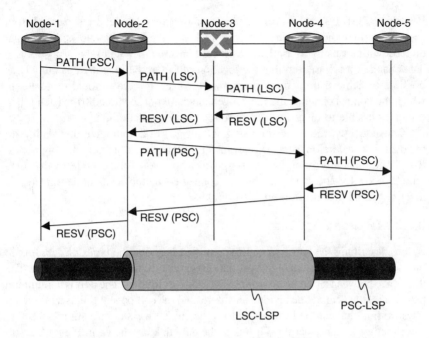

FIGURE 8.19 Hierarchization of LSP.

8.6 LINK MANAGEMENT PROTOCOL

8.6.1 NECESSITY OF LMP

To manage the link between neighboring nodes, a Link Management Protocol (LMP) has been introduced as a GMPLS protocol [17].

As described in Section 8.3, the data plane and the control plane are separated from each other in a GMPLS network. This is because there are interfaces that cannot identify and process such packets as TDM, LSC, and FSC. Two neighboring nodes need not have the control channel and the data link use the same physical media. This control channel is used to operate the routing protocol and the signaling protocol. As a result, it is not necessarily the case that the data transmission is valid on the data channel, even if a control signal is correctly transmitted on the control channel, and vice versa.

The concept of a TE link has been utilized in GMPLS networks. The TE link is an abstract link created by bundling multiple links and is used for the purpose of convenience in route computation of the LSP or to enhance the extensibility of the routing protocol.

In such a condition as described above, it is not possible to homologize the TE link to the data link that belongs to relevant TE link by using the conventional protocols. Also, in the data link that is connected to the local node, it is not possible to homologize the physical port of a local node to that of a remote node.

Further, in GMPLS, the control channel and the data channel are separated from each other, including the interface that receives and transmits an optical signal directly without using an optical-electrical conversion. When a fault occurs in a data channel, it is essential that the location of failure be identified as quickly as possible to achieve immediate recovery. However, in conventional protocols in which the control channel and the data channel are not separated from each other, it is not possible to identify the failure location on a data link.

Consequently, the main roles of LMP are (a) to homologize the TE link to the data link that belongs to the relevant TE link between the neighboring nodes, (b) to automatically homologize the physical port of a local node to that of a remote node for the data link that is connected to the local node, and (c) to identify the failure location on a data link.

8.6.2 Types of Data Link

Before describing the function of LMP, we first describe the types of data link in this section. There are two types of data link: a port and a component link. The difference between a port and a component link is as follows. The port is a minimum physical ink that cannot be divided any more, and the component link is a minimum physical link that can be further divided by using a time slot or a shim label.

Here, we assume that the TE link is the data link, as shown in Figure 8.20. In Figure 8.20(a), the TE link consists of bundled multiple OC-192c paths. In this case, the physical minimum unit is OC-192c that cannot be divided further into a smaller unit path than this. Therefore, the minimum logical link of this link is also OC-192c. There are two parameters that identify the minimum logical unit of link: a link identifier that indicates the TE link, and an interface identifier that indicates the physical port interface. In the case of a port, the interface identifier corresponds one-to-one to the label. In contrast, in Figure 8.20(b), the physical minimum unit is OC-192, which is a component link and can be further divided into four OC-48c paths. Therefore, in this case, the minimum physical unit of the link is OC-48c. There are three parameters necessary to identify the minimum logical unit of a link: a link identifier that indicates the TE link, an interface identifier that indicates the physical port interface, and a label that indicates the time slot of OC-48c.

8.6.3 Functions of LMP

LMP has four functions:

Control-channel management
Link-property correlation
Connectivity certification
Failure management

Control-channel management and link-property correlation are indispensable functions of LMP, and connectivity certification and failure management are optional functions.

GMPLS (Generalized Multiprotocol Label Switching)

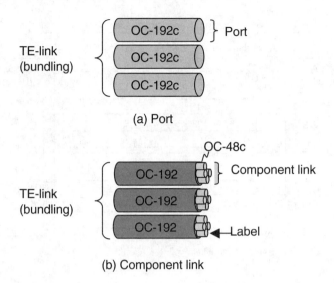

FIGURE 8.20 Types of data link.

8.6.3.1 Control-Channel Management

To make the LMP function operate, there must be at least one bidirectional control channel between two nodes that are connected by a TE link. When at least one bidirectional control channel has been established, the connected nodes are called "LMP neighbors." The function of control-channel management is to establish and manage the control channel.

Figure 8.21 shows an example of control channel between two nodes. In this example, although the control channel is using a different physical medium than the one used in the data link, it is also possible to use the same physical medium. Communication of control data between neighboring nodes is executed by IP protocol. All of the LMP packets operate as UDP (User Datagram Protocol) packets having an LMP port number. In communication by IP protocol, an IP router can be inserted between the neighboring nodes, and if communication by IP protocol is possible, there is no restriction in assembling the control channel. This control channel can also be used for protocol communication of the OSPF extension or the RSVP extension.

The control channel of each direction is identified by a control-channel identifier. The control channels for both directions can be homologized by using a Config message. The Config message is transmitted from local node to remote node to establish the control channel. At this time, the necessary IP address of the interface of the control channel in the remote node is obtained manually or automatically. The Config message includes a local control-channel identifier, a transmit-side node identifier, a message identifier, Config parameters, etc. An LMP message is transferred securely using a 32-bit message identifier and a

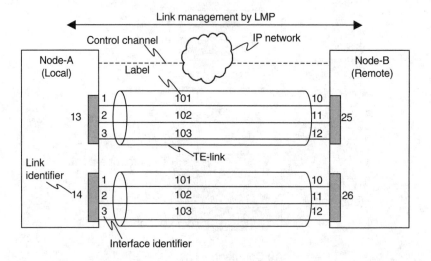

FIGURE 8.21 Link management protocol.

resending function. The number of the message identifier increases incrementally and returns to 0 when it reaches some maximum value.

The remote node that receives the Config message transmits a ConfigAck message, if there was no problem in the content of the Config message, to establish the control channel. If there is any problem in the content of the Config message, it transmits a ConfigNack message that includes the content of the problem. In this case, the control link channel is not established, and the remote node must modify the content of the Config message and try again to establish the control channel.

When the control channel is established, LMP Hello protocol is activated. The LMP Hello protocol exchanges a Hello message between two neighboring nodes and confirms the normalcy of the control channel. When the Hello message cannot be exchanged normally, the TE link enters a Degraded state in which the TE link operates normally, but the control channel does not operate normally. When the TE link returns to the normal state from the Degraded state, the control channel returns to the established state. However, if the TE link cannot not recover from the Degraded state, the TE link is disconnected.

8.6.3.2 Link-Property Correlation

Link-property correlation is executed to synchronize the properties of a TE link between the neighboring nodes by bundling multiple data links (ports or component links) to the TE link. This is executed after the TE link is manually established or after the TE link is automatically established by a connectivity-verification function, which is an optional function of LMP, by using a LinkSummary message, a LinkSummaryAck message, and a LinkSummaryNack message.

As shown in Figure 8.21, when there is one or more TE links between the neighboring nodes, each TE link has a link identifier at each node at both ends,

GMPLS (Generalized Multiprotocol Label Switching) 217

and each data link (port or component link) has an interface identifier at each node at both ends. The link identifier or interface identifier is expressed with IP v.4, IP v.6, or unnumbered.

It is possible to synchronize the attributes of a TE link between neighboring nodes by bundling multiple data links (ports or component links) to a TE link with a LinkSummary message. The LinkSummary message includes a TE-link object and a data-link object. The TE-link object distinguishes the local link identifier and the remote link identifier of the TE link, and indicates whether the connectivity-certification and the failure-management functions, which are optional functions of LMP, are supported or not. The data-link object includes the information of data link (interface identifier, interface switching type, etc.) that composes the TE link.

A local node that receives the LinkSummary message from the remote node checks the content of the LinkSummary message against the information of the TE link that the local node itself possesses. The TE-link information that the local node possesses includes the information that is used when the TE link is automatically established by the connectivity-verification function as an optional function of LMP or the configuration information that is used when the TE link is manually established.

A LinkSummaryAck message is transmitted to the initiator node of the Link-Summary message when the interface identifier of the data link that belongs to the TE link agrees with the definition of the link property. When the node receives the LinkSummaryAck message, the properties of the TE link between the neighboring nodes are verified. If the interface identifier does not agree with the content, it transmits a LinkSummaryNack message that includes the content of the problem. The node that receives the LinkSummaryNack message transmits a new LinkSummary message reflecting the content of the LinkSummaryNack message to the neighboring node.

The LinkSummary message may become large as the number of data-link objects increases. Thus, in transmitting and receiving the LMP message, the node that transmits the LMP packet must have a function to fragment the IP packet, and the node that receives the LMP packet must have a function to assemble the fragmented IP packet.

8.6.3.3 Connectivity Verification

There are two methods for establishing the TE link: manually and automatically. When it is done manually, the connectivity of the data link that belongs to the TE link is verified manually, and then the properties of the TE link are set up to the nodes at both ends as the information of configuration. The connectivity-verification function of LMP is a function for establishing the TE link automatically.

It is difficult to verify the connectivity of the data link between transparent interfaces that transmit and receive optical signals without converting them to or from an electrical signal. Therefore, the connectivity of the data link must be verified before the user traffic is transferred. To do this, the data link is switched to a device that can transmit and receive electrical signals in the node at both ends, and then a Test message is transmitted to the data link to verify connectivity. When the connectivity of multiple data links needs to be verified, it is not necessary to transmit simultaneously the Test message to multiple data links, but it is a good idea to verify the connectivity of each data link one-by-one in turn. The node that supports the connectivity-verification function must be able to connect at least one of the interfaces to be verified to a device that can transmit and receive a Test message. Here it should be noted that, in LMP, only the Test message is transferred on the data link, and all other LMP messages are transferred on the control channel.

The connectivity-verification function is executed using such messages as BeginVerify message, BeginVerifyAck message, BeginVerifyNack message, EndVerify message, EndVerifyAck message, Test message, TestStatusSuccess message, TestFailure message, and TestStatusAck message.

Here we describe the operation of connectivity verification using Figure 8.21 as an example. Let us consider connectivity verification of the TE link between the link identifier 13 of node A and the link identifier 25 of node B.

1. Node A transmits a BeginVerify message through the control channel. The BeginVerify message includes a local identifier* 13 of the TE link, and node A begins connectivity verification of the data link that belongs to the TE link against node B.
2. When node B receives the BeginVerify message, it creates a verification identifier and maps this verification identifier to the TE link from node A. This verification identifier is used when node B receives the Test message from node A. Node B recognizes the link identifier 13 of node A and transmits the local link identifier 25 of itself, a remote link identifier** 13, and a verification identifier by putting them on the BeginVerifyAck message to node A.
3. When node A receives the BeginVerifyAck message, a Test message is transmitted periodically on the data link with interface identifier 1. The Test message transmits the verification identifier given to node B and the local interface identifier 1 to node A. When the Test message is transmitted on the data link, node B connects the its data link interface to a device that can transmit and receive the electrical signals for testing. As a result, when the data link with an interface identifier of 10 is connected to the relevant device, node B can receive the Test message.
4. When node B receives the Test message, the interface identifier of node A and that of node B are mapped to each other, and the TestStatusSuccess message is transmitted to node A through the control channel. The TestStatusSuccess message includes the interface identifier and the verification identifier of both nodes. The verification identifier identifies the TE link or link identifier.
5. Node A transmits the TestStatusSuccess message to node B to notify that it has received the TestStatusAccess message.
6. These procedures are repeated in the remaining data links to verify each connectivity.

---

* In this example, this is an unnumbered case, and the ID router is omitted.
** Node A is local when we look at it from the node B side.

GMPLS (Generalized Multiprotocol Label Switching)

7. When all of the data links have been tested, node A transmits the EndVerify message to node B through the control channel.
8. Finally, node B transmits the EndVerifyAck message to node A to indicate that it has received the EndVerify message through the control channel.

Using these procedures, it is possible to verify the connectivity of data links automatically.

8.6.3.4 Failure Management

Failure management is an optional function of LMP that has the role of locating the failure points of a TE link. The ability to detect a failure of a TE link and locate the failure point makes it possible to protect the network immediately and to recover from a failed state by restoration. The failure-management function is executed using such messages as ChannelStatus message, ChannelStatusAck message, ChannelStatusRequest message, and ChannelStatusResponse message.

In the case of a transparent interface that transmits and receives optical signals without converting to or from electrical signals, detection of a data link failure is done by measuring the degradation of the physical layer signal, such as an extinction of optical signal, for example. However, the function to determine degradation of the physical layer is independent of the LMP's function.

When a failure occurs on a data link between two nodes, it may often happen that all the downstream-side nodes of the LSP detect the failure and generate multiple alarms without detecting the failure point. Upstream/downstream is defined by the direction of the data stream. To avoid generation of multiple alarms for the same failure, a failure-notification function is available using a Channel-Status message in LMP.

To identify the link failure between neighboring nodes, the downstream nodes that detect a failure of the LSP notify that they have detected a failure to the upstream nodes using a ChannelStatus message. The upstream node checks whether an upstream link failure has been detected or not against the downstream link failure. If the upstream link seen from the node itself is normal, it notifies that there is a failure in the link between neighboring nodes to the downstream nodes using a ChannelStatus message. If a failure is also detected in the upstream node seen from the node itself, it notifies that the link between neighboring nodes is normal using a ChannelStatus message. If a ChannelStatus message was not transmitted from the upstream node, it transmits a ChannelStatusRequest message to the upstream node. When a failure point on the data link is identified, a recovery process from failure is executed by signaling protocol.

8.7 PEER MODEL AND OVERLAY MODEL

From the viewpoint of network operation, GMPLS supports two models, a peer model and an overlay model. Figure 8.22 and Figure 8.23 show the peer model and the overlay model, respectively. In this section, we describe a network in which an IP router and OXC exist.

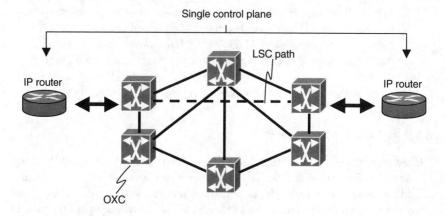

FIGURE 8.22 Peer model.

8.7.1 Peer Model

As shown in Figure 8.22, in the peer model, all of the IP routers and OXC within the network exist as "peers," that is, an equivalent thing, and they are controlled by a single control plane. The IP routers and OXC can obtain topology information of all the packet layers and λ-layers. The IP address that is set to the IP routers and OXC must be in the same address system.

Because the peer model operates by a single control plane, it is possible to draw out fully the merit of hierarchization of the LSP and of the multilayer traffic engineering that are characteristic features of GMPLS for a network with

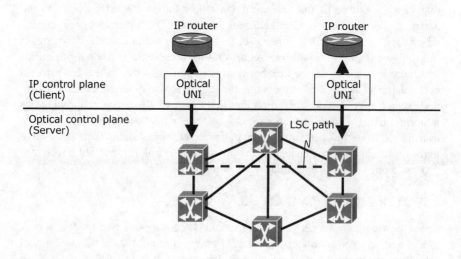

FIGURE 8.23 Overlay model.

a complex layer structure. Further, the peer model has the merit that it can execute end-to-end protection or restoration even if there are multiple mixed layers.

However, in the peer model, because the link states of all the layers are advertised to each node, vast amounts of information are transferred through the control plane. In addition, if multilayer traffic engineering is executed considering the link states of all the layers, the amount of computation required becomes vast. Given this scalability problem, it is difficult to realize a perfect peer model. In contrast, the overlay model can resolve this scalability problem. In the near future, it is expected that a more perfect peer model will be realized in stages by introducing the peer model into the overlay model.

8.7.2 Overlay Model

As shown in Figure 8.23, the overlay model is separated into an IP control plane and an optical control plane, both being connected by UNI (user network interface). The IP layer cannot know the topology of the λ-layer, and OXC cannot know the topology of the packet layer. Because the control planes are separated, the IP address that is set to the IP router and the IP address that is set to OXC can be from different address systems.

The overlay model can be considered as a client-server model in which the upper layer is seen as a client and the lower layer is seen as a server. When a request to set up the LSP (PSC-LSP) between IP routers is posed, the packet layer asks the λ-layer through the optical UNI whether it is possible to offer a desired network resource or not. If it is possible, it sets up the LSP in between IP routers using the resource of λ-layer, i.e., LSC-LSP.

The characteristic feature of the overlay model is that, because the control planes of the upper and lower layers are separated from each other, there is no scalability problem such as seen in the peer model. In addition, the upper layer's subnetwork, which is connected to the lower layer by an optical UNI, has no interaction with the upper layer's subnetwork that is connected by the other optical UNI. Therefore, even within the same upper layer, it is possible to update the control plane or the network elements, such as a node, independently of each other. This also applies to the relationship between the upper layer and the lower layer.

The peer model can fully exploit the advantages of GMPLS, such as multilayer traffic engineering, but in the interest of preserving scalability, the overlay model restricts full utilization of GMPLS functions by separating the control plane into two layers. Some consider the overlay model as an introductory step in the application of GMPLS networks. The advantages of multilayer traffic engineering in a peer model and the advantage of scalability with independent control planes in an overlay model are trade-offs. The choice of the best model for a GMPLS network — peer model or overlay model — is still being debated within the standardization organizations, including the IETF and the OIF (Optical Internetworking Forum).

REFERENCES

1. Awduche, D. and Rekhter, Y., Multiprotocol lambda switching: combining MPLS traffic engineering control with optical crossconnects, *IEEE Commun. Mag.*, Vol.39, Issue3, 111–116, 2001.
2. Awduche, D. et al., Requirements for Traffic Engineering over MPLS, RFC 2702, http://www.ietf.org/rfc/rfc2702.txt?number=2702 Sep. 1999.
3. Awduche, D., MPLS and traffic engineering in IP networks, *IEEE Commun. Mag.*, Vol.37, Issue12, 111–116, 1999.
4. Rosen, E., Viswanathan, A., and Callon, R., Multiprotocol Label Switching Architecture, RFC 3031. http://www.ietf.org/rfc/rfc3031.txt?number=3031
5. Banerjee, A. et al., Generalized multiprotocol label switching: an overview of routing and management enhancements, *IEEE Commun. Mag.*, Vol. 39, Issue1, 144–150, 2001.
6. Kompella, K. and Rekhter, Y., OSPF Extensions in Support of Generalized MPLS, IETF draft, http://www.ietf.org/internet-drafts/draft-ietf-ccamp-ospf-gmpls-extensions-12.txt, Oct. 2003 (work in progress).
7. Kompella, K. et al., Link Bundling in MPLS Traffic Engineering, IETF draft, http://www.ietf.org/internet-drafts/draft-ietf-mpls-bundle-06.txt, Dec. 2004.
8. Moy, J., OSPF Version 2, RFC 2328. http://www.ietf.org/rfc/rfc2328.txt?number=2328
9. Coltun, R., The OSPF Opaque LSA Option, RFC 2370. http://www.ietf.org/rfc/rfc2370.txt?number=2370
10. Katz, D. et al., Traffic Engineering Extensions to OSPF Version 2, RFC3630, http://www.ietf.org/rfc/rfc3630.txt?number=3630
11. Oki, E., Matsuura, N., Shiomoto, K., and Yamanaka, N., A disjoint path selection scheme with shared risk link groups in GMPLS networks, *IEEE Commun. Lett.*, 6, Vol. 6, Issue9, 406–408, 2002.
12. Okamoto, S., Oki, E., Shimano, K., Sahara, A., and Yamanaka, N., Demonstration of the highly reliable HIKARI router network based on a newly developed disjoint path selection scheme, *IEEE Commun. Mag.*, 40, 11, 52–59, 2002.
13. Oki, E., Matsuura, N., Shiomoto, K., and Yamanaka, N., A disjoint path selection scheme with shared risk link group constraints in GMPLS networks, *IEICE Trans. Commun.*, E86-B, no.8, pp. 2455–2462, 2003.
14. Banerjee, A. et al., Generalized multiprotocol label switching: an overview of signaling enhancements and recovery techniques, *IEEE Commun. Mag.*, Vol. 39, Issue7, 144–151, 2001.
15. Berger, L. et al., Generalized MPLS ó Signaling Functional Description, RFC 3471, http://www.ietf.org/rfc/rfc2702.txt?number=3471 Jan. 2003.
16. Ashwood-Smith, P. et al., Generalized MPLS Signaling ó RSVP-TE Extensions, RFC 3473, http://www.ietf.org/rfc/rfc2702.txt?number=3473 Jan. 2003.
17. Lang, J. et al., Link Management Protocol (LMP), IETF draft, http://www.ietf.org/internet-drafts/draft-ietf-ccamp-lmp-10.txt Oct. 2003 (work in progress).

9 Traffic Engineering in GMPLS Networks

This chapter describes several important studies on traffic engineering in Generalized Multiprotocol Label Switching (GMPLS) networks. These studies address multilayer traffic engineering, survivable networks, and wavelength-routed optical networks, and one presents a demonstration of a GMPLS-based router.

The first two sections discuss GMPLS-based multilayer traffic engineering technologies. Section 9.1 presents a distributed virtual-network topology (VNT) reconfiguration method for an IP-over-WDM (wavelength division multiplexing) network under dynamic traffic demand. The discussion includes a simple heuristic algorithm for calculating the virtual-network topology for distributed control [1]. Section 9.2 addresses a hierarchical cloud-router network (HCRN) to solve the problem of overcoming the scalability limit in a multilayer Generalized Multiprotocol Label Switching (GMPLS) network [2].

Section 9.3 describes wavelength-routed optical networks. Distributedly controlled dynamic wavelength-conversion (DDWC) optical networks based on a simple route and wavelength assignment (RWA) policy is introduced [3].

The next three sections address survivable GMPLS networks. Section 9.4 introduces a disjoint-path-selection scheme for survivable GMPLS networks with shared-risk link group (SRLG) constraints. In this so-called weighted-SRLG (WSRLG) scheme [4, 5], the number of SRLG members related to a link is treated as part of the link cost when the kth-shortest-path algorithm is executed. In WSRLG, a link that has many SRLG members is rarely selected as the shortest path. Section 9.5 presents a scalable shared-risk group (SRG)-based restoration method in shared-mesh wavelength-routed networks [6]. A backup-SRG concept to identify which SRG is backed up by individual links is introduced. The discussion is then extended to an admission control method for backup path using the backup-SRG concept to achieve 100% recovery performance.

Finally, Section 9.6 presents a demonstration of the photonic MPLS router, which was developed by NTT [7]. Integration of Multiprotocol Label Switching (MPLS) functions and Multiprotocol Lambda Switching (MPλS) functions can enhance the throughput of IP networks and remove bottlenecks that are derived from electrical packet processing. To enhance the packet-forwarding capability, NTT has proposed a photonic MPLS concept that includes MPλS, and NTT has demonstrated an IP, MPLS, and photonic MPLS integrated router system called the photonic MPLS router.

9.1 DISTRIBUTED VIRTUAL-NETWORK TOPOLOGY CONTROL

9.1.1 Virtual-Network Topology Design

The IP-over-WDM network consists of a set of WDM links and hybrid nodes, each of which consists of an electrical IP router part and an optical cross-connect path. A lightpath is established between the hybrid nodes. A set of lightpaths provides a virtual-network topology to carry IP packet traffic offered to the network. Adequate virtual-network topology is configured for a given traffic demand. The number of wavelengths per link and the number of transceivers per node is a limited resource. The adequate virtual-network topology is determined under the limited-resource constraint.

The virtual-network topology design problem has been studied extensively for a static traffic demand [8, 9]. The virtual-network topology can be designed for a given initial traffic demand, but as the network grows, the traffic demand can become significantly greater than the network was initially designed to handle. Reconfiguration of the virtual-network topology would be required to adapt to such a change in traffic demand.

Several methods for reconfiguring the virtual-network topology have been proposed [10, 11]. Those methods assume that the future traffic demand is given, and they are aimed at minimizing the topology change in the reconfiguration process. The new virtual-network topology is determined from the current one to adapt to the given traffic demand. The traffic demand is hard to anticipate accurately and fluctuates frequently in real networks. Thus traffic measurement and reconfiguration of the virtual-network topology should be orchestrated to cope with unpredictable traffic demand. A method for reconfiguring the virtual-network topology under dynamic changes in traffic demand would be required to cope with unpredictable traffic demand.

The problem of reconfiguring the topology of a virtual network under the load of dynamic traffic was studied in a recent work [12]. The proposed solution uses an off-line algorithm for time-variant traffic. It assumes that a set of traffic matrices at different instants is known *a priori*. Another work on virtual-network topology under dynamic traffic includes an on-line method of reconfiguring the virtual-network topology [13]. The method monitors the traffic instead of assuming future traffic pattern. A simple adjustment to the virtual-network topology is applied to mitigate congestion and reclaim network resources for underutilized lightpaths. The method is based on centralized control, which collects the traffic-demand measurement and calculates the new virtual-network topology for the obtained measurement of traffic demand. The centralized controller initiates a lightpath setup/teardown procedure. A heuristic algorithm is used to calculate the virtual-network topology. To mitigate the congestion, a new lightpath is established between the end nodes of multihop traffic with the highest load using the most congested lightpath.

This section presents a distributed VNT reconfiguration mechanism under dynamic unpredictable traffic [1]. In a distributed approach, each node decides whether it should initiate a lightpath setup/teardown procedure. The distributed approach requires a properly implemented mechanism for coordination between nodes, otherwise, a new virtual-network topology might be formed inconsistently. The proposed method uses a link-state routing protocol for each node to share the same virtual topology and the traffic demand over the individual lightpath, which is measured at the originating node. Each node calculates the new virtual-network topology using a simple heuristic algorithm, compares it with the old one, and initiates the lightpath setup/teardown procedure if necessary.

9.1.2 DISTRIBUTED NETWORK CONTROL APPROACH

9.1.2.1 Virtual-Network Topology

It is assumed that a hybrid node, consisting of an electrical IP router part and an optical cross-connect (OXC) part, is used (Figure 9.1). Every port of the electrical IP router is connected to the OXC port via internal fiber. All traffic between nodes is carried over the WDM link. Some of the traffic goes to the electrical IP router after passing through the OXC port, and some exits the node through the OXC port. A lightpath is established between nodes by setting up the cross-connects along the route between nodes. The lightpath is terminated at the transceiver of the electrical IP router part of the end nodes. The number of wavelengths per link and the number of transceiver ports per node is a limited resource in determining the virtual-network topology. In the following discussion, let N_p and N_l denote the number of transceiver ports per node (Figure 9.1), and let w denote the number of wavelengths WDM links.

The virtual-network topology should be designed so that a set of traffic demand is carried efficiently (Figure 9.2). Let Γ_{ij} denote a traffic-demand matrix whose (i,j) elements indicate traffic demand between node i to node j. An adequate virtual-network topology is determined for a given traffic-demand matrix under the constraint of the number of wavelengths and the number of transceiver ports.

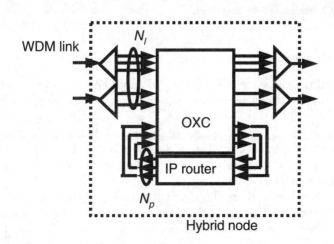

FIGURE 9.1 Hybrid node architecture (unfolded view). (Copyright 2003, IEEE.)

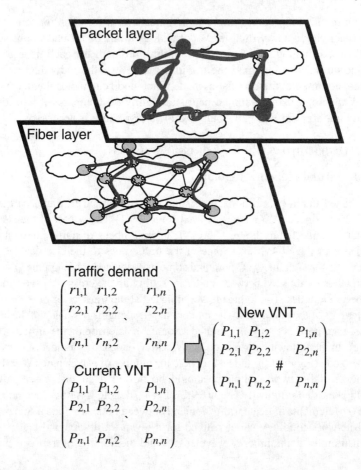

FIGURE 9.2 Virtual-network topology.

9.1.2.2 Design Goal

Our method is designed with three goals in mind. First, the method should be simple. A simple method means quick responsiveness and easier implementation. Second, the method should be efficient. The goal of traffic engineering is to optimize network resource utilization, so the method should be efficient in terms of using network resources. Third, the method should work as a distributed system because such systems are robust against failure and work automatically with minimum human intervention.

In the virtual-network topology reconfiguration process, centralized coordination is avoided by using a special class of virtual-network topology calculation algorithm. Two issues need to be addressed in designing the distributed method for the VNT reconfiguration. The first one is the order of lightpath setup or teardown, and the second one is the presence of conflicting requests on a lightpath, where one node requests teardown of a lightpath while another requests persistent use of the lightpath. These issues are easily resolved by a centralized method because the centralized method does not permit a change of lightpaths through reconfiguration unless it has resolved the issue beforehand using a centrally maintained network database.

The first issue stems from the fact that different virtual-network topologies are obtained if the order of lightpath setup/teardown is changed. If two nodes assume a different order of lightpath setup/teardown, they have inconsistent views of the new virtual-network topology. In calculating the virtual-network topology, the order of the lightpath setup/teardown is fixed: the new lightpath setup is tried in descending order of its traffic load to determine the route, and the existing lightpath is torn down in ascending order of its traffic load. Note that the "actual" order of setup/teardown is not fixed, even though the "calculating" order of setup/teardown is fixed.

The second issue stems from the fact that one node might request teardown for an underutilized lightpath while another might maintain it for QoS (quality of service) concerns. In this situation, these two nodes have inconsistent views of the virtual-network topology. This can occur when the two nodes are using different algorithms.

9.1.2.3 Overview of Distributed Reconfiguration Method

In the proposed method, each node uses the same virtual topology reconfiguration algorithm, so conflicting requests cannot occur. Each node periodically measures the traffic carried over the lightpath originating from the node. If the measured traffic changes in a measurement cycle, a link-state advertisement (LSA) packet carrying the traffic measurement data is flooded throughout the network. If the traffic carried over a lightpath exceeds the high threshold (T_H), each node initiates a VNT reconfiguration procedure by calculating a better virtual-network topology to quell the congestion. In addition, underutilized lightpaths should be torn down, if possible, for future lightpath setup requests. If the traffic carried over a lightpath falls below the lower threshold (T_L), the lightpath is torn down only if new congestion would not occur. This test is performed assuming that all underutilized lightpaths were removed. If the teardown of a lightpath could cause new congestion, it is not torn down. Comparing the better virtual-network topology and the existing one, each node identifies which lightpaths need to be set up or torn down. If the node is the one from which the new lightpath is originating, it initiates a lightpath setup procedure. On the other hand, if the node is the one from which the existing lightpath is originating, it initiates a lightpath teardown procedure. Each node performs these tasks independently, so no centralized coordination mechanism is required.

The proposed method measures traffic demand periodically. If the burst transfer period is much shorter than the measurement cycle, the traffic increase is not detected, and virtual topology reconfiguration is not initiated. For short burst transfer, optical burst switching is a promising approach. In optical burst switching, the ingress node has a burst-detection mechanism to initiate a lightpath setup procedure. Even though the virtual topology reconfiguration approach is more suited for timescales longer than mid-term, a burst-detection mechanism could be incorporated into the proposed virtual-network reconfiguration method to cope with short burst transfer. In addition, short-term traffic fluctuation need to be smoothed out for stable control. A low-pass filter is used for this purpose [14]. Maximum instantaneous utilization during a measurement cycle is regarded as a traffic demand.

9.1.2.4 Distributed Control Mechanism

Our method works as a distributed system. Each node in the network initiates the VNT reconfiguration procedure without centralized coordination. The key component of the distributed mechanism is to use a link-state routing protocol for each node to share the same virtual topology and the traffic demand over the individual lightpath, which is measured at the originating node. The link-state routing protocol is used to flood information on the virtual-network topology and traffic demand over the lightpath throughout all nodes in the network.

Once the information on the virtual-network topology and traffic demand over the lightpath is shared by all nodes, they all use an identical virtual-network topology calculation algorithm. After each node calculates the next VNT and compares it with the current one to identify which lightpaths should be set up and torn down, the originating node of the lightpath starts the procedure for lightpath setup/teardown (Figure 9.3). The VNT is reconfigured after each node finishes setting up and tearing down the lightpaths. Once the new VNT is achieved, the IP traffic is rerouted over the new VNT.

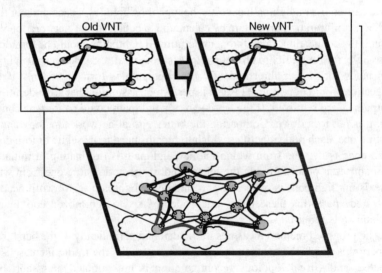

FIGURE 9.3

Traffic Engineering in GMPLS Networks

To minimize disruption in tearing down a lightpath, the lightpath is advertised as a dormant state by the originating node using the link-state routing protocol. When a node receives the link state of a dormant link, it detours all IP traffic around the dormant link. After the originating node confirms that no IP traffic is carried over the lightpath, it tears down the dormant lightpath.

The distributed method extends a link-state protocol. All link states in the network need to be shared by all nodes. The larger the network size, the larger the number of link states. For a large-scale network, a hierarchy could be used to achieve scalability. On the other hand, network utilization could be compromised due to the suboptimality of the hierarchical approach.

9.1.2.5 Heuristic Algorithm for VNT Calculation

A heuristic algorithm is used for calculating the VNT because it is simple enough to work quickly. To make the method work in a distributed manner, any order of setup/teardown of lightpaths is not assumed in designing the VNT calculation algorithm.

The algorithm adds new lightpaths to mitigate congestion and, when possible, removes existing underutilized lightpaths for reclamation. The VNT algorithm should not assume any order of setup/teardown of lightpaths initiated by individual originating nodes. Multiple new lightpaths might contend for the same resources, i.e., the number of wavelength links and the number of transceiver ports, before all underutilized lightpath candidates are removed. To avoid this situation, the heuristic algorithm adds new lightpath candidates first without relying on resources returned by removed lightpaths, and then removes existing underutilized lightpaths.

The heuristic algorithm uses two parameters to define congested and underutilized lightpaths: T_H and T_L denote, respectively, thresholds for congested (high) and underutilized (low) lightpaths. Pseudocode for the lightpath addition part of the heuristic algorithm is given in Figure 9.4. If traffic demand over the lightpath is greater than T_H, a new lightpath is set up to reroute traffic from the congested lightpath. The end nodes of the new lightpath are selected among the adjacent nodes of the end nodes of the congested link. The ingress node $v_{ingress}$ and egress node v_{egress} of the new lightpath are selected such that the sum of traffic demand from $v_{ingress}$ to the ingress node of the congested link and from the egress node of the congested link to v_{egress} is maximized. The ingress node of the congested link can be v_{egress}, and the egress node of the congested link can be $v_{ingress}$.

Pseudocode for the lightpath deletion part of the heuristic algorithm is given in Figure 9.5. If traffic demand over the lightpath is less than T_L, it is torn down for reclamation if possible. It is torn down only if new congestion would not occur after its removal. All underutilized lightpaths are tested in a deterministic order in calculating the new virtual-network topology. Note that even though the deterministic order of lightpath deletion in the virtual-network topology calculation is assumed, we do not have to care about the order of lightpath teardown after the new virtual-network topology is determined.

```
AddLightpath(){
Route(TrfMtx);
for (int i=0; i<size; i++) {
for (int j=0; j<size; j++) {
if (i!=j) {
if (uB[i][j]>TH*cap[i][j]) {
for (int k=0; k<size; k++) {
if ((PortAvailable(k)==true)
&& ((k!=i && LighPath(k,i))
|| (k==i))){

                for (int l=0; l<size; l++) {
if (((k!=i) || (l!=j)) && (l!=k)

                    && (PortAvailable(l))

                    && ((j!=l && LightPath(j,l))
|| (j==l))

                    && (NoLightPath(k,l))) {
if (uB[k][i]+uB[j][l]>=max) {
max = uB[k][i]+uB[j][l];
Src = k;
Dst = l;

                                        }

                    }
                                        }
    SetupLightPath(Src,Dst)
            }
}
}
```

FIGURE 9.4

9.1.3 Protocol Design

9.1.3.1 GMPLS Architecture

GMPLS has recently attracted much attention from industry and academia. Intelligent IP routing techniques are applied to various kinds of switching networks such as SDH/SONET (synchronous digital hierarchy/synchronous optical network), wavelength switch, and fiber switch. GMPLS provides a powerful blend of optical transport and IP routing (Figure 9.6) [15, 16] as well as a powerful set of control mechanisms for multilayer label-switched path (LSP) networks composed of IP, SDH/SONET, lambda, and fiber layers.

Traffic Engineering in GMPLS Networks

```
void removelink(){
route(TrfMtx);
for (int i=0; i < size; i++) {
for (int j=0; j < size; j++) {
if (i != j) {
if ((uB[i][j] < TL*cap[i][j])

                && isRemovable(i,j)) {

                int Src = i;

                int Dst = j;
        removeNominalLambdaPath(Src,Dst);

                if (checkRoute(TrfMtx)) {//
no congestion
        removeLambdaPath(Src,Dst);
                }
                }
}
}
}
}
```

FIGURE 9.5

GMPLS provides a framework for multilayer LSP network control. In the multilayer LSP network, a set of the lower-layer LSPs serves as the virtual-network topology (VNT) for the upper-layer LSPs. For example, a set of wavelength LSPs serves as the VNT for the LSPs in the IP/MPLS packet layer. The fiber network topology is used as the VNT for routing the WDM-layer LSP, which in turn serves as the VNT for routing the IP-layer LSP. The VNT is reconfigured by setting up or tearing down WDM-layer LSPs in the WDM-layer using the

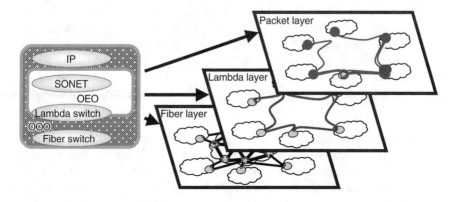

FIGURE 9.6 GMPLS-based multiregion network.

GMPLS signaling protocol. Traffic demand in the IP layer is disseminated by the GMPLS routing-protocol extensions. Each node calculates the adequate virtual topology to meet the IP-layer traffic demand in a distributed manner. Each node then initiates the path setup/teardown requests after comparing the new and old VNTs.

In a GMPLS architecture, a link-state routing protocol is used for topology discovery. Each node advertises a link state including all links originating from the node. The residual resource on the link is also carried in the link state, which is flooded throughout the entire network. When a node sets up the LSP to a destination, it calculates the feasible route for the LSP by running the constraint-based shortest-path-first (CSPF) algorithm. If the feasible route is found, the node initiates the LSP setup procedure.

9.1.3.2 Forwarding Adjacency in Multilayer Path Network

The multilayer path network control mechanism introduced by GMPLS plays a key role in the proposed distributed VNT reconfiguration. GMPLS introduces the concept of forwarding adjacency (FA) to control multilayer path networks [15]. In a multilayer network, the lower-layer LSP is used for forwarding the upper-layer LSP. If two nodes are directly connected by a lower-layer LSP, they are regarded as being in the state of "forwarding adjacency" in the upper layer. An optical LSP (i.e., lightpath) is set up first before an electrical packet LSP is set up. Once the optical LSP is established, it is advertised by routing protocol as FA-LSP to allow packet LSPs to use it for forwarding. Setup or teardown of LSPs triggers the change of virtual topology for the upper-layer LSP network. In this way, FA-LSP enables us to implement a multilayer LSP network control mechanism in a distributed manner. In multilayer LSP networks, the lower-layer LSPs form the virtual topology for the upper-layer LSP. The upper-layer LSP is routed over the virtual topology.

9.1.3.3 Switching Capability

A switching-capability concept is introduced in GMPLS protocol to support various kinds of switching technologies in a unified way. Five classes of switching capability are defined: PSC (packet-switch capable), L2SC (layer-2-switch capable), TDM (time-division-multiplex capable), LSC (lambda-switch capable), and FSC (fiber-switch capable) [15]. Each end of a link in a GMPLS network is associated with at least one switching capability. PSC is associated with an interface that can delineate IP/MPLS packets (e.g., LSR's interface), and LSC is associated with an interface that can switch individual wavelengths multiplexed in a fiber link (e.g., OXC's interface). Every link in the link-state database has switching capabilities on both ends.

Switching capability is used to implement multilayer routing. An optical LSP is routed over the topology composed of a link state, either of whose ends is LSC. In Figure 9.7, an optical LSP is routed from LSR B, to OXC A, to OXC C, to LSR E. The path for the optical LSP is composed of links, either of whose ends is LSC. Once the optical LSP is set up, it is advertised as FA-LSP [15], both ends of which are PSC. In calculating the path for packet LSP, the link-state

Traffic Engineering in GMPLS Networks

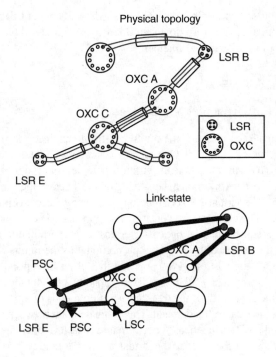

FIGURE 9.7

database is filtered to include the link, both ends of which include only PSC. In this way, hierarchical routing of packet LSP and optical LSP is done by using a link-state database filtered with respect to switching capability.

The lower-layer (optical LSP) network provides the virtual-network topology to the upper-layer (packet LSP) network. The virtual-network topology is configured by setting up or tearing down optical LSPs in the lower layer. By using GMPLS signaling and routing protocol, the virtual-network topology can be altered easily. By changing the virtual-network topology, traffic engineering can be implemented. The lightpath topology is altered to adjust the traffic demand of the IP layer.

9.1.3.4 Protocol Extensions

In our scheme, each node runs the VNT calculation algorithm, which requires information about the current virtual topology and the traffic demand of the packet LSPs. The information on traffic demand of the packet LSPs needs to be advertised. The GMPLS link-state routing protocol [17] is extended to include the traffic-demand information.

When the optical LSP is released, disruption must be avoided. The packet LSPs need to be rerouted before tearing down the underlying optical LSP. To advertise the optical LSP as dormant, the GMPLS routing protocol is extended. Once each node receives the dormant link state, it reroutes the packet LSPs over the dormant optical LSP to other nondormant optical LSPs. After all packet LSPs are rerouted, the dormant optical LSP is torn down. In this way, the graceful teardown of LSP is implemented in a distributed manner.

9.1.4 Performance Evaluation

9.1.4.1 Effect of Dynamic VNT Change

Here we investigate the effect of the multilayer traffic engineering method using a sample network model. The BXCQ (Branch Exchange with quality of service constraints) method is used in calculating the virtual topologies for the optical and electrical layers [18, 19]. The network model, consisting of 11 nodes, is the same as that used by Oki et al. [18]. The virtual-network topology producing the local optimum network cost is calculated. The network cost consists of an optical-layer cost and an electrical-layer cost. It is assumed that the optical-layer cost is proportional to the number of LSC ports and the electrical-layer cost is proportional to the number of PSC ports. Let x and y denote the cost of a PSC port and that of an LSC port, respectively. It is assumed that an optical LSP has a fixed capacity of 2.4 Gbps (this value is determined by the capability of the transceivers at both ends of an optical LSP) and that there is symmetrical traffic demand. The traffic demand between all source–destination (SD) node pairs is identical. A performance measure called the average nodal degree is introduced to characterize the virtual-network topology. The average nodal degree is defined as the average number of (virtual) links originating from the node. If there are 11 nodes in the virtual network, the fact that the average nodal degree is 10 implies that the virtual network is a full-mesh network.

9.1.4.1.1 Cost-Saving Effect

Here we evaluate the cost-saving effect of dynamic topology reconfiguration. The network cost is calculated for both a fixed topology and dynamic topology change, and these costs are compared. Figure 9.8 shows the network cost as a function of traffic demand of SD-pairs. It is assumed that the optical LSP has a fixed capacity of 2.4 Gbps, the same as OC48c/STS-16. The network topology optimized when the traffic demand is 150 Mbps is used for the fixed topology. As the traffic demand increases, the network cost is reduced with variable topology. As seen in Figure 9.8, the dynamic topology configuration reduces the network cost by half. These results confirm that the traffic-engineering method based on the virtual topology reconfiguration is effective.

9.1.4.1.2 Virtual Degree

Here we investigate the virtual-network topology for different traffic demands. Figure 9.9 shows the relationship between the optimum average nodal degree of an optical LSP's virtual topology and the traffic demand between SD node pairs. The average virtual degree is defined as the average number of links originating from a node in the virtual-network topology. A value of 2 means that the virtual-network topology is close to a ring, and 10 means that it is close to full mesh for a physical network consisting of 11 nodes.

Traffic Engineering in GMPLS Networks

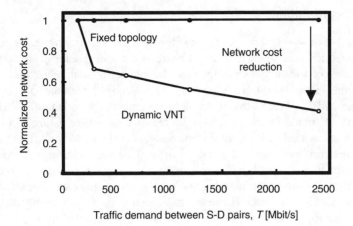

FIGURE 9.8 Network cost as a function of traffic demand of SD pairs. (Copyright 2003, IEEE.)

It is observed that the optimum average nodal degree increases with the traffic demand between SD node pairs. The average virtual degree is close to 2 when the traffic demand is small, and it is close to 10 when the traffic demand is large. This result agrees with the following observation. When the traffic demand is small, it is economical to aggregate the individual traffic demand between several SD node pairs into a single optical LSP. When the traffic demand increases, the number of SD pairs carried by a single optical LSP is reduced, and the direct optical LSP is used to carry the fat traffic demand between SD node pairs. From this result, it is confirmed that our approach, which uses the direct optical LSP for the heavier SD node pairs, is effective.

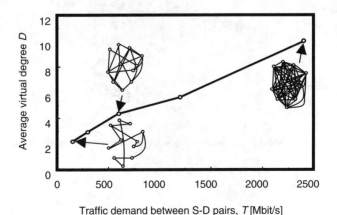

FIGURE 9.9 Average virtual degree as a function of traffic demands backed up by wavelength. (Copyright 2003, IEEE.)

9.1.4.2 Utilization

The number of wavelengths per link and the number of transceiver ports per node is a limited resource in determining the virtual-network topology. The impact of the number of wavelengths per link and the number of transceiver ports on the utilization is investigated. Recall that N_p and N_l denote the number of transceiver ports per node, and recall that w denotes the number of wavelength WDM links. The NSF (National Science Foundation) network model consisting of 14 nodes and 25 links as used by Ramaswami and Sivarajan [8] was adopted. The adequate virtual-network topology is determined for a given traffic-demand matrix under the constraint of the number of wavelengths and the number of transceiver ports. Recall that Γ_{ij} denotes a traffic-demand matrix whose (i,j) element indicates the traffic demand between node i to node j. It is assumed that i,j is uniformly distributed in the range $[0, r_1]$. Here we evaluate the maximum permitted load $r_{max} = \max r_1$ obtained by our virtual-network topology reconfiguration method using a bisection method. $T_H = 0.8$ and $T_L = 0.1$ are assumed.

Figure 9.10 shows the permitted load as a function of the number of wavelength links w (= 2, 4, 6, 8, 10, 12, 14, 16, 18, 20, 22, 24, 26, 28, 30, 32) and the PSC port ratio defined as N_p/N_l (= 0.1, 0.2, 0.4, 0.6, 0.8, 1.0). As shown in Figure 9.10, the permitted load increases as the number of wavelength links or the PSC port ratio increases. In this network model with the physical degree or a node port number and the average length of the shortest path between all nodes, when the number of wavelengths is moderate (between two and four), an increase of PSC port ratio does not improve the permitted load. A shortage of wavelength

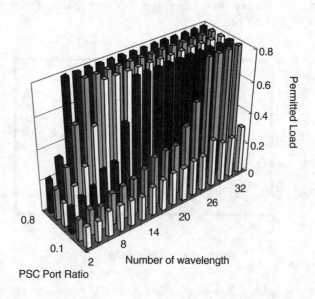

FIGURE 9.10 Permitted load as a function of the number of wavelength and PSC port ratio.

Traffic Engineering in GMPLS Networks

resources limits the utilization in this region. As the number of wavelengths increases, increasing the PSC port ratio significantly improves the permitted load.

9.1.4.3 Dynamic Traffic Change

Here we investigate how traffic fluctuation affects the performance of our method. It is assumed that $T_H = 0.8$ and $T_L = 0.4$, that the number of wavelengths $w = 24$, and that the PSC port ratio $N_p/N_1 = 0.2$. Three sets of traffic-demand matrices were used. Each set had 20 traffic-demand matrices randomly generated in sequence. It is assumed that i,j is uniformly distributed in the range $[r_0, r_1]$. The three sets of traffic-demand matrices ó $[r_0, r_1]$ = [0, 3], [0.1, 0.2], and [0.125, 0.175] ó had the same average load (= 0.15).

The number of added and deleted optical LSPs in each sequence was calculated, and the results in Figure 9.11, Figure 9.12, and Figure 9.13 show the respective number of existing and deleted optical LSPs as a function of time sequence for the three matrices. A comparison of these figures shows that the number of deleted optical LSPs is moderate for less variable traffic $[r_0, r_1]$ = [0.125, 0.175] (Figure 9.13) and increases for variable traffic $[r_0, r_1]$ = [0, 3] (Figure 9.11).

Figure 9.14, Figure 9.15, and Figure 9.16 show the respective maximum and average load of optical LSPs as a function of time sequence for the three matrices. A comparison of these figures shows that the maximum load of optical LSPs is controlled below the congestion threshold (T_H) of 0.8. These figures also show that the average load of the optical LSPs fluctuated around 0.3 to 0.4, but the fluctuation decreased with decreasing traffic variability, as seen by comparing the two extreme cases of $[r_0, r_1]$ = [0.125, 0.175] (Figure 9.16) and $[r_0, r_1]$ = [0, 0.3] (Figure 9.14).

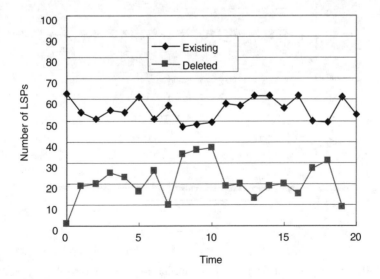

FIGURE 9.11 Number of lightpaths as a function of time sequence: $[r_0, r_1]$ = [0, 0.3]. (Copyright 2003, IEEE.)

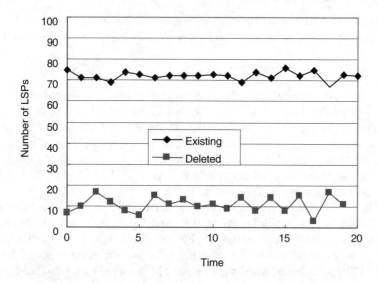

FIGURE 9.12 Number of lightpaths as a function of time sequence: $[r_0, r_1] = [0.1, 0.2]$. (Copyright 2003, IEEE.)

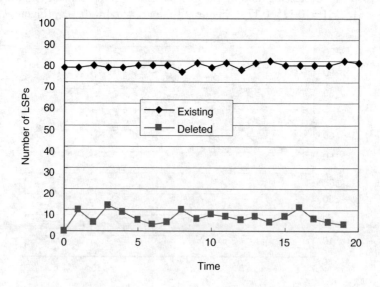

FIGURE 9.13 Number of lightpaths as a function of time sequence: $[r_0, r_1] = [0.125, 0.175]$. (Copyright 2003, IEEE.)

Traffic Engineering in GMPLS Networks

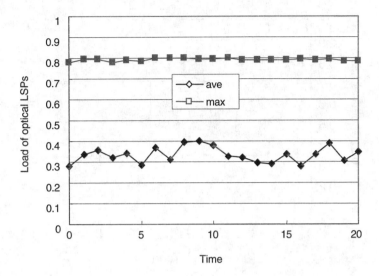

FIGURE 9.14 Load of optical LSPs as a function of time sequence: $[r_0, r_1] = [0, 0.3]$. (Copyright 2003, IEEE.)

The reconfiguration frequency is also affected by the high and low thresholds for detection of congestion and underutilization. The effect of these thresholds on the frequency of the reconfiguration of the virtual-network topology was evaluated by measuring the average number of optical LSPs under variable traffic

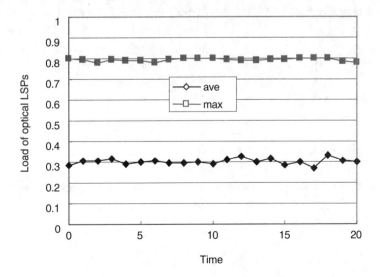

FIGURE 9.15 Load of optical LSPs as a function of time sequence: $[r_0, r_1] = [0.1, 0.2]$. (Copyright 2003, IEEE.)

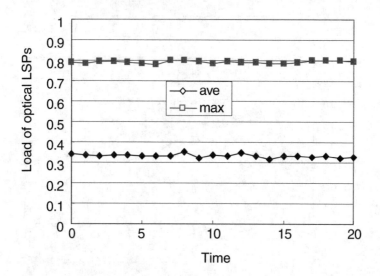

FIGURE 9.16 Load of optical LSPs as a function of time sequence: $[r_0, r_1] = [0.125, 0.175]$. (Copyright 2003, IEEE.)

conditions, where 100 traffic-demand matrices randomly generated in sequence for each element were randomly distributed in the range of $[r_0, r_1] = [0, 0.3]$.

Figure 9.17 shows the relationship between the threshold for congestion (T_H) and the number of optical LSPs. When the T_H is small, the number of existing optical LSPs is large because new optical LSPs are created to reduce the traffic load over the optical LSPs. As the T_H becomes larger, the number of existing

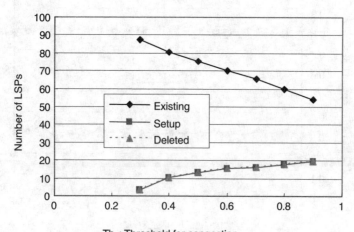

FIGURE 9.17 Relationship between the threshold for congestion and the number of optical LSPs. (Copyright 2003, IEEE.)

Traffic Engineering in GMPLS Networks

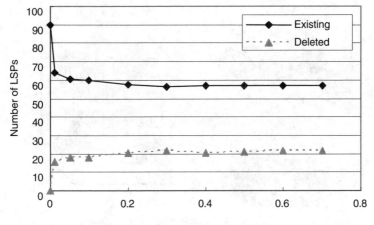

TL: threshold for under utilization

FIGURE 9.18 Relationship between the threshold for underutilization and the number of optical LSPs. (Copyright 2003, IEEE.)

optical LSPs becomes small because each optical LSP can accommodate much more traffic demand.

Figure 9.18 shows the relationship between the threshold for underutilization (T_L) and the number of optical LSPs. When the T_L is small, the number of deleted optical LSPs is small, and the number of existing optical LSPs is large because the underutilized optical LSP is not released. As the T_L becomes larger, the number of deleted optical LSPs becomes larger and the number of existing optical LSPs becomes smaller because the underutilized optical LSP is released.

In deleting optical LSPs, traffic could be disrupted. In the proposed method, when the optical LSP is torn down, it is advertised as dormant so that each node can have the electrical LSPs rerouted around the dormant optical LSP. Electrical LSPs are rerouted using a "make-before-break" technique to minimize packet loss during the switching process. When all electrical LSPs have been rerouted around the optical LSP, it is actually torn down. In this way, the disruption is minimized.

9.2 SCALABLE MULTILAYER GMPLS NETWORKS

9.2.1 SCALABILITY LIMIT OF GMPLS NETWORK

Networks with several layers include fiber networks, lambda networks, time-division multiplexing (TDM) networks, and packet networks. These networks are constructed using different technologies, and they are controlled and operated differently. Generalized Multiprotocol Label Switching (GMPLS) is now being standardized by the Internet Engineering Task Force (IETF). GMPLS enables us to control all different networks with one common protocol [20–22]. GMPLS expands the concept of MPLS, which is used for distributed traffic engineering

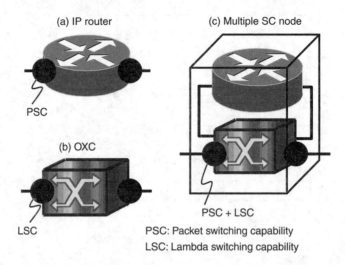

FIGURE 9.19 Switching capability.

in Internet Protocol (IP) packet-layer networks. GMPLS controls multilayer networks, including any of those mentioned above.

In a single-layer network, a node has only one kind of interface switching capability (SC) [15]. For example, an interface of an IP router has packet-switching capability (PSC), which provides the ability to switch packets in the packet layer (Figure 9.19(a)). An interface of an optical cross-connect (OXC) has lambda-switching capability (LSC), which provides the ability to switch wavelengths in the lambda layer (Figure 9.19(b)). In a GMPLS network, however, there are not only single-SC nodes but also multiple-SC nodes with more than one SC (Figure 9.19(c)) [23–24]. An example of a multiple-SC node is the so-called HIKARI router, which was developed by NTT [7, 25]. The HIKARI router has both LSC and PSC to support integrated packet- and lambda-layer networks. (Figure 9.19(c)).

GMPLS enables us to perform multilayer traffic engineering. The link-state information for all individual-layer networks is advertised distributedly to all nodes in a GMPLS network by a link-state routing protocol [17, 26]. All nodes utilize the same link-state database. Each node defines the shortest paths from itself to other nodes and a routing table. The required network resources, which include available bandwidth, wavelengths, and so forth, are efficiently reduced by GMPLS multilayer traffic engineering [18].

However, GMPLS networks have a scalability limit problem. In these GMPLS networks, all nodes support every layer's network topologies and calculate the shortest path to each destination node. All nodes advertise link states such as the presence or absence of a link and how many links are actually used. Routing protocols, such as Open Shortest Path First (OSPF), have this advertising function. In a network using OSPF, all nodes advertise link-state information whenever it changes.

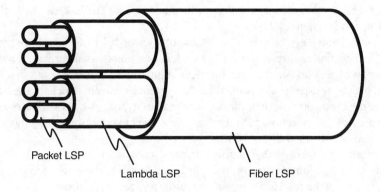

FIGURE 9.20 Hierarchial label switched path (LSP).

A multilayer GMPLS network applies the concept of a hierarchical label-switched path (LSP) (Figure 9.20) [27]. A lower-layer LSP is referred to as a link to establish a higher-layer LSP. In other words, a source node and a destination node connected via a lower-layer LSP are considered as adjacent nodes in the higher-layer network. This lower-layer LSP is called a forwarding-adjacency (FA)

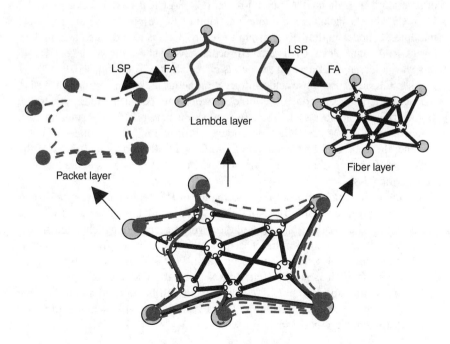

FIGURE 9.21 Label switched path (LSP) and Forwarding adjancency (FA) LSP.

LSP. An FA-LSP is recognized as a traffic-engineering (TE) link* in the higher-layer network (Figure 9.21).

Consequently, establishment or release of a lower-layer LSP is referred to as a link connection or disconnection and a network topology transformation by the higher-layer network. In the near future, in a multilayer GMPLS-controlled network, the shortest-path calculations caused by frequent advertisement of FA-LSPs by routing protocols will limit the network scalability, because network providers must provide for short hold path services and save network resources by dynamic rearrangement of established LSPs.

Another problem is that multilayer route computations do not have scalability for their complexity. In a single-layer network, the route computation volume is not large. However, in a multilayer network, route computation includes the decisions of whether to use cut-through [18] and which layer is selected. This computation is complex and time-consuming. The cut-through in the transit node may make the cost from the source node to the destination small if the requested bandwidth is large. If a route that utilizes cut-through in the packet layer is selected when the requested bandwidth is narrow, most of the lambda bandwidth will be idle, and the traffic allocation will be inefficient. These decisions thus depend on the requested bandwidth and available network resources.

This section presents a hierarchical cloud-router network (HCRN) to solve the problem of the scalability limit in terms of the multilayer network size [2]. A group of several nodes is defined as a cloud-router (CR). A CR is modeled as a multiple-SC node when it includes more than one kind of SC, even if there are no actual multiple-SC nodes in the CR. A group of several CRs is also defined as a higher-level CR. A higher-level CR is created recursively by several lower-level CRs. In this way, a hierarchical network, which is called HCRN, is established.

The CR advertises abstracted information about its internal structure. Using the abstracted information, the whole network can perform multilayer traffic engineering and achieve scalability. To abstract the network, the CR internal cost, which is part of the CR internal-structure information, is introduced. In this scheme, the ends of a link connecting two CRs are defined as interfaces of the CRs. The CR internal cost can be defined between CR interfaces and advertised outside the CRs.

In the following sections, the performance of HCRN in the case of abstracting a multilayer network by using the CR internal cost is evaluated. The reduction in the shortest-path calculation volume caused by changes in the network topology and optimum CR size is quantified. In addition, the effect of the CR internal-cost scheme on the probability of success in establishing an LSP is quantified. It is shown that HCRN using the CR internal-cost scheme can limit the probability that LSP establishment will be blocked and reduce the volume of shortest-path calculations, thus enabling a large-scale, multilayer network to achieve greater scalability.

---

* For brevity, a TE link is simply called a link.

9.2.2 Hierarchical Cloud-Router Network (HCRN)

9.2.2.1 HCRN Architecture

A hierarchical cloud-router network (HCRN) is proposed as a means to achieve large-scale, multilayer network scalability. In the HCRN scheme, a multilayer network is divided into groups that are treated as virtual nodes called cloud-routers (CRs), as shown in Figure 9.22. A network consisting of CRs is also divided into groups that are treated as higher-level CRs. An HCRN as a hierarchical network consisting of CRs can be developed in the same way that a higher-level CR- is created recursively by several lower-level CRs. Figure 9.23 shows a schematic of an HCRN.

The HCRN scheme's distinguishing characteristic is that a CR becomes equivalent to a multiple-SC node when it contains nodes with different SCs, even if there are no multiple-SC nodes within the CR. In Figure 9.22, interface a of the CR-containing node A can have a PSC + LSC interface, even if node A has only an LSC interface, because packet LSPs can be switched inside the CR (for example nodes B, E).

Every CR operates its link-state routing protocol independently. This link state is advertised inside the CR. The ends of a link connecting two CRs are defined as interfaces of the CRs. A CR internal cost is defined as the cost between two

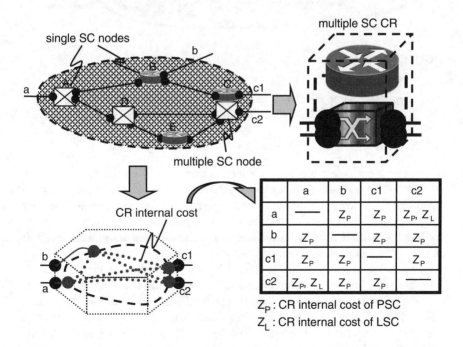

	a	b	c1	c2
a	—	Z_P	Z_P	Z_P, Z_L
b	Z_P	—	Z_P	Z_P
c1	Z_P	Z_P	—	Z_P
c2	Z_P, Z_L	Z_P	Z_P	—

Z_P : CR internal cost of PSC
Z_L : CR internal cost of LSC

FIGURE 9.22 Multiple SC cloud router (CR) and CR internal cost.

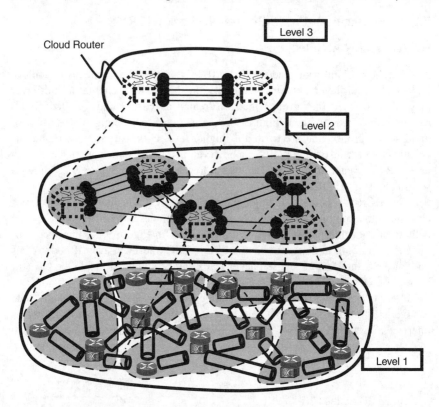

FIGURE 9.23 Schematic of HCRN.

interfaces of the same CR. CR internal cost is advertised outside the CR via the interface. CR internal cost is defined for all CR interface pairs that can be reached.

9.2.2.2 CR Internal-Cost Scheme and Network Topology

When a path can be established between two interfaces inside a CR, the interface pair is said to have reachability. An interface can thus know whether there is reachability between itself and other interfaces, because unreserved link resources, such as available bandwidth and wavelengths, are advertised by the routing protocol. If there is reachability for the interface pairs in the CR, a CR internal cost can be defined for the interface pairs.

Figure 9.24 shows an abstracted single-layer network based on the CR internal-cost scheme. The source node receives a link state advertised by a node in the same CR and a CR internal cost advertised by another CR's interface. The schematic in Figure 9.24 also shows the source node's network topology, which is made from the link state and the CR internal cost. The CR's interface receives link-state information of only the same or a higher-level CR. The outer network can recognize the reachability and cost between interfaces, but it cannot recognize the network

Traffic Engineering in GMPLS Networks

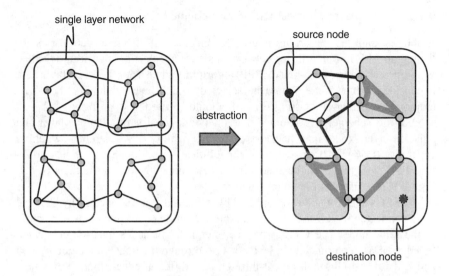

FIGURE 9.24 Graph of abstracted single-layer network.

topology inside the CR. In this way, the CR internal network topology is hidden, and other CRs are informed which interface is selected for LSP establishment.

By using this abstraction scheme, a multilayer network can be abstracted. Every SC in a CR is converted to a vertex. Figure 9.25 shows a multilayer abstraction. The interface in the CR calculates the shortest path and the CR internal cost, which are advertised outside the CR.

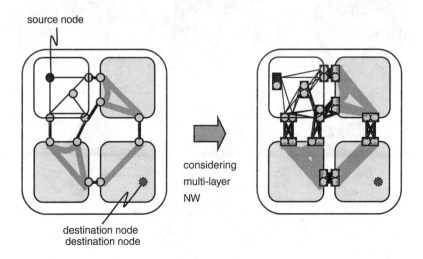

FIGURE 9.25 Graph of abstracted multilayer network.

9.2.2.3 Multilayer Shortest-Path-First Scheme

A shortest-path-first algorithm is used to select a cost-effective route for establishing an LSP in a multilayer network. This route-selection algorithm is called a multilayer shortest-path-first (MSPF) algorithm. When a new LSP is established, it considers which route should be selected, which layer switching should be employed, whether an existing FA-LSP should be used, and so forth.

For this scheme, it is assumed that a network consists of a lambda layer and a packet layer. Nodes have packet-switching capability (PSC), lambda-switching capability (LSC), or both switching capabilities. A multiple-SC node has trunk-type PSC parts [3, 25] and can adaptively determine whether to use them or not. Figure 9.26 illustrates the node structure.

The SCs in every node are replaced by vertices that are connected by physical links connected by edges in the graph, not including edges that connect vertices in the same node. The vertices between neighboring nodes must be connected by edges. Based on this graph, LSPs that go through the transit node hop by hop and LSPs that go through by cut-through take different edges, though they take the same fiber route. Different costs to these two LSPs can be allocated.

In the case of going through the PSC vertex P, both the PSC interface and the LSC interface are occupied. In the case of going through the LSC vertex L, only the LSC interface is occupied. The cost of the edge is expressed as the summation of the reciprocal number of available interfaces. For example, in Figure 9.26, when

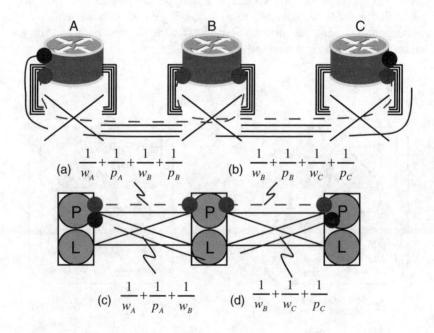

FIGURE 9.26 Calculation of CR internal cost.

Traffic Engineering in GMPLS Networks

a lambda LSP has been established, to establish the packet LSP and transfer the IP packet from node A to node B, the edge connecting vertex P on node A and vertex P on node B is selected. The cost is then expressed as (a) where w_A is the number of available LSC interfaces in node A and p_A is the number of available PSC interfaces in node A. In the case where a lambda LSP has been newly established by a cut-through packet layer on node B, to establish the packet LSP and transfer the IP packet from node A to node C, the edges connecting vertex P on node A with vertex L on node B and vertex L on node B with vertex P on node C are selected. The cost here is expressed as (c) + (d).

The cost of LSP establishment using the lower-layer FA-LSP is expressed as Z, which is formulated as follows when the packet LSP is established without establishing a new lambda LSP but using an existing lambda LSP instead.

$$Z = \varepsilon * \frac{BW_{req}}{BW_{res}} \qquad (9.1)$$

As described above, the minimal-cost LSP in a multilayer network can be determined by Dijkstra calculation [28], replacing SCs in every node with vertices and links with edges.

9.2.3 Performance Evaluation

9.2.3.1 Scalability

A higher-layer CR considers all lower-layer CRs interface within it as vertices. The scheme calculates the shortest-path by using all interfaces in its own CR. This has the advantages of using network resources effectively and providing a large traffic-intake capacity, because the shortest-path calculation engine can have an in-depth link state.

The parameters for the performance evaluation are defined as follows. N is the number of nodes in the network, and S is the number of nodes in the CR. S is called the CR size. If $N = S^p$ (p = variable), the network level increases by one. S and p determine the shortest-path calculation volume. When N is a constant number, a larger p reduces the shortest-path calculation volume. The volume in the $S^p < N \leq S^{p+1}$ case at every level and the largest volume, which determines the dominant level, are calculated.

Figure 9.27 shows the correlation between the number of levels and the order of the shortest-path calculation volume, p denotes the number of levels. CR size becomes small when N is constant and p becomes large, which means that the network level becomes deep. The calculation volumes of the dominant levels are highlighted by the heavy closed lines. The results show that the calculation volume becomes small when the number of levels becomes large, except in the $1 < p \leq 2$ range (Figure 9.28). However, the large number of

p	$0 < p \leq 1$	$1 < p \leq 1.5$	$1.5 < p \leq 2$
Level 1	$N^2 \log N$	$\frac{1}{p} N^{\frac{2}{p}} \log N$	$\frac{1}{p} N^{\frac{2}{p}} \log N$
Level 2	—	$\left(2 - \frac{2}{p}\right) N^{4 - \frac{4}{p}} \log N$	$\left(2 - \frac{2}{p}\right) N^{4 - \frac{4}{p}} \log N$
Level 3	—	—	—
Level 4	—	—	—
Level i+1	—	—	—
Calculation volume	$N^2 \log N$	$\frac{1}{p} N^{\frac{2}{p}} \log N$	$\left(2 - \frac{2}{p}\right) N^{4 - \frac{4}{p}} \log N$

p	$2 < p \leq 3$	$3 < p \leq 4$	$i < p \leq i+1$
Level 1	$\frac{1}{p} N^{\frac{2}{p}} \log N$	$\frac{1}{p} N^{\frac{2}{p}} \log N$	$\frac{1}{p} N^{\frac{2}{p}} \log N$
Level 2	$\frac{2}{p} N^{\frac{4}{p}} \log N$	$\frac{2}{p} N^{\frac{4}{p}} \log N$	$\frac{2}{p} N^{\frac{4}{p}} \log N$
Level 3	$\left(2 - \frac{4}{p}\right) N^{4 - \frac{8}{p}} \log N$	$\frac{2}{p} N^{\frac{4}{p}} \log N$	$\frac{2}{p} N^{\frac{4}{p}} \log N$
Level 4	—	$\left(2 - \frac{6}{p}\right) N^{4 - \frac{12}{p}} \log N$	$\frac{2}{p} N^{\frac{4}{p}} \log N$
Level i+1	—	—	$\left(2 - \frac{2i}{p}\right) N^{4 - \frac{4i}{p}} \log N$
Calculation volume	$\frac{2}{p} N^{\frac{4}{p}} \log N$	$\frac{2}{p} N^{\frac{4}{p}} \log N$	$\frac{2}{p} N^{\frac{4}{p}} \log N$

FIGURE 9.27 Order of shortest-path calculation volume.

levels causes a long transmission delay in the link state from the lowest level to a higher level.

Larger numbers of nodes can be operated with HCRN than in a normal GMPLS network. Figure 9.29 shows the correlation between the number of nodes and the order of the shortest-path calculation volume when A is 30 min and $p = 5$. The figure shows that a multilayer network has scalability up to 200 nodes by using HCRN, while it has a scalability limit of 20 nodes without HCRN.

A shorter advertisement interval period for LSP establishment or release can converge faster than with a normal GMPLS network. Figure 9.30 shows the correlation between the advertisement interval period and the order of the shortest-path calculation volume. The shortest-path calculation volume for a single-layer network, V_s, is given by Equation 9.2, and that for a multilayer network, V_m, is given by Equation 9.3. Here, N is the number of nodes, L is the number of links, A is the advertisement interval period for LSP establishment or release, and SC is the number of switching capabilities. Figure 9.30 shows

Traffic Engineering in GMPLS Networks

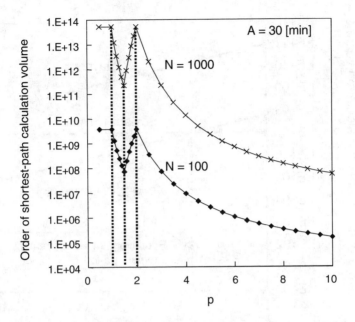

FIGURE 9.28 Relation between the number of levels and the order of the calculation volume.

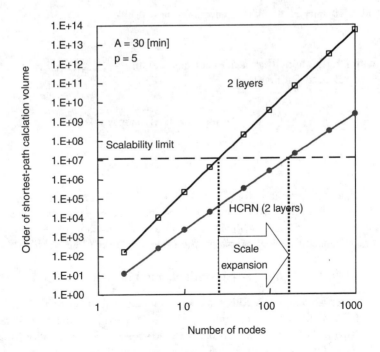

FIGURE 9.29 Network size scalability.

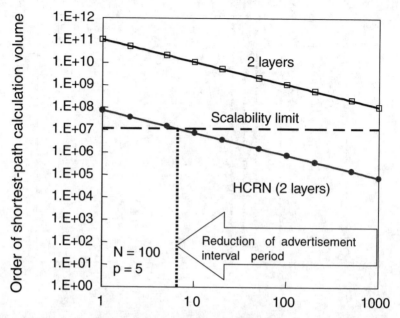

FIGURE 9.30 Interval of link-state advertisement scalability.

that with HCRN, a multilayer network has scalability if each FA-LSP is established or released within 10 min.

$$V_s = N^2 \log N (= \max\{L\} \log N) \qquad (9.2)$$

$$V_m = (1 + 30A^{-1}(N \cdot SC)^2)(N \cdot SC)^2 \log(N \cdot SC) \qquad (9.3)$$

The calculation volume for N layers ($n > 3$) is approximately $(n - 1) \times$ (the calculation volume for two layers) and is of the same order as that for a two-layer network. This is because the increase in calculation volume in the multilayer network is caused by the conversion of the topology in the lower network.

9.2.3.2 Effect of Multilayer Network Hierarchization

The CR internal-cost and MSPF schemes, as mentioned in Section 9.2.2.2 and Section 9.2.2.3, allow that the shortest path is calculated in an abstracted hierarchical network. There is a trade-off between the shortest-path calculation volume and the blocking probability for LSP establishment. When p is large and the network is deep, the CR size and the shortest-path calculation volume become small.

Traffic Engineering in GMPLS Networks

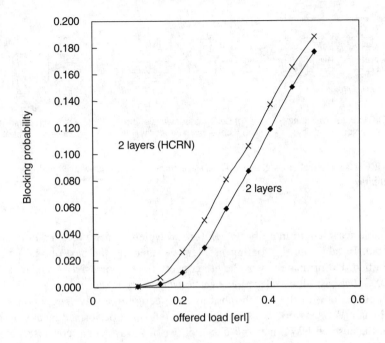

FIGURE 9.31

However, the link-state information that a node has also becomes small, as does the portion of the network that the node can recognize. In this case, many network resources may be used wastefully when an LSP is established. Figure 9.31 shows the blocking probabilities in a hierarchical network and a nonhierarchical network. The results show that the blocking probability for LSP establishment in the hierarchical multilayer network is only 10% higher than that in the nonhierarchical multilayer network.

9.3 WAVELENGTH-ROUTED NETWORKS

9.3.1 ROUTING AND WAVELENGTH ASSIGNMENT (RWA) PROBLEM

Wavelength-division multiplexing (WDM) optical networks are mainly categorized into three types in terms of wavelength-conversion capability. The first type covers wavelength-convertible (WC) networks. The second type covers nonwavelength-convertible (NWC) networks. The third covers limited-wavelength-convertible (LWC) networks.

In a WC network, any wavelength can be converted into any other wavelength at optical cross-connects (OXCs) on an optical path. When an optical path is set up, any wavelength can be used as long as there is at least one

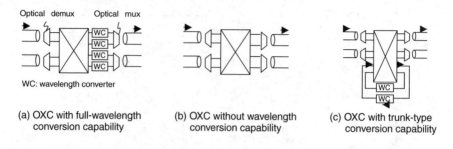

FIGURE 9.32 Optical cross-connect (OXC) structures with different wavelength-conversion capabilities.

available wavelength in a fiber. Therefore, wavelength resources are efficiently utilized. In addition, the routing problem of selecting the best optical path is simplified. Unfortunately, wavelength converters are expensive. An example of an OXC that has full wavelength-conversion capability is shown in Figure 9.32(a). Each output port of the switch fabric is equipped with a wavelength converter.

In an NWC network, wavelength conversion is not performed on any optical path. Because an NWC network does not use any wavelength converters at the OXCs, no wavelength converters are needed.* However, an NWC network utilizes wavelength resources less efficiently than a WC network. When an optical path is set up, the routing and wavelength assignment (RWA) problem has to be solved. In general, the blocking probability, the probability that an optical setup request will be rejected, is much higher in an NWC network than that in a WC network. Figure 9.32(b) shows a nonwavelength convertible OXC.

An LWC network is an intermediate solution between WC and NWC networks. An LWC network offers the advantage of WC networks (better wavelength resource utilization) as well as that of NWC networks (lower wavelength-conversion cost). Figure 9.32(c) shows an OXC with trunk-type wavelength converters. Such converters are used only when needed. Optical paths in an LWC network experience few or no conversions in wavelength.

Several studies have addressed the effectiveness of an LWC network with trunk-type wavelength converters [29, 30]. They showed that trunk-type wavelength conversion is sufficient for achieving wavelength utilization levels close to those of networks with full wavelength conversion [31]. However, these studies assumed that complete knowledge about the optical-link states in the network was held by a centralized-control operating system that collected the optical-link states of each OXC. This centralized approach is rather impractical when the network size is large.

Distributedly controlled traffic engineering (TE) is considered to be a promising way of realizing cost-effective and flexible optical WDM networks. In Generalized

---

* Although 3R functions (i.e., reshaping, retiming, and regeneration) are required in NWC, they are not considered here to simplify the discussion on the wavelength conversion cost.

Multiprotocol Label Switching (GMPLS) networks [15, 20], a source node finds an appropriate route and wavelength based on the optical link-state information collected with a routing protocol such as Open Shortest Path First (OSPF) [17], and sets up an optical path by using a signaling protocol such as RSVP-TE [22].

Two studies [25, 32] have introduced extensions of the RSVP-TE signaling protocol to better utilize network resources in NWC and LWC networks based on distributedly controlled traffic engineering. In the proposed extensions, after the route of an optical path is chosen by using the link-state information advertised by OSPF, the wavelength is adaptively set by using extended RSVP-TE. However, in these two studies [25, 32], performance evaluations were left as remaining issues. Although several papers [33, 34] addressed the performance of an NWC network with distributed control, no study has investigated the performance of an LWC network with trunk-type wavelength converters when distributed control is adopted.

Several questions are obvious. When distributedly controlled traffic engineering is adopted in an LWC network, how much is the blocking probability improved? How many wavelength-conversion resources are needed in an LWC network to emulate a WC network? How does the hop length of optical paths impact the blocking probability? This section provides answers to these questions.

This section describes a performance analysis of distributedly controlled dynamic wavelength-conversion (DDWC) optical networks that use the extended signaling protocol of RSVP-TE [25, 32] to better utilize the network resources of an LWC network [3]. A simple routing and wavelength assignment (RWA) policy is introduced. In the RWA policy, a source node first chooses an optical-path route by using aggregated optical link-state information and then sends an RSVP-TE Path message on the selected route; wavelength assignment is then performed. If wavelength conversion is needed at a transit node, the wavelength is adaptively converted by using trunk-type wavelength converters. The advertised information overhead imposed by OSPF and the processing capability needed to solve the complicated RWA problem are saved. It is shown via simulations that the performance of the DDWC network with our simple RWA policy emulates that of an equivalent WC network but with many fewer trunk-type wavelength converters.

9.3.2 Distributedly Controlled Dynamic Wavelength-Conversion (DDWC) Network

9.3.2.1 DDWC Network with Simple RWA Policy

Figure 9.33 shows a DDWC network. It consists of OXCs with trunk-type wavelength converters. Each node (OXC) advertises its own optical-link states to other nodes and collects optical-link states from other nodes by using the extended OSPF described in the literature [17, 32].

To set up an optical path, a source node has to determine which route and wavelength should be chosen. The problem is the so-called route and wavelength

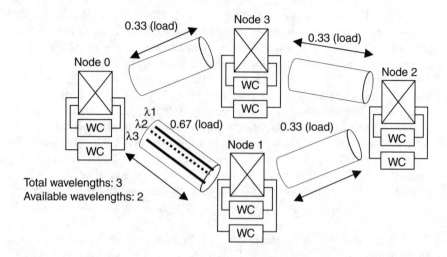

FIGURE 9.33 Distributedly controlled dynamic wavelength-conversion network.

assignment (RWA) problem [35]. If the availability of each wavelength is advertised to all links in the network, it is difficult to scale the network in terms of the number of nodes and the number of wavelengths because the amount of advertised information rapidly becomes excessive. In addition, even if a source node knows all link information, the processing power required by the source node to solve the complicated RWA problem can become excessive.

To reduce the amount of advertised information and the required processing power, a simple RWA policy is introduced. A source node first chooses an optical-path route by using aggregated optical link-state information, and then sends an extended RSVP-TE signaling Path message on the selected route to the destination node. The Path message collects available wavelengths and wavelength converters, if needed, on the route. Finally, the destination node determines which wavelengths should be adopted.

9.3.2.2 Optical Route Selection

Each optical route is selected based on a least-loaded shortest routing policy and the use of aggregated optical link-state information. The load l_i of fiber link i is defined as the ratio of the number of reserved wavelengths (NRW_i) in the fiber link i between neighboring nodes to the number of accommodated wavelengths (NW_i) in the fiber link i (i.e., $l_i = NRW_i/NW_i$).

In the least-loaded routing policy, the link load is taken as a link cost. Considering link cost, the shortest route is determined so that $\sum_{i \in route} l_i$ is minimized.

In OSPF extensions, wavelengths in a fiber link between neighboring nodes are bundled into a traffic-engineering (TE) link. The optical-link information in a TE link between two neighboring nodes is advertised to other nodes in the network.

Traffic Engineering in GMPLS Networks 257

Therefore, multiple ports of a node, each of which corresponds to a wavelength, can be combined into a TE link. In OSPF extensions, opaque LSA sub-TLVs include maximum reservable bandwidth and unreserved bandwidth types, which sub-TLV types are 7 and 8, respectively [26]. NW_i and NRW_i are expressed using maximum reservable bandwidth and unreserved bandwidth for a bundled TE link between neighboring nodes as described below.

$$\begin{aligned} NW_i &= \text{maximum reservable bandwidth for a bundled TE-link} \\ &= \text{sum of maximum reservable bandwidth of all component links} \end{aligned} \quad (9.4)$$

$$\begin{aligned} NRW_i &= NW_i - \text{unreservable bandwidth for a bundled TE-link} \\ &= NW_i - \text{sum of unreservable bandwidth of all component links} \end{aligned} \quad (9.5)$$

Note that, each component link corresponds to a different wavelength. The units of the maximum reservable bandwidth and unreservable bandwidth are defined as bytes per second by Kompella and Rekhter [17]. Because the values of NRW_i and NW_i are independent of bytes per second, modification of the units or a sub-TLV definition is needed. This modifications was proposed by Oki et al. [32].

9.3.2.3 Extended Signaling Protocol of RSVP-TE

9.3.2.3.1 AND Scheme

After an optical route is determined, the source node sends an extended RSVP-TE Path message over the selected route to the destination node. When the Path message attempts to set up an optical path at each source/transit node, it carries a set of available labels, i.e., labels that are unreserved on all TE links from the source node to the transit node. The label set is called the AND label set. The source node sends the AND label set, which is a set of the available labels of it own egress link to the next-hop node. The transit node creates a new AND set by using the transmitted AND set that was sent by the previous node and the available labels of the egress link. If there is no available label in the created AND set, the transit node must perform wavelength conversion. When the wavelength is converted at the transit node, the AND set is reset to be transmitted to the next-hop node on the selected route. If such conversion is not possible, the request for optical-path setup is rejected.

Figure 9.34 shows two examples of AND usage. In case 1, the available-label set in link A is $\{\lambda 1, \lambda 3, \lambda 4\}$. The available-label set in links A and B is $\{\lambda 3, \lambda 4\}$, as $\lambda 1$ is not available in link B. The available-label set that the Path message carries from the source node to the destination node is thus $\{\lambda 3\}$, as $\lambda 4$ in link C is not available. In this case, no wavelength conversion is required for optical-path setup. The destination node chooses $\lambda 3$ to set up the optical path and sends back the reservation message to the source node over the selected route.

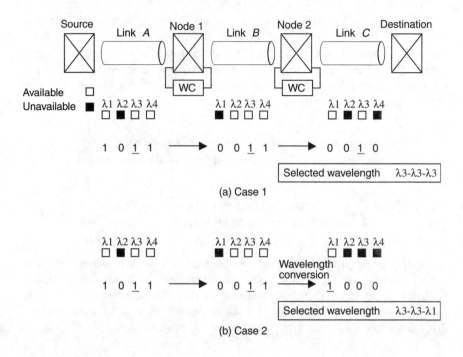

FIGURE 9.34 Examples of AND usage.

In the case of 2, the available-label set in links A and B is {λ3, λ4}. However, because only λ1 is available in link C, there is no available label in the AND label set that is created by node 2. Therefore, node 2 decides to convert either λ3 or λ4 to λ1 by using a trunk-type wavelength converter. Note that, according to the first-fit policy [35], λ3 is chosen.

9.3.2.3.2 ALL Scheme

In the AND scheme, wavelength conversion is performed at the transit node only when the label set is empty. On the other hand, in the ALL scheme, the destination node decides at which transit node the wavelength conversion is performed after considering the available labels and wavelength converters at each transit node. For this purpose, the Path message carries all information on availability of labels and wavelength converters on the selected route instead of the aggregated information, as used in the AND scheme.

When a Path message attempts to set up an optical path between each source/transit node pair, it carries a set of all available labels for all TE links on the optical-path route. The label set is called an ALL label set. A destination node receives the ALL label set and decides which labels should be used on the optical path. In the ALL scheme, the destination node has several options when deciding which node should perform wavelength conversion.

Traffic Engineering in GMPLS Networks

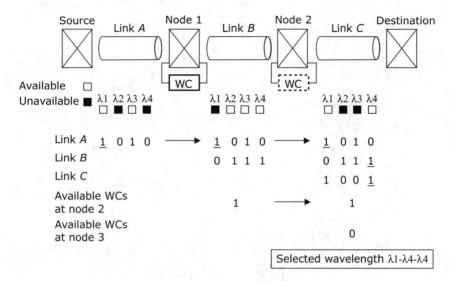

FIGURE 9.35 Example of ALL scheme.

Figure 9.35 shows an example of the ALL scheme in use. The Path message carries all information on the wavelengths available on the links and wavelength converters on the selected route. The example shows that a wavelength converter is available at node 1 but not at node 2. In this example, the AND scheme is not able to set up an optical path. This is because node 2 cannot perform wavelength conversion, although it is needed. The ALL scheme, on the other hand, can set up an optical path due to the available wavelength converter at node 1. Therefore, the ALL scheme offers improved blocking performance over the AND scheme at the cost of an increase in information transfer overhead and the cost of performing the sophisticated wavelength-selection algorithm at the destination node.

9.3.3 Performance of DDWC

The performance of the DDWC network was evaluated by using computer event-driven simulations. The impact on the blocking probability of the DDWC network is investigated using several important parameters such as the number of trunk-type converters at each node, x, the optical-path hop length, h, and the number of wavelengths in a fiber, w. For this purpose, the simple network model shown in Figure 9.36, which has 11 nodes ($N = 11$), was assumed. Bidirectional fibers are set between neighboring nodes. The routing and wavelength assignment policy described in Section 9.3.2 is employed. Although the AND and ALL schemes were introduced, we focus on the AND scheme in the evaluation to simplify the discussion of DDWC performance. This is because the performance

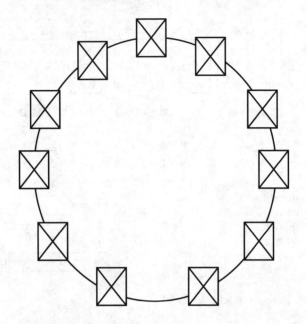

FIGURE 9.36 Simple network model with 11 nodes.

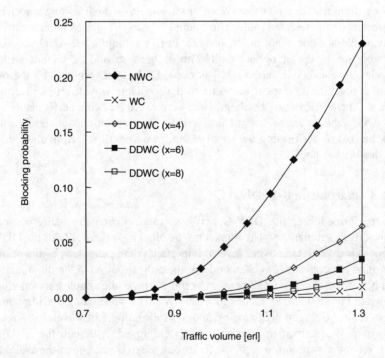

FIGURE 9.37 Blocking probability vs. traffic demand ($h = 5$, $w = 32$).

Traffic Engineering in GMPLS Networks

of the ALL scheme may be superior to that of the AND scheme, so the AND scheme provides conservative numerical results. The simulations assume that traffic demand between all source and destination nodes is the same. Requests for optical-path setup follow a Poisson distribution. The holding time of each source and destination node pair is considered to follow an exponential distribution.

Figure 9.37 shows that as the number of trunk-type wavelength converters, x, of the DDWC network increases, the blocking probability approaches that of the WC network. The number of wavelengths in a fiber $w = 32$ is set. The traffic demand (horizontal axis) is defined as that between each source and destination node pair. In Figure 9.37, the blocking probability of optical-path requests whose optical-path hop length h was 5 was focused. The results show that the DDWC network combined with our RWA policy dramatically reduces the wavelength-conversion cost compared with the WC network.

Figure 9.38 shows the blocking probability of optical-path requests with $h = 3$. The difference in the blocking probability between the NWC and WC networks is much smaller than that with $h = 5$. Figure 9.39 shows the blocking probability

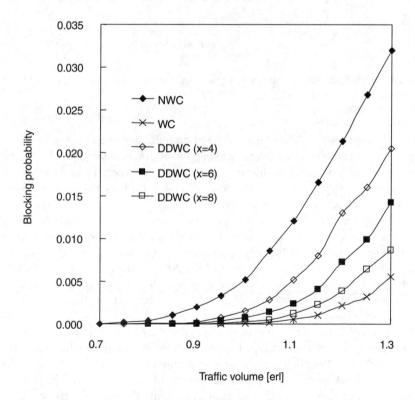

FIGURE 9.38 Blocking probability vs. traffic demand ($h = 3$, $w = 32$).

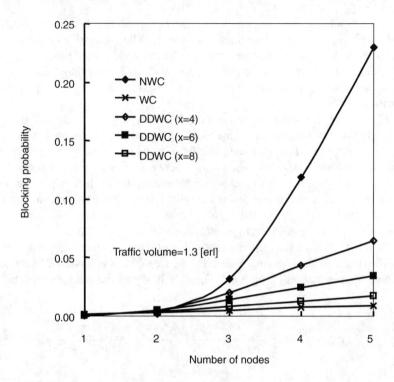

FIGURE 9.39 Blocking probability vs. optical-path hop length (traffic demand = 1.3 erl, $h = 5$, $w = 32$).

versus the optical-path hop length for the given traffic volume of 1.3 (erlang). As the number of hops increases, the blocking probabilities become larger. Therefore, the impact of x becomes significant. This means that when the cost-effective x to approximately emulate the WC network is designed, blocking probabilities with large enough h values should be considered.

As the number of trunk-type converters increases, the blocking probability decreases, as shown in Figure 9.40. The minimum x is defined as the point at which the difference in blocking probabilities between the DDWC and WC networks is less than 30%. This is called the 30%-emulation point e. The arrow in Figure 9.40 indicates the 30%-emulation point for the conditions considered. For example, when the traffic volume is 1.3 erl, where the blocking probability of the WC network is about 0.01, the 30%-emulation point is $e = 10$. This means that the DDWC network achieves $(pw - e)/pw \times 100 = 69\%$ cost reduction, in terms of wavelength converters, compared with the WC network while achieving 30% emulation. Note that p is the number of ingress fiber ports in the node. In the ring network model, p is set to 1, as shown in Figure 9.41. The WC network needs w wavelength converters for each fiber, while the DDWC network needs e trunk-type wavelength converters.

Traffic Engineering in GMPLS Networks

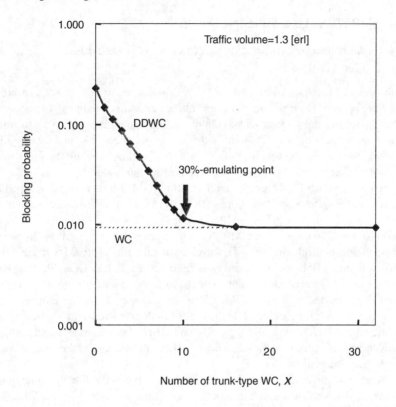

FIGURE 9.40 Blocking probability vs. number of trunk-type wavelength converters ($h = 5$, $w = 32$).

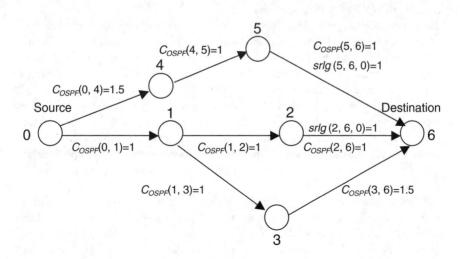

FIGURE 9.41 Network example. (Copyright 2002, IEEE.)

9.4 SURVIVABLE GMPLS NETWORKS

9.4.1 A Disjoint-Path-Selection Scheme with Shared-Risk Link Groups

With optical-fiber bandwidth and node capacity increasing explosively, a break in a fiber span or node failure can cause a huge amount of damage to customers [36]. Therefore, network providers should design survivable networks to minimize communication loss. Disjoint-path routing enhances the survivability of a network. Several disjoint paths, which are routed without sharing the same links or nodes, must be set between source and destination nodes.

Generalized Multiprotocol Label Switching (GMPLS) is being developed by the Internet Engineering Task Force (IETF) [15]. It is an extended version of Multiprotocol Label Switching (MPLS). While MPLS was originally developed to control the IP-packet layer, GMPLS controls several layers, such as IP-packet, time-division-multiplexing (TDM), wavelength, and optical-fiber layers, in a distributed manner. IP link-state-based routing protocols such as Open Shortest Path First (OSPF) [37] and Intermediate System to Intermediate System (IS-IS) [38] are being extended to support GMPLS by advertising TE (traffic engineering)-link states defined for each layer. In the OSPF/IS-IS extensions, information about the shared-risk link group (SRLG) is also advertised. For example, when multiple wavelengths are carried on the same fiber, SRLG information is taken into account for the disjoint-path routing.

Several algorithms have been presented to efficiently find the maximum number of disjoint paths between source and destination nodes [39–42]. Bhandari [43] described several algorithms to take account of the equivalence of SRLGs. However, no efficient algorithm has addressed generally finding the maximum number of disjoint paths taking into consideration the SRLG information advertised by the OSPF/IS-IS extensions.

A kth-shortest-path algorithm is widely used to find disjoint paths because of its simplicity [43]. First, the shortest-path algorithm provided by Dijkstra [28] is used to find the first path for a given network topology. Node disjoint paths are considered as disjoint paths here. Next, the links and nodes that are used by the first route are pruned from the given network topology. Note that pruning a node means that links that are connected to it are also pruned. Then, the second-shortest path is searched for in the pruned network topology. In the same way, the kth-shortest disjoint path is searched for. Although the kth-shortest-path algorithm does not find the *maximum* number of disjoint paths, Dunn et al. [40] showed that its results are very nearly equal to the maximum-flow solution.

Let us explain how the kth-shortest-path algorithm is performed for a network with SRLG in a conventional manner. The first path is found based on link costs advertised by OSPF, which is regarded as a link-state-based routing protocol here. Then, the kth-shortest-path algorithm prunes the links and nodes on the first route and also the links that belong to the same groups as the links and nodes on the first route. To find more disjoint paths, the same procedure

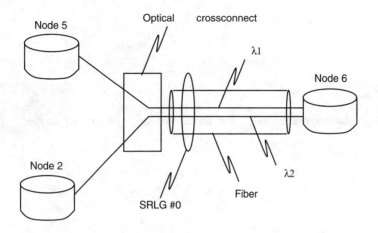

FIGURE 9.42 SRLG example. (Copyright 2002, IEEE.)

is followed. However, when a link that shares many risks with other links is used as a disjoint route, a problem occurs in that a sufficient number of disjoint paths cannot be obtained.

Figure 9.41 shows a network example. Consider the solution of disjoint paths between source node 0 and destination node 6. Link cost $C_{OSPF}(i,j)$ for link $L(i,j)$* is advertised by OSPF. $SRLG(i,j,g)$ indicates the SRLG information, where $SRLG(i,j,g) = 1$ or 0 means that $L(i,j)$ does or does not belong to SRLG g, respectively. $L(5,6)$ and $L(2,6)$ belong to SRLG 0. Figure 9.42 shows an SRLG example for $L(5,6)$ and $L(2,6)$. Both links use the same fiber by using wavelengths $\lambda 1$ and $\lambda 2$, respectively, through an optical cross-connect (OCX). When the kth-shortest algorithm is used by taking C_{OSPF} as the link cost, route 0-1-2-6 is selected as the shortest path. To find the second disjoint path, links and nodes on the first selected path, which are $L(0,1)$, node 1, $L(1,2)$, node 2, and $L(2,6)$, are pruned. In addition, because $L(5,6)$ belongs to SRLG 0, to which $L(2,6)$ belongs, $L(5,6)$ is also pruned. As a result, we cannot find more than one disjoint path between nodes 0 and 6, as shown in Figure 9.43.

Although $L(5,6)$ and $L(2,6)$ belong to only SRLG 0 in Figure 9.42, a link may generally belong to more than one SRLG. For example, a link could belong to two SRLGs when one SRLG is a fiber and the other is a conduit.

This section describes a disjoint-path-selection scheme for GMPLS networks with SRLG constraints [4, 5]. It is called the weighted-SRLG (WSRLG) scheme. It treats the number of all SRLG members related to a link as part of the link cost when the kth-shortest-path algorithm is performed. In WSRLG, a link that has many SRLG members is rarely selected as the shortest path. Simulation results

---

* In GMPLS, link $L(i,j)$ is referred to as a "TE link." For simplicity, we call it a "link" here.

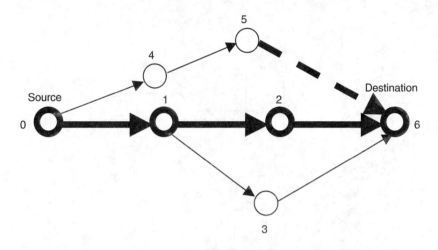

FIGURE 9.43 Example of conventional disjoint-path-selection algorithm.

show that WSRLG finds more disjoint paths than the conventional kth-shortest-path algorithm.

9.4.2 WEIGHTED-SRLG PATH SELECTION ALGORITHM

The terminology used in this section is as follows:

$C_{\text{OSPF}}(i,j)$ = Link cost for $L(i,j)$ that is advertised by OSPF
$\text{SRLG}(i,j,g)$ = SRLG information; $\text{SRLG}(i,j,g) = 1$ means that $L(i,j)$ belongs to SRLG g, and $\text{SRLG}(i,j,g) = 0$ means that it does not
$C_{\text{comp}}(i,j)$ = Link cost for $L(i,j)$ that is used to find kth-shortest paths
α Weight factor for SRLG
$D_{\text{req}}(s,d)$ = Required number of disjoint paths between nodes s and d
$C_{\text{path}}(s,d)$ = Sum of costs for all disjoint paths between nodes s and d
$K(s,d)$ = Number of obtained disjoint paths between nodes s and d
G = Number of SRLGs in a network
$M(g)$ = Number of members of SRLG g; $M(g) = \sum_i \sum_j \text{SRLG}(i,j,g)$

$C_{\text{comp}}(i,j)$ is defined by

$$C_{\text{comp}(i,j)} = \frac{1-\alpha}{C_{\text{OSPF}}^{\max}} C_{\text{OSPF}(i,j)}$$

$$= \frac{\alpha}{\text{SRLG}^{\max}} \max\{\text{SRLG}(i,j),1\}$$

(9.6)

where

$$\text{SRLG}(i,j) = \sum_{g}^{G} M(g) sr\log(i,j,g) \qquad (9.7)$$

$$C_{\text{OSPF}}^{\max} = \max_{i,j} C_{\text{OSPF}(i,j)} \qquad (9.8)$$

and

$$\text{SRLG}^{\max} = \max\{\max_{i,j} \text{SRLG}(i,j), 1\}, \qquad (9.9)$$

When $\text{SRLG}(i,j) = 0$, $\max\{\text{SRLG}(i,j), 1\}$ in the second term of Equation 9.6 is set to 1 so that the hop count for the link can be considered. Therefore, the value of the second term of Equation 9.6 is affected by the hop count even when $L(i,j)$ does not belong to any SRLG group, or $\text{SRLG}(i,j) = 0$. When SRLG g has no members, $M(g) = 0$. Because SRLG g has some members that share the same risk, $M(g) \geq 2$ is meaningful. Therefore, $M(g) = 1$ is not considered. $\text{SRLG}^{\max}$ determines the sensitivity of α. The smaller the value of $\text{SRLG}^{\max}$ is, the more sensitive to α is the value of $C_{\text{comp}}(i,j)$.

$C_{\text{path}}(s,d)$ is expressed by

$$C_{\text{path}(s,d)} = \sum_{k=1}^{k(s,d)} \sum_{(i,j) \in \text{path}_k} C_{\text{OSPF}(i,j)} \qquad (9.10)$$

where path k is the kth-shortest path found by the kth-shortest-path algorithm.

In WSRLG, α is set to an appropriate value by using a modified version of the well-known binary search method [44] so that $C_{\text{path}}(s,d)$ can be reduced as much as possible under the condition that $K(s,d)$ is equal to or larger than $D_{\text{req}}(s,d)$. Next, we explain why the conventional binary search method was modified. In our estimate, $K(s,d)$ and $C_{\text{path}}(s,d)$ mostly increase with α. However, this estimate is not always true. $K(s,d)$ and $C_{\text{path}}(s,d)$ do not always increase monotonically with α. Therefore, the conventional binary search method may miss an appropriate α that satisfies the required number of disjoint paths. As a result, the α that is finally obtained by the conventional binary search method does not satisfy the required number of disjoint paths. To avoid this problem, the modified method searches for α while remembering the most appropriate candidate in the regular binary search process.

The initial values are set as $\alpha_{\min} = 0$, $\alpha_{\max} = 1.0$, $K_{\text{temp}} = \infty$, and $C_{\text{temp}} = \infty$. Here, ε is used as a parameter to judge whether the modified binary search method converges. It should be set considering the value of $\text{SRLG}^{\max}$, which determines the sensitivity of α.

WSRLG is described as:

Step 1: $\alpha = \dfrac{\alpha_{min} + \alpha_{max}}{2}$.

Step 2: $K(s,d)$ is calculated by the following kth-shortest-path algorithm by taking SRLG into account. In this process, $C_{path}(s,d)$ is also obtained.

Step 3: If $K(s,d) \geq D_{req}(s,d)$, then $\alpha_{max} = \alpha$ is set. Otherwise, $\alpha_{min} = \alpha$ is set.

Step 4: If $K(s,d) \geq D_{req}(s,d)$ and $C_{path}(s,d) < C_{temp}$, then $\alpha_{temp} = \alpha$, $K_{temp} = K(s,d)$, and $C_{temp} = C_{path}(s,d)$ are set.

Step 5: If $\alpha_{max} - \alpha_{min} > \varepsilon$, go to Step 1. Otherwise, go to Step 6.

Step 6: If $K_{temp} \geq D_{req}(s,d)$, then $\alpha = \alpha_{temp}$. A set of disjoint paths obtained with this α is considered to be a solution. Otherwise, no set of disjoint paths that satisfies the required conditions is found.

The kth-shortest-path algorithm with SRLG is described below. First, $k = 1$ is set as the initial value.

Step 1: The kth-shortest path between source and destination nodes is searched for based on link cost $C_{comp}(i,j)$. If the path is found, go to Step 2. Otherwise, $K(s,d) = k$ is set, and the kth-shortest algorithm is ended.

Step 2: Prune links $L(i,j)$ and nodes that are on the kth-shortest path. For all gs, if $SRLG(i,j,g) = 1$, all links $L(i',j') = 1$, where $SRLG(i',j',g) = 1$, are also pruned.

Step 3: $k = k + 1$ is set and go to Step 1.

Consider WSRLG applied to the network model in Figure 9.41. Here, α is set to 1.0, as a special case. Figure 9.44 shows that WSRLG finds two disjoint paths between nodes 0 and 6, while the conventional k-shortest scheme finds only one path. $C_{comp}(2,6)$ and $C_{comp}(5,6)$ are twice as large as other $C_{comp}(i,j)$. Route 0-1-3-6 is selected as the shortest path. Then, route 0-4-5-6 is found as the second-shortest path. Thus, WSRLG is able to find more disjoint paths.

9.4.3 Performance Evaluation

To evaluate the performance of WSRLG, we use network topologies generated in a random manner under the condition that average node degree D is satisfied for a given number of nodes N and at least one path exists between every source–destination node pair. D is the average number of other nodes to which individual nodes are connected by links. D is defined as

$$D = \dfrac{\sum_{i}^{N}\sum_{j}^{N} a(i,j)}{N} \qquad (9.11)$$

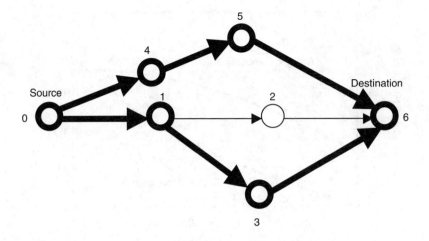

FIGURE 9.44 Example of WSRLG path-selection algorithm.

where $a(i,j)$ is element (i,j) of the network adjacency matrix:

$$a(i,j) = \begin{cases} 0 & \text{when there is no link from node } i \text{ to node } j \\ 1 & \text{when there is a link from node } i \text{ to node } j \end{cases} \quad (9.12)$$

Other evaluation conditions are assumed to be as follows. $M(g)$ is set to m, which is independent of g to simplify the discussion. SRLG(i,j,g) is set randomly under the condition that $\sum_i \sum_j \text{SRLG}(i,j,g) = m$ is set randomly between 0 and 1, unless otherwise stated. The number of sample network topologies for each evaluation is more than 100.

In Section 9.4.3.1, α is manually set to show its dependence without considering link-capacity constraints. In Section 9.4.3.2, the effects of adaptive selection by the modified binary search method are shown without considering link-capacity constraints. In Section 9.4.3.3, link-capacity constraints are considered.

9.4.3.1 Fixed α

The value for α is manually set to show its dependence. Figure 9.45 shows the average number of disjoint paths between source and destination nodes and the normalized average path cost of $C_{\text{path}}(i,j)$, where the path cost for $\alpha = 0$ is assumed to be 1.0. As α increases, the average number of disjoint paths becomes large. When $\alpha = 1$, the average number of disjoint paths is 19% more than for $\alpha = 0$. Note that $\alpha = 0$ means that the conventional disjoint-path-selection algorithm is used. $C_{\text{path}}(i,j)$ also increases with α. When α is large, the second term in Equation 9.6 is taken into account more than the first term in finding the shortest paths. By using

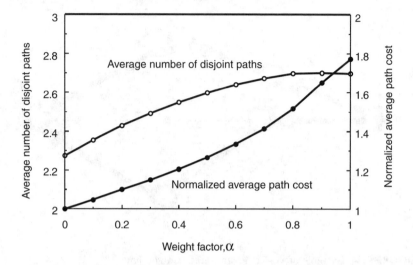

FIGURE 9.45 Number of disjoint paths and path cost $C_{path}(s,d)$ vs. α ($N = 20$, $D = 6$, $G = 16$, $m = 14$). (Copyright 2002, IEEE.)

WSRLG, α is automatically determined so that the required number of disjoint paths can be found while suppressing the increase in path cost.

Figure 9.46 shows how many source–destination node pairs have more disjoint paths for $\alpha > 0$ than for $\alpha = 0$, divided by the total number of source–destination node pairs, expressed as a ratio. WSRLG with $\alpha = 1$ dramatically increases the ratios to 41%.

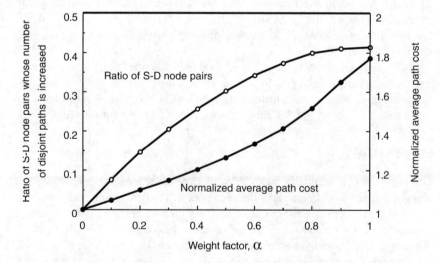

FIGURE 9.46 Ratio of SD node pairs whose number of disjoint paths is increased and path cost $C_{path}(s,d)$ vs. α ($N = 20$, $D = 6$, $G = 16$, $m = 14$). (Copyright 2002, IEEE.)

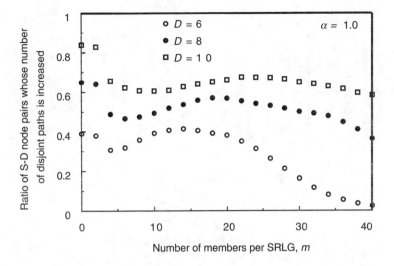

FIGURE 9.47 Ratio of SD node pairs whose number of disjoint paths is increased vs. number of members per SRLG ($N = 20$, $G = 16$).

Figure 9.47 shows the impact of the number of SRLG members m ($= M(g)$) on this ratio. When m is around 0, the ratio is high. As m increases, the ratio at first decreases, then increases, and finally decreases. The turning point depends on D. In Figure 9.47, we observe under which topology ó i.e., under which conditions of D and m ó WSRLG is the most effective compared with the conventional scheme.

Note that the highest ratios are obtained at around $m = 0$. Because the SRLG factor is not considered with $m = 0$, the effect is not directly due to the WSRLG algorithm. In this case, the number of hops, which is implied in Equation 9.6, is taken into account. To make the effectiveness of WSRLG even clearer, we consider the condition of $C_{OSPF}(i,j) =$ constant instead of variable values. Figure 9.47 shows the ratios of source–destination node pairs whose number of disjoint paths is increased under $C_{OSPF}(i,j) =$ constant. When $m = 0$, there is no gain by WSRLG, because the conventional scheme also finds a disjoint path to minimize the number of hops by the kth-shortest-path algorithm. As m increases, maximal points, which depend on D, appear. These maximal points correspond to those in Figure 9.47.

Figure 9.47 and Figure 9.48 also show the dependency of D. When N is constant, WSRLG provides better performance with higher D. In other words, WSRLG solves the conventional problem even more effectively in dense networks. This is because, once a link bottleneck is solved by WSRLG, more alternative routes are available in dense networks than in the conventional scheme.

Different Ds are considered with a fixed N in Figure 9.47 and Figure 9.48 to reveal the dependency of D, and different Ns with a fixed D are considered in Figure 9.49 and Figure 9.50 to observe the dependency of N. In these latter

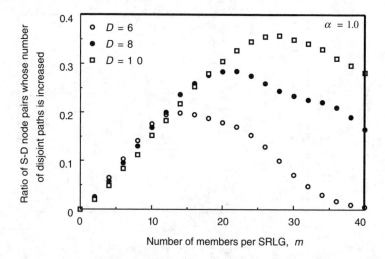

FIGURE 9.48 Ratio of SD node pairs whose number of disjoint paths is increased vs. number of members per SRLG ($N = 20$, $G = 16$, $C_{OSPF}(i,j)$ = constant).

figures, the WSRLG performance tendency is similar to that observed in Figure 9.47 and Figure 9.48, even when N is changed. WSRLG provides better performance with larger N with constant D. When D is given, larger N gives more alternate available routes. This results in better performance by WSRLG, the same behavior as observed in Figure 9.47 and Figure 9.48.

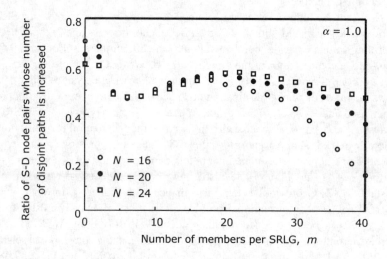

FIGURE 9.49 Ratio of SD node pairs whose number of disjoint paths is increased vs. number of members per SRLG ($D = 8$, $G = 16$).

Traffic Engineering in GMPLS Networks

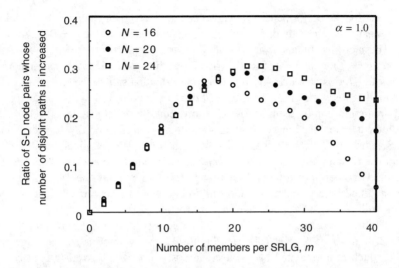

FIGURE 9.50 Ratio of SD node pairs whose number of disjoint paths is increased vs. number of members per SRLG ($D = 8$, $G = 16$, $C_{OSPF}(i,j)$ = constant).

9.4.3.2 Adaptive α

Figure 9.51 shows how many source–destination node pairs are satisfied with $K(s,d) \geq D_{req}(s,d)$, divided by the total number of source–destination node pairs, expressed as a ratio. $K(s,d)$ is set to 2 for all source–destination node pairs. Three values of

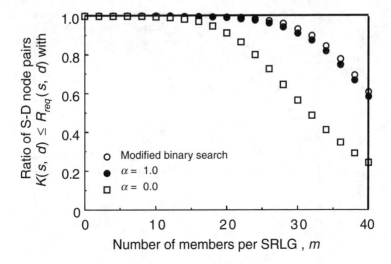

FIGURE 9.51 Ratio of SD node pairs with [trial mode] vs. number of members per SRLG ($N = 20$, $D = 8$, $D_{req}(s,d) = 2$, $G = 16$).

α are set for the performance comparison. One is adaptively selected by the modified binary search method. The other two α values are fixed at $\alpha = 0$ and $\alpha = 1$.

The modified binary search method clearly outperforms $\alpha = 0$. Even when the number of members per SRLG m increases, the modified binary search method keeps high ratios of source–destination node pairs that satisfy the conditions. This means that network providers can further relax topology designs to satisfy the required condition by using WSRLG compared with the conventional scheme. In addition, the modified binary search method slightly outperforms $\alpha = 1$. As explained in Section 9.4.2, $K(s,d)$ does not always increase monotonically with α, so adaptive α with the modified binary search method sometimes satisfies the required number of disjoint paths, although $\alpha = 1$ sometimes does not. As a result, the modified binary search method can find a few more source–destination node pairs with $K(s,d) \geq D_{req}(s,d)$ than the fixed value of $\alpha = 1$.

The modified binary search method reduces path cost $C_{path}(s,d)$ and finds a greater number of disjoint-path sets. Figure 9.52 shows normalized average path cost, which is normalized by the average cost with $\alpha = 0$. Average path cost is obtained as an average value of path costs among the source–destination pairs that satisfy the required conditions. The modified binary search method suppresses the path cost more than the fixed value of $\alpha = 1$ does, because it keeps searching for a path whose cost is as low as possible under the required conditions.

9.4.3.3 Link-Capacity Constraints

Section 9.4.3.1 and Section 9.4.3.2 showed that WSRLG enables network providers greater freedom to relax topology designs to satisfy the required conditions

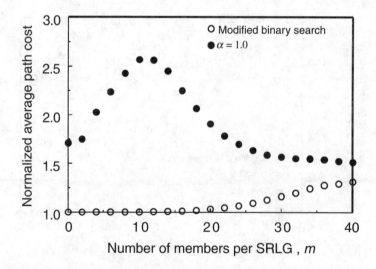

FIGURE 9.52 Normalized path cost vs. number of members per SRLG ($N = 20$, $D = 8$, $D_{req}(s,d) = 2$, $G = 16$).

Traffic Engineering in GMPLS Networks

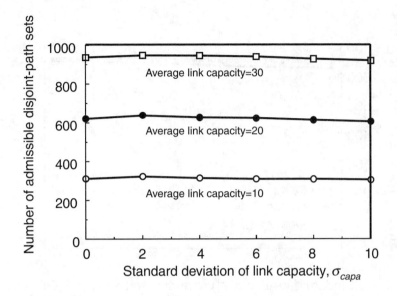

FIGURE 9.53 Dependence of deviation of link-capacity deviation ($N = 20$, $D = 8$, $D_{req}(s,d) = 2$, $G = 16$, $m = 20$).

compared with the conventional scheme. This is effective whether or not link-capacity constraints are considered.

Figure 9.53 shows the dependence of link capacity. The capacity of each link is given a value that follows a normal distribution specified by the average value and standard deviation, σ_{capa}. The average value of the link capacity is set to 10, 20, and 30. Link capacity is normalized by the path bandwidth requested by each source node. If a value given by a normal distribution is less than 0, it is set to 0. All the path bandwidths are assumed to be the same. The modified binary search method is adopted. $D_{req}(s,d)$ is set to 2 for all source–destination node pairs. To evaluate the number of admissible disjoint-path sets, each source–destination node pair that is randomly selected continues to assign its disjoint-path set until the number of admissible disjoint-path sets does not increase any more due to the link-capacity constraints. Figure 9.53 demonstrates that the number of admissible disjoint-path sets is not much affected by σ_{capa} at various average link-capacity values in our tested models. WSRLG is effective with various link-capacity constraints.

Figure 9.54 shows the dependency of the number of members per SRLG, $M(g)$, which is given a value that follows a normal distribution specified by the average value and standard deviation, σ_{mem}. The average value of $M(g)$ is set to 10, 15, and 20. If a value given by a normal distribution is less than 2, it is set to 0. Note that we do not consider $M(g) = 1$, as described in Section 9.4.2. Otherwise, we use a floor value of $M(g)$. Figure 9.54 demonstrates that the number of admissible disjoint-path sets is not much affected by σ_{mem} at various average $M(g)$ in our tested models, either. WSRLG is also effective with link-capacity constraints when $M(g)$ is varied.

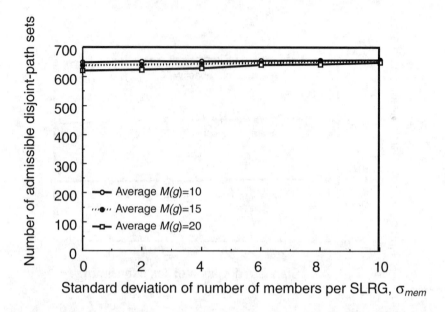

FIGURE 9.54 Dependence of deviation of number of SRLG members ($N = 20$, $D = 8$, $D_{req}(s,d) = 2$, $G = 16$).

9.5 SCALABLE SHARED-RISK-GROUP MANAGEMENT

Wavelength-routed networks have attracted much attention as a means of meeting the insatiable growth of the Internet's traffic demand [18, 25, 45]. In the wavelength-routed networks, traffic is carried via a wavelength path, which is called a label-switched path (LSP) in the IETF GMPLS terminology [15, 16]. Availability is a big concern because a single-wavelength LSP carries a huge amount of data (at 10 Gbps or more), and a lot of information is lost if the wavelength LSP fails. Quick LSP backup methods need to be developed for high availability [46–48]. LSP backup methods are classified with respect to backup-path calculation and resource reservation epochs, as shown in Table 9.1.

**TABLE 9.1
Classification of LSP Backup Methods**

	Backup Route Calculation Epoch	Backup Resource Reservation Epoch
Protection	Precalculated	Prereserved
Restoration		
Reactive	After failure	Not reserved
Preplanned	Precalculated	Not reserved

The protection method sets up working and backup LSPs in advance. The same data is copied into both working and backup LSPs at the sending node, and the data from the working LSP is selected by the receiving node. The LSP is switched from the working to the backup immediately after a failure is detected. Even though this method quickly recovers from failure, it requires twice the resources because the wavelength resources for both working and backup LSPs are occupied in parallel.

The restoration method sets up only the working LSP, allowing the backup LSP to be set up only after the working LSP fails. This method does not require twice the resources. However, the recovery time is slower than that of the protection method because it takes time to invoke the signaling/routing procedure to set up the backup LSP after the failure. The restoration methods are further classified into two subclasses: the reactive and the preplanned methods. The reactive method calculates the route for the backup LSP after a failure and therefore takes a longer recovery time than that of the preplanned method. The preplanned method calculates the route for the backup LSP in advance and therefore takes less recovery time. In this section, the preplanned restoration method is addressed because it is quick and consumes fewer wavelength resources.

Another viewpoint of classification of LSP-backup methods is the protected span of the LSP: link-backup and path-backup methods. The link-backup method prepares backup LSPs for individual links of the LSP. The LSP is repaired locally after a failure. This method needs to calculate multiple routes for the backup LSPs. The path-backup method prepares the backup LSP for the whole portion from the ingress to the egress nodes of the working LSP. This method calculates a single route for the backup LSP. In this section, the path-backup is considered because it does not have to calculate multiple routes for the working LSP.

Five goals in designing the restoration method were raised. The goal of the proposed method includes (1) 100% recovery performance, (2) efficient bandwidth utilization, (3) scalability, (4) less computation complexity, and (5) autonomous control [6]. To this end, a backup-SRG concept to guarantee 100% recovery performance is introduced that allows backup wavelength resources to be shared. A routing method in favor of working LSP is developed to use wavelength resources efficiently. Hierarchical SRG assignment is developed to reduce the number of backup SRGs to be used in admission control at each node.

9.5.1 SRG Concept

The route of the backup LSP and that of the working LSP should be disjoint to avoid a service outage when a single failure event occurs. For example, if working and backup LSPs are assigned to different wavelengths in the same fiber, both working and backup LSPs would simultaneously fail if the fiber were cut. Thus the backup LSP should be assigned to a wavelength in a different fiber that cannot fail when the fiber of the working LSP fails. Note, however, that the working and backup LSPs could simultaneously fail, even if they are accommodated in different fibers, if they are assigned to the same conduit.

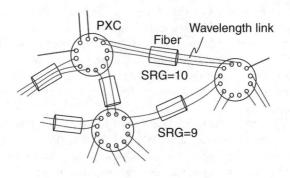

(a) Node is duplex

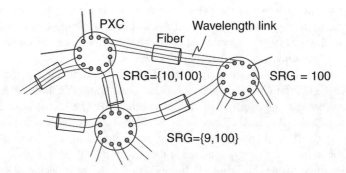

(b) Node is not duplex (node protection is required)

FIGURE 9.55 SRG.

The concept of a shared-risk link group (SRLG) attached to the link facilitates implementation of "disjointedness" in routing (Figure 9.55(a)) [49–51]. SRLG identifies a failure cause (e.g., fiber cut, laser down, etc.). For instance, an SRLG can be assigned to a single fiber, indicating that all wavelength paths routed over that fiber will be affected if the fiber is cut. An SRLG can also be assigned to a conduit in which multiple fibers are bundled. Multiple SRLGs (for conduit and fiber) can be attached to a wavelength link. An SRLG associated with a wavelength link is a set of SRLG tags, the failure of which causes failure of the link. Links with the same SRLG can fail simultaneously if the SRLG event occurs. An SRLG can also be attached to an LSP as a set of SRLGs of links that constitute the LSP. The SRLGs attached to the link (or LSP) are advertised by a routing protocol so that all nodes share the SRLGs attached to all links [23, 17]. Each node calculates the route disjoint with the working LSP for the backup LSP.

The SRLG concept was developed in the course of designing a link-failure recovery mechanism [49–51]. In designing a node-failure recovery mechanism,

Traffic Engineering in GMPLS Networks

the SRLG concept was naturally applied to node failure. The SRG (shared-risk group) concept is used to address the cause of both link failure and node failure. If a node is a duplex system, an SRG cannot be assigned to the node if a single failure event is considered. If a node is not a duplex system, we need to protect node failure by assigning SRG to the node. In Figure 9.55, we compare two cases: (a) duplex node and (b) nonduplex node. If the node is a duplex system, SRG is not assigned to the node, and SRGs assigned to the links are not affected. On the other hand, if the node is a nonduplex system, SRG (= 100) is assigned to the node, and this value is also assigned to the links originating from the node. By incorporating the SRG assigned to the node into the SRG assigned to the links originating from the node, the constraint-based routing first (CSPF) calculation (e.g., Dijkstra algorithm over the pruned link state) is simplified.

By using the CSPF algorithm combined with the SRG attribute attached to the link, a route for the backup LSP can be calculated such that it is disjoint with the route for the working LSP. Thereby, the backup LSP cannot fail when any links in the working LSP fail. However, there may be competition for the wavelength resource from different backup LSPs unless the wavelength resource is reserved in advance. If conflict occurs, one of conflicting backup LSPs is saved, and the others are lost. Such loss can be avoided by anticipating potential interference between all backup LSPs that would be activated in response to the same failure. Therefore we need to take into account how different backup LSPs share the wavelength resources to achieve 100% recovery performance.

9.5.2 SRG-Constraint-Based Routing (SCBR)

There are two conditions for achieving 100% recovery performance under the assumption of a single failure-event occurrence. The first one is for SRG-disjointedness between working and backup LSPs. If we set up working and backup LSPs, they should be SRG-disjoint. SRG-disjoint means that there are no common elements between SRGs for working and backup LSPs. Unless they are SRG-disjoint, a single failure event can cause outage of both working and backup LSPs. The second condition is for SRG disjointedness between different working LSPs. If two working LSPs are SRG-disjoint, their backup LSPs can share the same network resource while maintaining 100% recovery performance under the assumption of a single failure-event occurrence (Figure 9.56). If working LSPs associated with any pair of backup LSPs sharing the same SRG maintain the second condition, 100% recovery performance is achieved.

It is assumed that the working LSP is established to meet 100% recovery performance if and only if both working and backup LSPs are feasible. Even if the route for the backup LSP is not found in the SRG-constraint-based routing calculation, we can set up the backup LSP to accommodate additional working and backup LSPs to achieve 100% recovery performance. However, there is a trade-off between throughput gain and outage risk. The risk of an LSP outage can be estimated by multiplying the probability of failure of links shared by both working and backup LSPs [52].

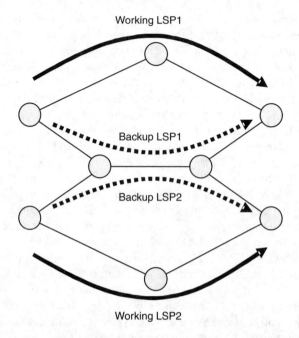

FIGURE 9.56 Backup LSPs can share the same resource if working LSPs are SRG-disjoint.

9.5.2.1 Admission Control at Link Using Backup-SRG Concept

We introduce a concept of "backup SRG" to identify which SRGs are backed up by a wavelength link (or LSP). It indicates a set of SRGs that the link (or LSP) protects. In Figure 9.57, the working LSP traverses links 1 and 2, whose respective SRGs are a and b. It follows that the SRG for the working LSP is a, b. The backup LSP is routed along with the SRG-disjoint route of the working path. It traverses links 3, 4, and 5, with the result that backup SRG for those links is set to a, b. The route for the backup LSP is selected to avoid resource contention between other backup LSPs. The links whose backup SRG coincide with that of the backup LSP are excluded before we calculate the route for the backup LSP. The route, which is both "SRG-disjoint" with the working LSP and "backup-SRG-disjoint" with other backup LSPs, is selected for the backup LSP. In this way, the backup LSP safely shares the backup resource (wavelength link) with other backup LSPs, thus excluding any risks of resource contention unless multiple link failures occur simultaneously. (Note that multiple link failures simultaneously occur at extremely low probability if each link failure occurs at very low probability.)

Each fiber link keeps track of wavelength links usage for working and backup LSPs. Wavelength links usage for a backup LSP is broken down into SRGs in the backup SRG. The number of backup LSPs is kept in SRG piecewise. The backup LSP carried over the link accounts for the number of used wavelength links associated

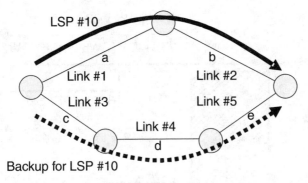

FIGURE 9.57 Backup-SRLG concept.

with individual SRGs included in its backup SRG. To cover failure of any SRG, the wavelength links usage is administered so that the sum of the maximum wavelength links for SRG and the wavelength links for the working LSPs does not exceed the number of wavelength links in the fiber links (Figure 9.58).

The number of available wavelength links for the working LSP is obtained by subtracting the number of wavelength links for the existing working LSPs from the number of total wavelength links in the fiber. Wavelength links allocated for backup LSPs can be potentially preempted by a new working LSP. When a new working LSP request arrives, it is accepted if the link has available resource for the new working LSP. Some of the existing backup LSP could be bumped out if necessary.

The number of available wavelength links for the backup LSP is calculated in an SRG-piecewise manner. The number of available wavelength links for an individual SRG in a backup SRG is obtained by subtracting the number of wavelength links for the existing working LSPs and the existing backup LSPs that have the SRG in their backup SRGs from the number of total wavelength links in the fiber. When the new backup LSP request arrives, it is accepted if the link has available resources for all SRGs included in the backup SRG. This bandwidth-management concept can be described with the backup-capacity model developed by Datta et al. [53].

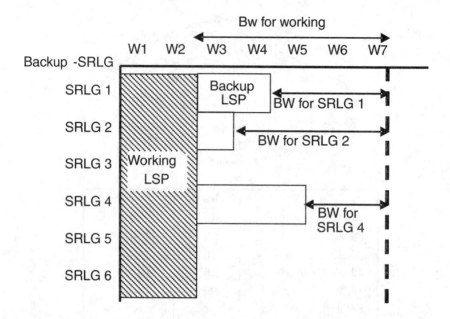

FIGURE 9.58 Resource management at link for working and backup LSPs with SRG breakdown.

9.5.2.2 SCBR

9.5.2.2.1 Link Protection Case

We use the constraint-based shortest-path-first (CSPF) routing for both working and backup LSPs. The working LSP has a higher priority over the backup one because the working LSP occupies wavelength resources while the backup LSP shares wavelength resources with other backup LSPs.

First, the route for the working LSP is calculated using Dijkstra's algorithm on the link-state database after pruning the links whose residual bandwidth is less than requested. Then the SRG is obtained for the working LSP by summing the SRGs of links on the route. Second, the route for the backup LSP is calculated using Dijkstra's algorithm on the link-state database after pruning the links whose residual bandwidth for the backup SRG is less than requested to back up the SRG of the working LSP as well as the links including SRG assigned to the working LSP. To guarantee 100% recovery performance, the LSP is admitted only if routes for both working and backup LSPs are found.

The backup LSP could be bumped out to leave enough wavelength resource to admit a new working LSP. In this case, the bumped backup LSP is rerouted over the second-constraint shortest path. The route for the bumped backup LSP is recalculated. Again, the working LSP has a higher priority over the backup LSP that protects other working LSPs because the working LSP occupies the wavelength resources while the backup LSP shares the wavelength resources

Traffic Engineering in GMPLS Networks

with other backup LSPs. The new working LSP is admitted only if the route for the bumped backup LSP is found to maintain 100% recovery performance of existing LSPs.

9.5.2.2.2 Node Protection Case
If an SRG is assigned to a node, it is assigned to the links originating from the node to simplify calculation of an SRG-constraint-based routing algorithm. We can calculate the SRG-constraint-based route using Dijkstra's algorithm on the link-state database after pruning the links that do not satisfy the SRG constraint, except for the ones assigned to the source and the destination nodes.

9.5.2.2.3 Hierarchical SRG Assignment
As described previously, each link maintains residual bandwidth for each backup SRG. As the number of SRGs increases, the storage to reserve residual bandwidth grows. To maintain scalability, we introduce a hierarchy of SRG. We define an area concept such that SRG has only to be unique within an area. The SRGs assigned to all links and nodes within the area are obscured outside the area. Only the SRG assigned to the area is used to calculate the route's portion outside the area (Figure 9.59). Note that the link state is visible but that the SRG of the link and node is not visible outside the area to which they belong. In Figure 9.59, the NSF network model, which consists of 14 nodes and 25 links, is used as an example. In this model 39 (= 14 + 25) SRG tags are required in the original setting. We divide the network into four areas. Area 1 consists of three nodes (nodes 1, 2, and 3); area 2 consists of three nodes (nodes 4, 5, and 7); area 3 consists of three nodes (nodes 6, 10, and 14); and area 4 consists of five nodes (nodes 8, 9, 11, 12, and 13). Nine links are used to connect these four areas. Three internal SRGs are visible in area 1. Two internal SRGs are visible in area 2. Two internal SRGs are visible in area 3. Five internal SRGs are vi sible in area 4. In addition to the internal SRGs, all external SRGs need to be kept at each node. External SRGs are assigned to the areas and the

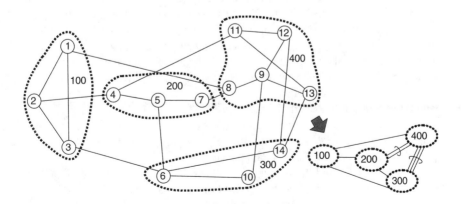

FIGURE 9.59

links between areas. The number of SRGs that need to be kept depends on the area. In this example, area 4 requires the maximum number of SRGs (17 = (4 − 1) + 9 + max(5,3,3,2)), and areas 2 and 3 require the minimum number of SRGs (14 = (4 − 1) + 9 + min(5,3,3,2)). In this way, the number of backup SRGs that need to be kept at each link is reduced.

9.5.3 Distributed Routing Calculation

The bandwidth-management concept can be implemented in either a centralized or distributed way [54]. In a distributed scenario, the source node of the wavelength path calculates the route for both working paths [54] and backup paths [55]. The backup SRG attached to the link (or LSP) is advertised by a routing protocol so that all nodes share the backup SRGs attached to all links, thus allowing any node to calculate the route for the backup LSP to achieve 100% recovery performance in a shared-mesh restorable network. A simple probabilistic routing method has been developed by Sengupta and Ramamurthy [54]. Sharable bandwidth for each SRG is flooded throughout the network using link-state routing extensions [55].

The routes for both working and backup LSPs are calculated by the node at the ingress of the LSP [54]. A link-state routing protocol is used for routing calculation [17, 23]. The link state, which includes various attributes associated with the link, is flooded through the entire network in the link-state routing protocol so that each node in the network can calculate the route for the LSP. Every time the link-state changes, a new link-state update packet is generated to notify the change.

The link is associated with attributes such as SRG and backup SRG. The SRG is a set of SRG tags attached to the link. The SRG of the link is advertised because it is included in the link state. The SRG is a static attribute. The backup SRG is a set of SRG tags that the link backs up in the event of failure. The backup SRG is a dynamic attribute. Every time the backup LSP is routed over the link, the backup SRG of the link could be updated to include the backup SRG of the backup LSP. However, if the backup SRG is included in the link-state update packet, a new link-state update packet is flooded every time the backup LSP is routed over the link. A surge of link-state update packets might be generated if multiple backup LSPs were set up at once. Instead, we propose that a full-booked backup SRG be included in the link-state update packet to reduce the number of link-state update packets. The full-booked backup SRG is a set of SRGs showing which links do not have sufficient resources to back up LSPs. In the example of Figure 9.58, if the wavelength resource for SRG 4 is fully booked (for example, two more working LSPs are set up), then SRG 4 is advertised as the full-booked backup SRG in the link-state update packet.

Once the SRGs and the full-booked backup SRGs are advertised in the link-state update packets, all nodes in the network can calculate the routes for both working and backup LSPs. For the working LSP, the ingress node calculates the route by running Dijkstra's algorithm on the graph consisting of the links with sufficient wavelength resources (constraint-based shortest path first). The working LSP is set up along with the route by the signaling protocol [21, 22]. The SRGs for the working LSP are collected along with the route. The ingress node then calculates the route for the backup LSP by running Dijkstra's algorithm on the graph consisting of the links, which are SRG-disjoint with the working LSP and backup-SRG-disjoint with all other backup LSPs. The former condition is applied on the original graph by removing the links having common SRG tags with the backup SRG of the backup LSP. The latter condition is applied on the original graph by removing the links whose full-booked-backup-SRG attribute has common SRG tags with the backup SRG of the backup LSP. The backup LSP is set up along with the route. In this way the routes for both the working and the backup LSPs are calculated in a distributed manner.

9.5.4 Performance Evaluation

9.5.4.1 Shared Restoration versus Protection

We investigated how many working wavelength LSPs can be established while maintaining a 100% recovery ratio by setting up the associated backup LSPs. The NSF net model consisting of 14 nodes and 21 links is used as a model network (Figure 9.60). We assume that the number of wavelengths in each fiber link is identical. Let w denote the number of wavelengths in each fiber link. The effect of w on the number of admissible working LSPs is evaluated through computer simulation. New LSP establishment requests, whose source–destination pair is randomly determined, are generated consecutively in the simulation. The maximum number of wavelength LSPs is 91 in a full-mesh configuration. We assume that a wavelength LSP consists of bidirectional working and backup LSPs. As previously

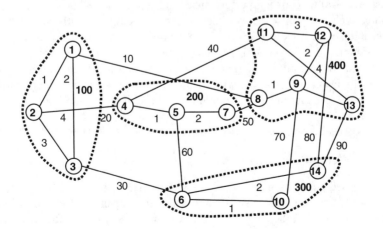

FIGURE 9.60

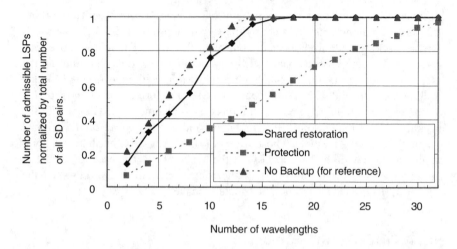

FIGURE 9.61 Relationship between the number of wavelengths and the number of admissible working LSPs.

mentioned, a wavelength LSP is set up only if both working and backup LSPs are admitted. Figure 9.61 shows the relationship between the number of wavelengths and the number of admissible LSPs normalized by the total number of all source–destination combinations.

When the number of wavelengths exceeds 18, all of the SD pairs are connected by wavelength LSP while maintaining 100% recovery performance. Compared with the number of wavelengths achieved by the method without backup LSPs, in which all SD pairs are connected with 14 wavelengths per link, the proposed shared-mesh restoration method requires just four extra wavelengths to achieve 100% recovery performance (i.e., an extra 28.6% in wavelength resources achieves 100% recovery performance). Compared with the number of wavelengths with backup LSPs by the 1+1 protection method, in which backup LSP occupies the wavelength resources and more than 32 wavelengths are required to connect all SD pairs, the proposed shared-mesh restoration method requires 14 fewer wavelengths while maintaining 100% recovery performance (i.e., a 56.3% reduction in wavelength resources achieves the same recovery performance).

9.5.4.2 Effect of Bumping of Existing Backup LSPs

The proposed restoration method bumps the existing backup LSPs if a new working LSP requests the wavelength to which those backup LSPs are assigned. The rationale behind the rule is that the working LSP occupies the wavelength resource while the backup LSPs can share the wavelength resource as long as

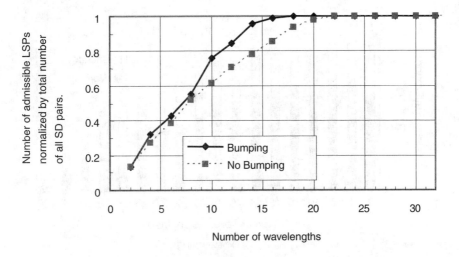

FIGURE 9.62 Number of admissible LSPs by bumping and nonbumping methods.

there are no conflicts between the identical SRGs. If the backup LSPs are not bumped, the new working LSP is routed over a suboptimal route, consuming extra wavelength resource.

Figure 9.62 shows the number of admissible wavelength LSPs achieved by the bumping as well as the nonbumping methods. The bumping method outperforms the nonbumping method. The bumping method uses the shortest path for working LSP, which occupies wavelength resources, at the expense of the second-shortest path for backup LSP, which allows resource sharing between other backup LSPs. Therefore the bumping method makes better use of wavelength resources compared with the nonbumping method.

Figure 9.63 shows the wavelength resource utilization as a function of the number of wavelengths per fiber link for both the bumping and nonbumping methods. The bumping method allocates more wavelength resources to working LSPs than to backup LSPs compared with the nonbumping method. We confirmed that the bandwidth-sharing effect between backup LSPs in the bumping method improves network throughput.

9.5.4.3 Link Protection versus Node Protection

So far we have assumed that only link failure is protected in the performance evaluation. If we need also to protect against node failure, the network resource utilization could be reduced because it is less likely that an SRG-constraint-based route will be found. We evaluated how the number of admitted LSPs is reduced if node-failure protection is introduced in addition to link-failure protection.

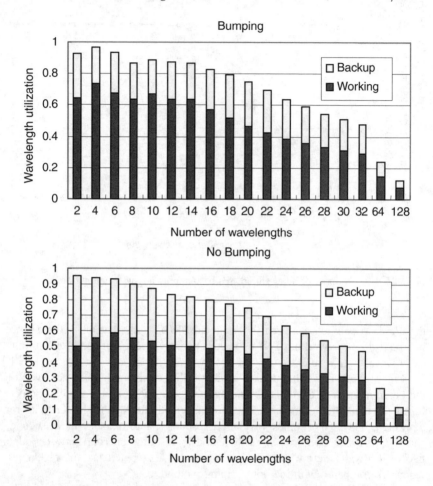

FIGURE 9.63 Wavelength resource usage by working and backup LSPs.

We assigned different SRGs to all nodes, as shown in Figure 9.60. Figure 9.64 shows the relationship between the number of the admitted LSPs normalized by the number of the tried LSPs. A slight difference is observed if we protect against node failure.

9.5.4.4 Hierarchy

Each link maintains backup-SRG piecewise bandwidth information. As the network size grows, the number of SRGs increases as well. To maintain scalability and reduce the amount of backup-SRG piecewise bandwidth information that must be stored, a hierarchy in SRG is introduced. There is, however, a trade-off between the reduction of required storage and the networkís resource utilization. If we introduce hierarchy in SRGs, the network resource utilization could be reduced because it is less likely that an SRG-constraint-based route will be found. The number of admitted LSPs may also be reduced.

Traffic Engineering in GMPLS Networks 289

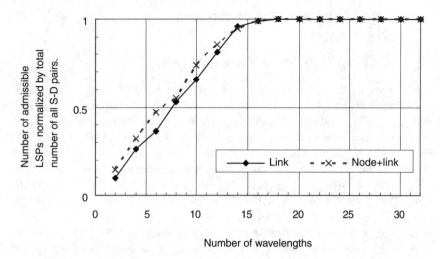

FIGURE 9.64 Performance comparison between link failure protection and node-link failure protection.

We evaluated the number of admitted LSPs assuming the SRG hierarchy as shown in Figure 9.60. Figure 9.65 shows the number of admitted LSPs for both single-area SRG and multiarea SRG. In this example, little difference is observed, but the amount of storage for backup-SRG piecewise bandwidth information is reduced. As noted previously for this model, the maximum number of backup SRGs is 17 (area 4) and the minimum is 14 (areas 2 and 3), and the total number of backup SRGs is 39.

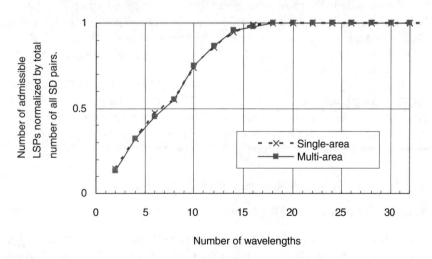

FIGURE 9.65

9.6 DEMONSTRATION OF PHOTONIC MPLS ROUTER

9.6.1 Integration of IP and Optical Networks

The amount of IP data traffic has grown remarkably over the past few years. Massive routers and flexible route-control mechanisms are required to help deal with this growth, and optical technology must catch up with this trend.

From the viewpoint of network control, Multiprotocol Label Switching (MPLS) [56] has shown great promise as a solution to the growth in traffic. MPLS was initiated on the control plane of cell- or frame-based systems such as Asynchronous Transfer Mode (ATM), frame relay, Ethernet, and Point-to-Point Protocol/High-Level Data Link Control (PPP/HDLC). A few years ago, some proposals were submitted pertaining to the MPLS control capability on optical networks, for instance Multiprotocol Lambda Switching (MPLambdaS or MPλS) [57] and Photonic MPLS [58]. These issues have become topics of intense discussion and have been broadened to Generalized MPLS (GMPLS) [20], which is still under discussion in many standardization bodies, such as the Internet Engineering Task Force (IETF), International Telecommunications Union-Telecommunication standardization sector (ITU-T), and Optical Internetworking Forum (OIF).

To realize an MPλS prototype, two fundamental modifications are required between MPλS and MPLS. The first is that the relationships between the label values and wavelength numbers must be defined. This will lead to fewer labels used in MPλS than in MPLS due to the limitations in physical wavelength-division multiplexing (WDM) transmission characteristics. In the MPλS case, there is a one-to-one mapping between wavelengths and labels; therefore, the number of wavelength labels is limited because the state-of-the-art technology allows only a limited number of wavelengths.

The second modification is that an optical label-switched path (OLSP) must be associated with a pair of conventional label-switched paths (LSPs) because OLSPs are bidirectional and, conversely, LSPs are basically unidirectional. These differences severely affect the path-control protocols. MPλS was implemented while taking these issues into account.

This section reports on the concept of a photonic MPLS router (HIKARI router) and describes the usage of a kind of MPλS, which is implemented as a modified MPLS protocol [7]. A newly developed efficient disjoint-path-selection scheme with shared-risk link-group (SRLG) constraints, called a weighted-SRLG (WSRLG) scheme, is presented. Finally, some examples of HIKARI router networks are presented.

9.6.2 Photonic MPLS Router (HIKARI Router)

9.6.2.1 Concept of HIKARI Router

The HIKARI router incorporates not only the lambda-switching capability (LSC), but also the packet-switching capability (PSC) within one router box. Conventional intelligent optical cross-connect systems (OXCs) have only the lambda-switching capability or a fiber-switching capability (FSC). Conventional MPLS routers have only the PSC function. The HIKARI router combines both LSC and PSC functions, thus the advantages of both the packet-switching technology and the circuit-switching technology can be utilized with suitable router-management systems.

The HIKARI routers establish IP networks over photonic networks with distributed autonomous control. By communicating with other routers, each HIKARI router controls the OLSP. According to the virtual-wavelength-path (VWP) concept [59], wavelength conversion is available in optical networks, and in the OLSP, there is a wavelength label for each section. It is difficult to establish VWP networks; however, VWP achieves flexible selection of a label at each node to compensate for the small number of labels. The introduction of wavelength conversion can reduce the required wavelength resources by approximately 20% [59]; however, it increases the node cost. A reduction in the number of wavelength converters and regenerators is required to realize cost-effective HIKARI router networks. This topics is addressed in a paper by Sato et al. [25].

The HIKARI router can handle both LSPs and OLSPs simultaneously because LSP is handled by PSC and OLSP is handled by LSC. LSPs are seamlessly connected by OLSPs, and LSPs are aggregated on OLSPs. Therefore, a controller can manage OLSPs as a layer in the LSP hierarchy. Deploying OLSPs improves the throughput of networks by taking advantage of the cut-through effect. In many cases, HIKARI routers reduce the electrical forwarding fabric by approximately one-third to one-half [25]. This is one advantage of the combined switching capabilities.

As shown in Figure 9.66, the HIKARI router comprises five functional blocks: a WDM function unit, an optical switch unit, an L1 trunk unit, an L2/L3 trunk unit, and a network element manager (NE manager). A lambda-routing unit (LRU) is defined as an integrated unit comprising the WDM unit, the optical switch unit, and the L1 trunk unit. The LRU can switch and add/drop OLSPs, and wavelength conversion is performed at the λ-conv in L1 trunks. The HIKARI routers can exchange messages through the optical supervisory channel (OSC). The OSC and optical paths are multiplexed at WDM units. The NE manager monitors all the circuit elements in the node, supervises remote optical repeaters, performs restoration, monitors the quality levels of the signals (L1 level and L2 level), and controls the LSPs and OLSPs.

The proposed HIKARI router is based on a universal photonic platform or optical backplane with the addition of many trunks. L1 trunks can be replaced with λ-conv functions, signal-regeneration functions (i.e., reshaping, retiming, and regeneration [3R] functions), and optical-amplifier functions. L2/L3 trunks can be replaced with IP-router functions, MPLS-router functions, and layer-2 switch functions. These functions are used adaptively. In other words, if a signal is degraded by fiber loss as well as nonlinear effects such as PMD (polarization mode dispersion) or ASE (amplified spontaneous emission), the 3R function is activated. In addition, wavelength conversion is also used when signaling is blocked by wavelength overbooking. Of course, L3 packet forwarding is used adaptively when L3 packet forwarding is performed.

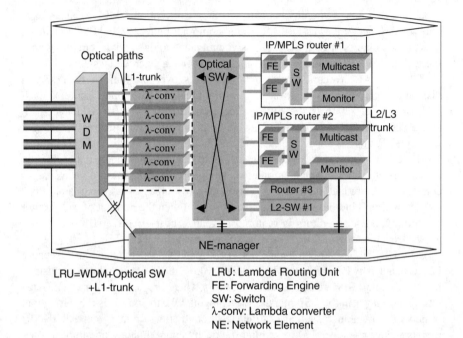

FIGURE 9.66 HIKARI router functional configuration. (Copyright 2002, IEEE.)

Note that the HIKARI router can transfer three types of signals [26]. Type A involves L3 switching functions for operations such as MPLS path aggregation and L3-packet-level forwarding. This is the same as the conventional MPLS-router signal flow. Type B represents relay connections that need wavelength conversion. Type-B connections are also supported by the adaptive use of the 3R function. Finally, type C represents transparent transfer or bit-rate restriction-free signals. The HIKARI router supports all transfer capabilities. Type C is not implemented in the prototype system described in this section.

9.6.2.2 HIKARI Router Characteristics

Table 9.2 gives the characteristics of the HIKARI router. The HIKARI router is an IP router system. Therefore, the packet-processing capability is the most important. The developed HIKARI router system was designed to achieve correspondence to a maximum 5 gigapackets per second (5 Gpps) capability. Some portion of the input packets are processed at L3 trunks on a packet-switching basis, and others are processed at the LRU on a lambda-switching basis.

An optical-switch-based (planar light-wave circuit thermooptics switch) integrated OXC was implemented as the LRU, which establishes both the optical-path cross-connect and long-haul WDM transmission functionalities. Hence, this

TABLE 9.2
HIKARI Router Characteristics

Items	Characteristics
Packet-processing capability	Corresponds to maximum 5 Gpps (shared with L3 trunk and LRU)
System throughput	Maximum 2.56 Tbps
Interface for L3 trunk	Packet over SONET/SDH (POS), ATM, GbE, etc.
Interface between L3 trunk and LRU	POS (2.5 Gbps or 10 Gbps), 10 GbE-WAN PHY
Optical switches	Planer light-wave circuit thermooptical switches
Operating wavelength range	1550-nm band (C band)
Optical channel speed	2.5 Gbps or 10 Gbps
Maximum number of wavelengths	32 per fiber
Number of WDM fiber ports	Maximum of eight
Total switch scale	256 × 256 channels
Optical supervisory channel (OSC)	OC-3 (156 Mbps)
MPS signaling channel (trial mode)	IP over ATM over OSC
MPS protocol (trial mode)	Proprietary CR-LDP extension [49]

prototype system includes two optical network element (ONE) functions in one box. Most of the conventional OXCs require a WDM transmission system outside the OXC box. In contrast, the LRU is equipped with the all-in-one type ONE. This feature aids in reducing operational costs, maintenance costs, device costs, office-space requirements, electrical energy, and the rate of failure. In the next generation, networks will be enormous, and these aforementioned features will provide great benefits.

The LRU has a maximum switching capacity of 256 × 256. Thus, 128 bidirectional optical paths can be accommodated. The optical-path adaptation interfaces implemented in this system are SONET OC-48c. Thus, the maximum total capacity is 640 Gbps. If the interfaces were upgraded to OC-192c or 10 GbE, the capacity would become 2.56 Tbps. Actually, the optical-loss budget design was constructed to support 10-Gbps optical-path signals, making expansion possible without serious difficulties.

9.6.2.3 Optical-Layer Management Characteristics

This section describes the operation, administration, and maintenance (OA&M) aspects regarding the optical layer of the HIKARI router. The implemented optical-channel (OCh) frame format was a modified SDH G.707 frame. This frame format will be changed to the OTN G.709 frame. The optical supervisory channel (OSC) carries the optical transmission section overhead (OTS-OH), the optical multiplexing section overhead (OMS-OH), and a high-speed data communication channel (DCC).

The OSC was established by modification of the SONET OC-3 signal format. The overhead region of the OSC was rearranged for transmission in OTS-OH and OMS-OH. The payload region of the OSC is used for DCC. OTS-OH and OMS-OH are used to manage WDM links. The DCC is used for the network-management channel and the MPλS signaling channel. Data communication networks (DCN) that employ ATM are built based on DCC.

The NE manager controls and monitors the corresponding hardware elements and enables communication among the other NEs (e.g., other HIKARI routers) to exchange MPλS signaling protocol messages via a modified Constraint-Based Routing Label Distribution Protocol (CR-LDP) and exchange OA&M messages. A preliminary console in the network operation center (NOC) was also developed. Operators can control and supervise remote NEs using the console. The Simple Network Management Protocol (SNMP) is used to manage HIKARI routers. The original LRU management information bases (MIBs) and OLSP MIBs were developed. About 50 MIBs were newly implemented to the NE manager.

9.6.2.4 Implementation of MPλS Signaling Protocol

The OLSP control plane of the HIKARI router is operated by the CR-LDP protocol extended for photonic network control. In the extended protocol, LSPs are handled as bidirectional paths for smooth connection with OLSPs defined as bidirectional paths, while conventional MPLS handles LSPs as unidirectional paths. This implementation of the protocol was developed with the fewest possible modifications. In fact, the same protocol core can be adapted to the standard and extended protocols. Some modifications have been implemented in the path connection-point configuration and path termination-point configuration. In the standard CR-LDP, there is only one required connection in the NE (e.g., MPLS router) configuration for a unidirectional LSP. Contrarily, the extended CR-LDP must simultaneously handle a downstream connection and an upstream connection. The path connection/termination-point configurations are performed consecutively. When bidirectional path configurations are established in a distributed network management system (NMS), conflicts occasionally occur such as when two or more resource reservation requests are sent for the same resource at the same time. This conflict is avoided by using a priority control method implemented based on OLSP-IDs. The signaling channel for CR-LDP is made with IP over ATM over OSC links. This IP network is also used as a data communication network (DCN) for network management. An NMS can communicate with NEs via DCN using SNMP.

9.6.3 PHOTONIC NETWORK PROTECTION CONFIGURATION

The HIKARI router provides many types of protection and restoration configurations utilizing the combination of 1+1 protection, 1:1 protection, restoration, or no protection on the photonic network side. In the 1+1 protection, both working and protection OLSPs are set between source and destination HIKARI router pairs. Traffic is bridged to both OLSPs. On the other hand, in the case of 1:1

protection, a protection OLSP can carry extra traffic. This extra traffic is not protected when the working OLSP is experiencing failure. In case of the restoration, protection OLSPs are not actually set; rather, the resources are reserved. These reserved resources can share with many working OLSPs; therefore, the required network resources can be reduced.

To realize protection and restoration functions, an efficient disjoint-path-selection algorithm and fast-restoration signaling should be developed. The HIKARI router adopts the WSRLG scheme as an efficient disjoint-path-selection algorithm, as described in Section 9.4. It also adopts the protection and restoration functions as proposed by Shimano et al. [60, 61].

The implemented protection and restoration functions can be described as follows. Automatic protection switching (APS)-based protection (1+1 and 1:1) was implemented in the HIKARI router (photonic MPLS router [PR]). A protection time of less than 20 msec was achieved. A new and fast restoration scheme was also implemented on the HIKARI routers. Figure 9.67 shows the restoration procedures. The sample configuration is shown in Figure 9.67(a). An OLSP is established between PR1 and PR3 via PR2, with λ_a assigned to the link between PR1 and PR2 and λ assigned to the link between PR2 and PR3. On the other hand, the restoration path is calculated before any fault occurs, and the resources are reserved in the new method. Reservation for the restoration path corresponding to the OLSP is performed on a DCN over OSC DCC channels. As mentioned in Section 9.6.2.3, the OSC carries ATM traffic. Some ATM virtual circuits (VCs) are set as reserved on the OSC. In this case, the VCs are allocated to three sections: PR1-PR4, PR4-PR5, and PR5-PR3. When a failure occurs in the link between PR1 and PR2 (Figure 9.67(b)), optical-path-setup signaling is transported to related PRs through the VCs over OSC (Figure 9.67(c)). The restoration path is established on the same route after the signaling is successfully completed. All the nodes are connected to new optical paths according to the VC identifiers on the route of the restoration path. The mapping scheme is such that VC_a is mapped to λ_a, and so on. Therefore, the wavelength assignments are simultaneously completed (Fig. 9.67(d)). The detailed experimental results of the developed restoration scheme are described in Section 9.6.4.

9.6.4 Demonstration of HIKARI Router

The functions of the developed HIKARI routers were demonstrated at SuperComm2001 (June 2001) [60]. OLSP setup/teardown and OLSP protection were successfully performed. Figure 9.68 shows the configuration of the demonstration. As mentioned in Section 9.6.3, less than 20 msec OLSP protection time was achieved. For the digital video stream demonstration (demo. 2), 20 msec was sufficient to prevent a video stream frame misalignment. Another demonstration that was conducted was the automatic OLSP setup/teardown procedure (demo. 1). An operator initiated the OLSP setup command to PR 1 via the NOC console. PR 1 determined the OLSP route according to the command. PR 1 first sent a Label Request message to PR 2. PR 2 received the message and

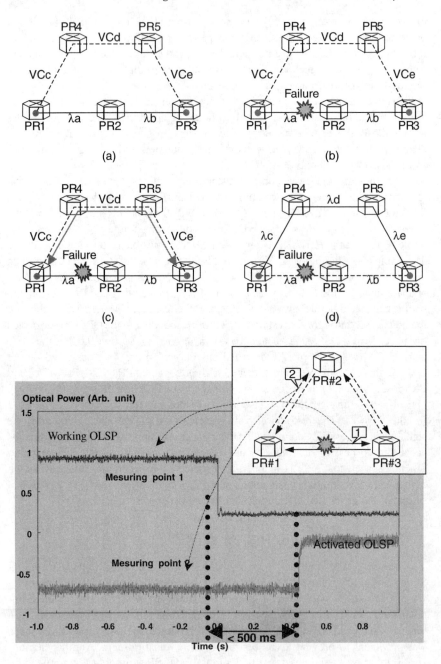

FIGURE 9.67 Fast restoration procedure: (a) restoration configuration, (b) link failure, (c) signaling for restoration-path configuration, (d) restoration completion, (e) experimental results of restoration demonstration. (Copyright 2002, IEEE.)

Traffic Engineering in GMPLS Networks

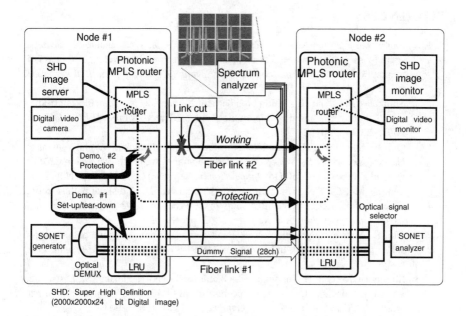

FIGURE 9.68 Configuration of the photonic MPLS router demonstration. (Copyright 2002, IEEE.)

replied to it by sending a Label Mapping message back to PR 1. After this signaling process, the OLSP was successfully established in less than 1 sec. In this demonstration, the OLSP setup/teardown procedure was repeated in a 5-sec cycle time.

Next, the restoration function was examined. A triangle network was constructed with three Photonic MPLS routers (Figure 9.67(e)). A working path was established between PR 1 and PR 3. A restoration path was reserved between PR 1 and PR 3 via PR 2. The working path was intentionally cut. PR 3 detected signal failures as restoration triggers in the path termination points (TPs). Then PR 3 initiated a restoration request to the other end of the corresponding path. Figure 9.67(e) shows the preliminary results of the demonstration. In the figure, the number in the balloons represents the measuring point. Measuring point 1 was set on the working path between PR 1 and PR 3. Measuring point 2 was set on the route of the restoration path. The restoration procedures were completed within 500 msec. The service downtime in the SONET layer was evaluated, and the time was approximately 650 msec. The service downtime in the PPP layer was evaluated, and the time was approximately 700 msec. The restoration time was sufficiently short to restore IP over photonic networks without initiating any recovery mechanism on the IP layer.

REFERENCES

1. Shiomoto, K., Imajuku, W., Oki, E., Okamoto, S., and Yamanaka, N., Scalable shared-risk group management in shared mesh restorable wavelength routed networks, Proc. *2003 Workshop on High Performance Switching and Routing,* 2003, pp. 189–194.
2. Shimazaki, D., Oki, E., Shiomoto, K., and Yamanaka, N., Scalable multi-layer GMPLS networks based on hierarchical cloud-routers, in Proc. *IEEE Globecom 2003,* 2003, pp. 3792–3796.
3. Oki, E., Shimazaki, D., Shiomoto, K., Matsuura, N., Imajuku, W., and Yamanaka, N., Performance of distributed-controlled dynamic wavelength-conversion GMPLS networks, Proc. Int. Conf. On *Optical Commun. Networks,* 1, 355–358, Nov. 2002.
4. Oki, E., Matsuura, N., Shiomoto, K., and Yamanaka, N., A disjoint path selection scheme with SRLG in GMPLS networks, Proc. in *2002 Workshop on High Performance Switching and Routing,* 2002, pp. 88–92.
5. Oki, E., Matsuura, N., Shiomoto, K., and Yamanaka, N., A disjoint path selection scheme with shared risk link groups in GMPLS networks, *IEEE Commun. Lett.,* 6, 406–408, 2002.
6. Shiomoto, K., Oki, E., Imajuku, W., Okamoto, S., and Yamanaka, N., Distributed virtual network topology control mechanism in GMPLS-based multi-region networks, *IEEE J. Selected Areas Commun.,* 21, 1254–1262, 2003.
7. Okamoto, S., Oki, E., Shiomoto, K., Sahara, A., and Yamanaka, N., Demonstration of the highly reliable HIKARI router network based on a newly developed disjoint path selection scheme, *IEEE Commun. Mag.,* 40, 11, 52–59, 2002.
8. Ramaswami, R. and Sivarajan, K.N., Design of logical topologies for wavelength-routed optical networks, *IEEE J. Selected Areas Commun.,* 14, 840–851, 1996.
9. Mukherjee, B., Banerjee, D., Ramamurthy, S., and Mukherjee, A., Some principles for designing a wide-area WDM optical network, *IEEE/ACM Trans. Networking,* 4, 684–696, 1996.
10. Labourdette, J.-F.P., Hart, G.W., and Acampora, A.S., Logically rearrangeable multihop lightwave networks, *IEEE Trans. Commun.,* 39, 1223–1230, 1991.
11. Banerjee, D. and Mukherjee, B., Wavelength-routed optical networks: linear formulation, resource budgeting tradeoffs, and a reconfiguration study, *IEEE/ACM Trans. Networking,* 8, 598–607, 2000.
12. Ricciato, F., Salsano, S., Belmonte, A., and Listanti, M., Off-line configuration of a MPLS over WDM network under time-varying offered traffic, in *Proc. IEEE Infocom 02,* Vol.1, pp. 57–65, 2002.
13. Gencata, A. and Mukherjee, B., Virtual-topology adaptation for WDM mesh networks under dynamic traffic, in *Proc. IEEE Infocom 02,* Vol.1, pp. 48–56, 2002.
14. Shiomoto, K., A simple bandwidth management strategy based on measurements of instantaneous virtual path utilization in ATM networks, *IEEE/ACM Trans. Networking,* 6, Issue 5, 625–634, 1998.
15. Banerjee, A., Drake, J., Lang, J.P., Turner, B., Kompella, K., and Rekhter, Y., Generalized multiprotocol label switching: an overview of routing and management enhancements, *IEEE Commun. Mag.,* 39, 1, 144–150, 2001.
16. Banerjee, A., Drake, J., Lang, J., Turner, B., Awduche, D., Berger, L., Kompella, K., and Rekhter, Y., Generalized multiprotocol label switching: an overview of signaling enhancements and recovery techniques, *IEEE Commun. Mag.,* 39, 7, 144–151, 2001.

17. Kompella, K. and Rekhter, Y., OSPF Extensions in Support of Generalized MPLS, IETF draft, draft-ietf-ccamp-ospf-gmpls-extensions-12.txt, Oct. 2003 (work in progress). http://www.ietf.org/internet-drafts/draft-ietf-ccamp-ospf-gmpls-extensions-12.txt
18. Oki, E., Shiomoto, K., Okamoto, S., Imajuku, W., and Yamanaka, N., A heuristic multilayer optimum topology design scheme based on traffic measurement for IP.photonic networks, in *Proc. Optical Fiber Communication Conference and Exhibit*, Mar. 2002, pp. 17–22.
19. Oki, E. and Yamanaka, N., Impact of multimedia traffic characteristic on ATM network configuration, *J. Network Syst. Manage.*, 6, 377–397, 1998.
20. Mannie, E. et al., Generalized Multi-Protocol Label Switching Architecture, RFC3945, Oct.2004 http://www.ietf.org/rfc/rfc3945.txt.
21. Berger, L., Ed., Generalized Multi-Protocol Label Switching (GMPLS) Signaling Functional Description, RFC3471, Jan. 2003. http://www.ietf.org/rfc/rfc2702.txt?number=3471.
22. Berger, L., Ed., Generalized Multi-Protocol Label Switching (GMPLS) Signaling Resource ReserVation Protocol-Traffic Engineering (RSVP-TE) Extensions, RFC 3473, Jan. 2003. http://www.ietf.org/rfc/rfc2702.txt?number=3473.
23. K. Shiomoto, Requirements for GMPLS-based multi-region networks, IETF draft, draft-shiomoto-ccamp-gmpls-mrn-reqs-01.txt, Feb. 2005.
24. Imajuku, W., Oki, E., Shiomoto, K., and Okamoto, S., Multilayer Routing Using Multilayer Switch Capable LSRs, IETF draft, draft-imajuku-ml-routing-02.txt, Jun. 2002.
25. Sato, K., Yamanaka, N., Takigawa, Y., Koga, M., Okamoto, S., Shiomoto, K., Oki, E., and Imajuku, W., GMPLS-based photonic multilayer router (Hikari router) architecture: an overview of traffic engineering and signaling technology, *IEEE Commun. Mag.*, 40, 3, 96–101, 2002.
26. Katz, D., Kompella, K., and Yeung, D., Traffic Engineering (TE) Extensions to OSPF Version 2, RFC 3473, Sep. 2003. http://www.ietf.org/rfc/rfc3473.txt
27. Kompella, K. and Rekhter, Y., LSP Hierarchy with Generalized MPLS TE, IETF draft, draft-ietf-mpls-lsp-hierarchy-08.txt, Mar. 2003 (work in progress). http://www.ietf.org/internet-drafts/draft-ietf-mpls-lsp-hierarchy-08.txt
28. Dijkstra, E.W., A Note on Two Problems in Connection with Graphs, *Numeriche Mathematik*, 1, 269–271, 1959.
29. Gerstel, O., Ramaswami, R., and Foster, S., Merits of hybrid optical networking, in *Proc. Optical Fiber Commun. Conf. (OFC) 2002 Technical Digest*, 2002, pp. 33–34.
30. Meyer, S., Quantification of wavelength contention in photonic networks with reach variation, in *Optical Fiber Commun. Conf. (OFC) 2002 Technical Digest*, 2002, pp. 36–37.
31. Ramaswami, R. and Siarajan, K., *Optical Networks: a Practical Perspective*, Morgan Kaufmann Publishers, San Francisco, 1998.
32. Oki, E., Matsuura, N., Imajuku, W., Shiomoto, K., and Yamanaka, N., Requirements of Optical Link-State Information for Traffic Engineering, IETF draft, draft-oki-ipo-optlink-req-00.txt, Feb. 2002.
33. Ho, P.-H. and Mouftah, H.T., Network planning algorithms for the optical Internet based on the Generalized MPLS architecture, in *Proc. IEEE Globecom 2001*, Vol. 4 pp. 2150–2154, 2001.

34. Ho, P.-H. and Mouftah, H.T., A novel routing protocol for WDM mesh networks, in *Proc. Optical Fiber Commun. Conf. (OFC) 2002 Technical Digest*, 2002, pp. 38–39.
35. Zang, H., Jue, J.P., and Mukherjee, B., A review of routing and wavelength assignment approaches for wavelength-routed optical WDM networks, *SPIE Optical Networks Mag.*, 1, 1, pp. 73–89, 2000.
36. Oki, E. and Yamanaka, N., A recursive matrix-calculation method for disjoint path search with hop link number constraints, *IEICE Trans. Commun.*, E78-B, 769–774, 1995.
37. Moy, J., OSPF Version 2, RFC 2328. Apr.1998, http://www.ietf.org/rfc/rfc2328.txt
38. Oran, D., OSI IS-IS Intra-Domain Routing Protocol, RFC 1142. Feb.1990, http://www.ietf.org/rfc/rfc1142.txt
39. Suurballe, J.W., Disjoint paths in a network, *Networks*, 4, 125–145, 1974.
40. Dunn, D.A., Grover, W.D., and MacGregor, M.H., Comparison of k-shortest paths and maximum flow routing for network facility restoration, *IEEE J. Selected Areas Commun.*, 12, 88–99, 1994.
41. Pollack, M. and Wiebenson, W., Solution of the shortest-route problem ó a review, *Operations Res.*, 8, 224–230, 1960.
42. Itai, A., Perl, Y., and Shiloach, Y., The complexity of finding maximum disjoint paths with length constraints, *Networks*, 12, 277–286, 1982.
43. Bhandari, R., *Survivable Networks Algorithms for Diverse Routing*, Kluwer, Dordrecht, Netherlands, 1999.
44. Aho, A.V., Hopcroft, J.E., and Ullman, J.D., *The Design and Analysis Computer Algorithms*, Addison-Wesley Series in Computer Science, Addison-Wesley, Reading, MA, 1974.
45. Shiomoto, K., Oki, E., Katayama, M., Imajuku, W., and Yamanaka, N., Dynamic multi-layer traffic engineering in GMPLS networks, in *Proc. WTC/ISS 2002*, Paris France, 2002.
46. Ye, Y., Dixit, S., and Ali, M., On joint protection/restoration in IP-centric DWDMbased optical transport networks, *IEEE Commun. Mag.*, 38, 6, 174–183, 2000.
47. Ghani, N. and Dixit, S., On IP-over-WDM integration, *IEEE Commun. Mag.*, 38, 3, 72–84, 2000.
48. Doverspike, R. and Yates, J., Challenges for MPLS in optical network restoration, *IEEE Commun. Mag.*, 29, 2, pp. 89–96, 2001.
49. Strand, J. and Chiu, A.L., Issues for routing in the optical layer, *IEEE Commun. Mag.*, 39, 2, 81–87, 2001.
50. Rajagopalan, B., Pendarakis, D., Saha, D., Ramamoorthy, R.S., and Bala, K., IP over optical networks: architectural aspects, *IEEE Commun. Mag.*, 38, 9, 94–102, 2000.
51. Xin, C., Ye, Y., Wang, T.-S., Dixit, S., Qiao, Ch., and Yoo, M., On an IP-centric optical control plane, *IEEE Commun. Mag.*, 39, 9, 88–93, 2001.
52. Papadimitriou, D., Poppe, F., Dharanikota, S., Hartani, R., Jain, R., Jones, J., Venkatachalam, S., and Xue, Y., Shared Risk Link Groups Encoding and Processing, IETF draft, draft-papadimitriou-ccamp-srlg-processing-01.txt, Nov. 2002.
53. Datta, S., Sengupta, S., Biswas, S., and Datta, S., Efficient channel reservation for backup paths in optical mesh networks, in *Proc. IEEE Globecom 2001*, pp. 2104–2108, 2001.

54. Sengupta, S. and Ramamurthy, R., Capacity efficient distributed routing of meshrestored lightpaths in optical networks, in *Proc. IEEE Globecom 2001*, pp. 2129–2133, 2001.
55. Yagyu, T., Suemura, Y., and Kolarov, A., Extensions to OSPF-TE for Supporting Shared Mesh Restoration, IETF draft, draft-yagyu-gmpls-shared-restorationrouting-00.txt, Jun. 2002.
56. Rosen, E. et al., Multiprotocol Label Switching Architecture, RFC 3031, Jan. 2001. http://www.ietf.org/rfc/rfc3031.txt
57. Awduche, D.O. et al., Multi-Protocol Lambda Switching: Combining MPLS Traffic Engineering Control with Optical Crossconnects, IETF draft, draft-awduchempls-te-optical-03.txt, Apr. 2001.
58. Okamoto, S., A proposal of the Photonic MPLS Network, OIF contribution OIF2000.017, Jan. 2000.
59. Sato, K. et al., Network performance and integrity enhancement with optical path layer technologies, *IEEE-JSAC*, SAC-12, 159–170, 1994.
60. Shimano, K. et al., MPLambdaS demonstration employing photonic routers (256 × 256 OLSPS) to integrate optical and IP networks, in *Proc. National Fiber Optic Engineers Conf. 2001 Tech. Proc.*, 2001, p. 5.
61. Shimano, K. et al., Implementation and demonstration of new fast restoration scheme for distributed photonic network control plane, in *Proc. Eur. Conf. Optical Commun.*, 2002.

10 Standardization

10.1 ITU-T (INTERNATIONAL TELECOMMUNICATION UNION-T)

The International Telecommunication Union (ITU) [1, 2] was formed in Madrid in 1932, the result of a coalition of the Universal Telegraph Union (founded in Paris in 1865) and the International Radio Telegraph Union (founded in Berlin in 1906). ITU is one of the specialized institutions of the United Nations (UN), and its purpose is to promote international cooperation for improving the electrical communications environment and its rational utilization, to enhance the efficiency of electrical communications operation, and to accelerate the adoption of effective technological measures. The number of affiliate countries is 189 (as of April 2005), and the headquarters is located in Geneva, Switzerland.

As shown in Figure 10.1, the UN telecommunication activity consists of four major sections: Telecom Standardization Sector (ITU-T), Radiocommunication Sector (ITU-R), Telecom Development Sector (ITU-D), and the Office of the Secretary General (GS). In more detail, it is composed of a Plenipotentiary Conference (PP), Council (C), World Conferences on International Telecommunications (WCIT), World/Regional Radiocommunication Conference (WRRC), Radiocommunication Assemblies (RA), Radiocommunication Study Group (ITU-R SG), World Telecommunication Standardization Assemblies (WTSA), Telecom Standardization Study Group (ITU-T SG), World/Regional Telecommunication Development Conference (WRTDC), Telecom Development Study Group (ITU-D SG), and four standing committees consisting of General Secretariat (GS), Telecommunication Standardization Bureau (TSB), Radiocommunication Bureau (RB), and Telecommunication Development Bureau (TDB).

The Telecom Standardization Study Group (ITU-T) comprises multiple study groups (SGs), and the chair of each study group guides discussions about the standardization of various telecommunication technologies in each session. There are 12 study groups at present, as shown in Table 10.1, plus a special study group (SSG) devoted to the topic of International Mobile Telecommunication-2000 (IMT-2000) "IMT-2000 and Beyond." Two of the study groups — SG13 and SG15 — have a special relationship with this SSG document.

The SG13 study group is related to multiprotocol- and IP-based networks, and it deals with matters relating to IP, B-ISDN, the global information infrastructure, and communication via satellite. At present, SG13 is working to establish a highly reliable and high-quality network by standardizing the operating technology

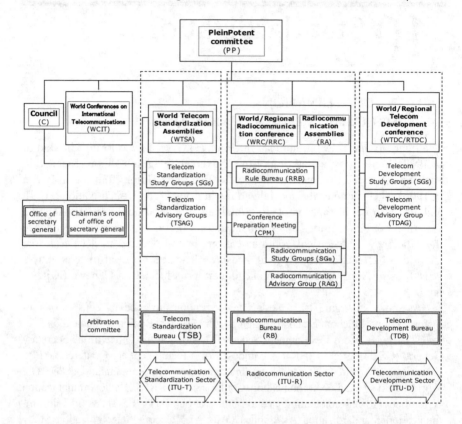

FIGURE 10.1 International Telecommunication Union (ITU) organizational chart.

that enables communication between networks or domains constructed using different technologies and operating with different protocols (SDH, ATM, MPLS, etc.). The SG13 group is now mainly discussing system architectures, interworking conditions, end-to-end performance, requirements for routing, and so on.

A working group (WG) is composed of multiple working parties (WP). At present, the following working parties are operating within the SG13 study group.

WP 1/13: Project Management and Coordination
WP 2/13: Architectures and Internetworking Principles
WP 3/13: Multiprotocol Networks and Mechanisms
WP 4/14: Network Performance and Resource Management

The discussion topics of each WP are organized as "questions." Examples of such "questions" are shown in Appendix 10.1.

TABLE 10.1
Structure of ITU-T Study Groups (Study Period 2001–2004)

Study Group	Topic
2	Operational aspects of service provision, networks, and performance
3	Tariff and accounting principles including related telecommunications economic and policy issues
4	Telecommunication management, including TMN
5	Protection against electromagnetic environment effects
6	Outside plant
9	Integrated broadband cable networks and television and sound transmission
11	Signaling requirements and protocols
12	End-to-end transmission performance of networks and terminals
13	Multiprotocol and IP-based networks and their internetworking
15	Optical and other transport networks
16	Multimedia services, systems, and terminals
17	Data networks and telecommunications software
SSG	Special study group: IMT-2000 and beyond

The SG15 study group is focused on the standardization of optical and other transport networks, systems, and equipment. The access system and metropolitan/long-distance systems are within the purview of the SG15 study group. At present, the following five WPs are operating under SG15.

WP 1/15: Network Access
WP 2/15: Network Signal Processing
WP 3/15: OTN (optical network termination) Structure
WP 4/15: OTN Technology
WP 5/15: Projects and Promotion

10.2 IETF (INTERNET ENGINEERING TASK FORCE)

The IETF [3, 4] was formed by computer science engineers with the aim of creating an Internet protocol that facilitates communication between computers with different operating systems and data formats.

The Internet's origin can be traced back to the early 1960s, when the Pentagon established a network called ARPANet (Advanced Research Projects Agency Network) to link the military, defense contractors, and university researchers. The Internet is currently operating based on a protocol group called TCP-IP, which was developed by Bob Kahn of DARPA and Vinton Cerf of Stanford University. The TCP-IP protocol supports autonomous distribution, which makes it highly robust, and provides end-to-end communication capability by supporting interoperability between different systems and different network media.

The first IETF conference, held in San Diego, California, in 1986, had only 15 attendees. Today, the membership exceeds 2000, and e-mail discussions among these members are carried out 24 hours a day. The IETF is a group within the Internet Society (ISOC), whose hierarchical structure is shown in Figure 10.2. ISOC comprises more than 150 member organizations with 8900 members from more than 170 countries. Developmental operations are being promoted under the hierarchy of the Internet Architecture Board (IAB), IETF, the Internet Engineering Steering Group (IESG), and the general area. IESG takes responsibility for standardizing Internet technologies discussed in IETF and has a role in guiding the direction of discussions within the IETF. There are currently eight main areas of discussion:

Applications Area (app)
General (gen)
Internet Area (int)
Operations & Management Area (ops)
Routing Area (rtg)
Security Area (sec)
Sub-IP Area (sub-ip)
Transport Area (tsv)

A director is appointed for each technological area, and there are multiple working groups (WGs) within each area. These WGs are listed in detail in Appendix 10.2. As for MPLS and GMPLS, they are discussed within the newly established sub-IP area. A new working group is organized whenever a new need for standardization is proposed, and the working group is terminated when

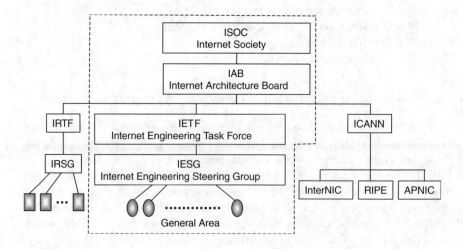

FIGURE 10.2 IETF organizational chart.

Standardization

its purpose has been achieved. The operation of a working group is based on a memorandum of intent (charter) that promotes active, focused, and useful discussion in an attempt to derive a speedy conclusion. Topics that are not specifically included in the charter are excluded from the discussion. Specifications of technology standards by IETF are written in a document called an RFC (request for comments) that is posted on a Web site for review by members through the Internet. IETF places importance on a "rough consensus and running code" rather than determining a tight specification, and once a specification has been accepted, it is considered to be a de facto standard in industry. The RFC process is a call for comments from Internet researchers as a means of improving the proposed specifications.

Figure 10.3 shows the RFC process in IETF. Any member can freely propose an Internet draft, which is then placed on a server for 6 months for public review. If the Internet draft is judged to be useful by the relevant working group, it becomes an official RFC. There are four types of RFC documents.

Standard-track RFC: This is a document that contains the specifications that a working group has judged should be the international standard in the industry. A standard-track RFC becomes a standard (S) when it passes through the phase of being a proposed standard (PS) and then a draft standard (DS). A PS is based on the premise that independent assembly tests have been performed and interoperability has been confirmed by multiple organizations. A DS is based on the premise that operational tests in a wide range of actual environments have been executed. When the proposal became a standard (S), the document receives an STD number.

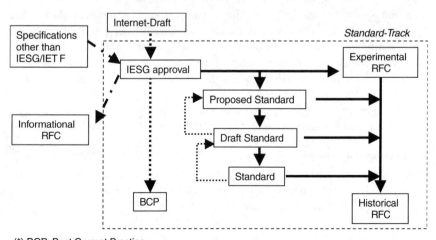

(*) BCP; Best Current Practice

FIGURE 10.3 RFC process in IETF.

Informational RFC: This is a document that contains useful information for industry even though it may not be a standard. This method is often used by companies to promote deployment of a new product without waiting for standardization. The specifications for that product are presented as an Informational RFC in an effort to position the specifications as a de facto standard.

Experimental RFC: This is a document relating to the discussion of technical specifications, not for the purpose of standardization but for the purpose of research. Although this document is often utilized for the purpose of pure research, it is also used by companies that are intent on positioning their own special specifications as a de facto standard.

Historical RFC: This is a document relating to information that should be reserved as a record of discussions in the standardization process. An example of an historical RFC would be the document recounting the discussions regarding implementation IP v.6 technology.

10.3 OIF (OPTICAL INTERNETWORKING FORUM)

The Optical Internetworking Forum (OIF) [5, 6] was established in 1998 in response to a proposal by Cisco Systems and Ciena Co. The purpose of OIF is to specify the physical (layer 1) and logical (layer 2) interfaces between the optical network and the data network (represented by the IP network) and to establish a standard operating method for optimum management and control of both networks. OIF is a representative forum to decide the industrial standards in North America. Besides the two companies mentioned here, the funding members are AT&T, Bellcore (now Telcordia), HP (now Agilent), Qwest, Sprint, and WorldCom. OIF has expanded its scope to include the interface specifications within devices, a multisource agreement (MSA) for optical transceivers, a MSA for variable-wavelength lasers, and the interface specifications between optical networks. The standards that are specified in OIF correspond to an implementation agreement (IA), which defines multiple specifications and then hands these off for the ultimate judgment of user selection in the marketplace.

OIF is an organization that specifies the requirements of ITU, etc. and standardizes the interface and protocol. Figure 10.4 shows the organizational chart for OIF. In 1999. There were three working groups (architecture, physical-link layer, and OAM&P) under the Technical Committee (TC). In February 2003, the number of working groups under the TC had grown to six. To illustrate the assigned roles of each WG, we will see how they interacted to arrive at the standards for OUNI 1.0 (Optical User Network Interface Version 1.0).

WG1 (Architecture): Described the function assignment between a data network and an optical network with an overlay model and created a reference model.

Standardization

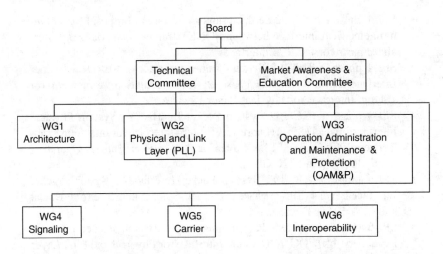

FIGURE 10.4 OIF organizational chart.

WG2 (Physical and Link Layer (PLL)): Adopted SDH/SONET (transmission rate 2.5 Gbps and 10 Gbps) as the layer-1 protocol, i.e., decided to adopt POS-IF as the physical-layer interface of OIF OUNI 1.0. Established the VSR (very short reach) interface to reduce costs for transmission interfaces faster than 10 Gbps.

WG3 (Operation, Administration and Maintenance & Protection, (OAM&P)): Made up the requirements for network protection and restoration, decided to have the layer-1 protocol be compliant with SDH/SONET, and denied SONET-Lite.

WG4 (Signaling): Established the OUNI signaling protocol.

WG5 (Carrier): Made up requirements from services.

WG6 (Interoperability): This WG did not exist at the time when OUNI 1.0. started. At present, it is executing planning and operation of interconnectivity demonstrations.

As described previously, standardization is processed in the form of an implementation agreement. Standards have been established yearly as follows:

Electrical Interfaces

- CEI-01.0 : Common Electrical I/O (CEI)- Electrical and Jitter Interoperability Agreements for 6G+bps and 11G+bps I/O
- CEI-02.0 : Common Electrical I/O (CEI) - Electrical and Jitter Interoperability agreements for 6G+ bps and 11G+ bps I/O
- CEI-P-01.0 : Common Electrical I/O - Protocol (CEI-P)- Implementation Agreement
- SPI-3 (OC-48 System Packet Interface) : Packet Interface for Physical and Link Layers for OC-48.

- SFI-4 phase 1 (OC-192 Serdes-Framer Interface) : Proposal for a common electrical interface between SONET framer and serializer/deserializer parts for OC-192 interfaces.
- SFI-4 phase 2 (OC-192 Serdes-Framer Interface) : SERDES Framer Interface Level 4 (SFI-4) Phase 2: Implementation Agreement for 10Gb/s Interface for Physical Layer Devices.
- SPI-4 phase 1 (OC-192 System Packet Interface) : System Physical Interface Level 4 (SPI-4) Phase 1: A System Interface for Interconnection Between Physical and Link Layer, or Peer-to-Peer Entities Operating at an OC-192 Rate (10 Gb/s).
- SPI-4 phase 2 (OC-192 System Packet Interface) : System Packet Interface Level 4 (SPI-4) Phase 2: OC-192 System Interface for Physical and Link Layer Devices.
- SPI-5 (OC-768 System Packet Interface) : System Packet Interface Level 5 (SPI-5) : OC-768 System Interface for Physical and Link Layer Devices.
- SFI-5 (40Gb/s Serdes Framer Interface) : Serdes Framer Interface Level 5 (SFI-5): 40Gb/s Interface for Physical Layer Devices.
- SxI-5 (System Interface Level 5) : System Interface Level 5 (SxI-5): Common Electrical Characteristics for 2.488 - 3.125Gbps Parallel Interfaces.
- TFI-5 (TDM Fabric to Framer Interface) : TDM Fabric to Framer Interface (TFI5)

Optical Transponder Interoperability

- LRI - (Long Reach Interop) : Interoperability for Long Reach and Extended Reach 10 Gb/s Transponders and Transceivers

Tunable Laser

- TL - Common Software Protocol, Control Syntax, and Physical (Electrical and Mechanical) Interfaces for Tunable Laser Modules.
- TLMSA - Multi-Source Agreement for CW Tunable Lasers.
- ITLA-MSA - Integratable Tunable Laser Assembly Multi-Source Agreement.

UNI - NNI

- UNI 1.0 Signaling Specification : User Network Interface (UNI) 1.0 Signaling Specification.
- UNI 1.0 Signaling Specification, Release 2
  -Common : - User Network Interface (UNI) 1.0 Signaling Specification, Release 2: Common Part
  -RSVP : - RSVP Extensions for User Network Interface (UNI) 1.0 Signaling, Release 2

Standardization

- CDR-01 : - Call Detail Records for OIF UNI 1.0 Billing.
- SEP-01.1 : - Security Extension for UNI and NNI
- SMI-01.0 : - Security Management Interfaces to Network Elements
- E-NNI-01.0 : - Intra-Carrier E-NNI Signaling Specification

Very Short Reach Interface

- VSR4-01 (OC-192 Very Short Reach Interface, 12 fiber 850nm): - Very Short Reach (VSR) OC-192 Interface for Parallel Optics.
- VSR4-02 (OC-192 Very Short Reach Interface, 1 fiber 1310nm): - Very Short Reach (VSR) OC-192 Interface for single fiber included as the 4dB link.
- VSR4-03.1 (OC-192 Very Short Reach Interface, 4 fiber 850nm): - Very Short Reach (VSR) OC-192 Four Fiber Interface Based on Parallel Optics.
- VSR4-04 (OC-192 Very Short Reach Interface, 1 fiber 850nm) : - Serial Shortwave Very Short Reach (VSR) OC-192 Interface for Multimode Fiber.
- VSR4-05 (OC-192 Very Short Reach Interface, OXC 1310nm) : - Very Short Reach (VSR) OC-192 Interface Using 1310 Wavelength and 4 and 11 dB Link Budgets.
- VSR5-01 (OC-768 Very Short Reach Interface) : - Very Short Reach Interface Level 5 (VSR-5): SONET/SDH OC-768 Interface for Very Short Reach (VSR) Applications.

As described later, as for the short-range optical parallel transmission method between IP routers, the standardizing procedure on the physical layer as a low-cost standard is under way [7] (see Figure 10.5).

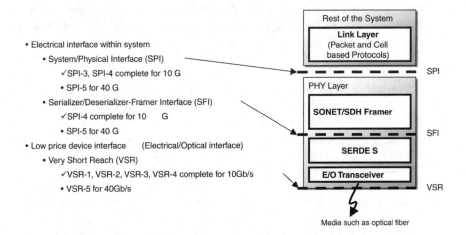

FIGURE 10.5 Standard of physical layer in OIF.

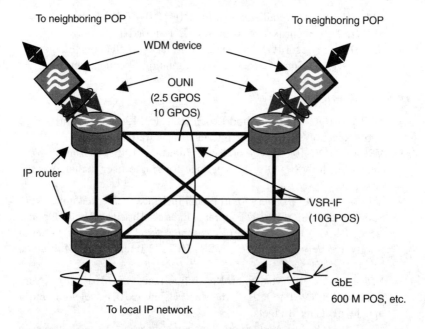

FIGURE 10.6 Reference model under discussion in OIF.

Figure 10.6 shows a network model that is under discussion in the OIF. This is an IP over a wavelength-division multiplexing (WDM) model in which an IP router and a WDM device are provided in each POP (point of presence) and the POPs are interconnected with point-to-point WDM transmission. This is an overlay model in which the IP device exists as a client of the WDM network, and the IP network exists on this WDM network. The section between an IP router and the optical WDM network is defined as an OUNI (optical user network interface), which is an important interface for controlling the optical network by OIF.

This model assumes that there is a WDM device as an edge device of the optical network. But it is also assumed that we cannot see inside the structure of the optical network, including whether or not there is an OXC (optical cross-connect), from the data-network side. In fact, if all of the traffic is processed by the IP router of the neighboring POP, there is no room for an OXC to exist within the optical network.

Because the IP routers in each POP are placed within the same building or on the same floor, it is enough that most interfaces (IF) are capable of transmitting within the range of a few hundred meters. Therefore, standardization of VSR-IF (very short reach interface) technology, which supports transmission ranges up to 300 m and greatly reduces the IF cost, was committed to OIF. Although VSR-IF was initially intended to be applied only to connections between IP routers, it was later assumed that it could also be applied to connections between the IP router and the WDM device as well as between the IP router and the OXC device

Standardization

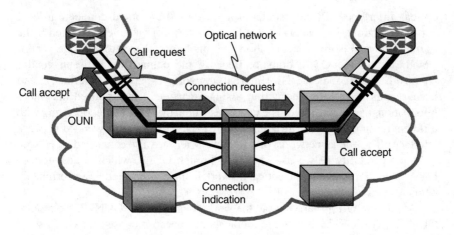

FIGURE 10.7 Control system in switched optical network.

that will be later described. Meanwhile, network structure has been developing from a simple point-to-point connection by a WDM terminal device to a dynamic WDM network such as optical cross-connect. Thus, the framework to set up a connection between IP routers through OUNI as the IP over a switched optical network is going to be intensively investigated.

Figure 10.7 shows the control system in a switched optical network. The optical network is seen here as a sort of black box through which it is possible to connect the IP router to other desired IP routers with OUNI.

The ONNI (optical network-to-network interface) model depicted in Figure 10.8 is a more complex model in that the optical network is constructed crossing over multiple carriers, and the optical devices consist of products from different vendors. This model is currently under discussion. The model in Figure 10.8 is an interface between carriers and within a carrier. Because the high-speed

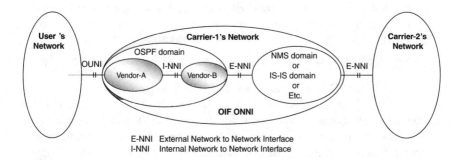

FIGURE 10.8 Interface allowing connections between carriers and vendors.

optical interface still has some problems in the optical-connection-level between vendors, it creates an islandlike network for each vendor. Even if the routing protocol is the same, as shown in this figure, there are multiple vendor islands within the OSPF domain. Because the connection between vendor islands is defined as I-NNI (internal network-to-network interface), and because this is an interface within a single carrier, it is supposed that more information may be disclosed. The interfaces within the same carrier operating with different routing protocols are connected by E-NNI (external network-to-network interface). The optical networks of different carriers are also connected by E-NNI.

Interoperability between 25 vendors through OUNI, which is an interface between the IP router and the optical network in the overlay model, was originally demonstrated at Supercom 2001.

At present, OIF is carrying out the work of establishing OUNI 2.0 to replace OUNI 1.0. Various functional requirements have been proposed for OUNI 2.0:

Nondisruptive connection modification: This would allow the network to change bandwidth or a call's service class without breaking the connection. In OUNI 1.0, these activities require the system to release the call and set it up again.

Multiple homing for diverse routing: This would make it possible to set up two connections that do not share the same node or link.

Transport of Ethernet interface: This would extend support for Fast Ethernet, Gigabit Ethernet, and 10-Gbit Ethernet, allowing the network to behave like a bridge or switch.

Mandatory discovery procedures: This would make neighbor-discovery and service-discovery functions mandatory. In OUNI 1.0, these functions are optional.

Enhanced security requirements: This would enhance security by transferring the encrypted protocol.

Separation of call and connection control: This would enhance the service function by separating call control from connection control.

Protection: This would support 1:N protection when an IP router and OXC are working together.

Sub-STS-1 rate interface: This would extend support to low-speed interfaces such as 1.5 Mbps (VC-11), 2 Mbps (VC-12), or 6 Mbps (VC-2). In OUNI 1.0, speeds of 2.5 Gbps (OC-48c) to 40 Gbps (OC-768c) have been officially supported.

IETF, OIF, and ITU share the mission of issuing implementation agreements, as shown in Figure 10.9. IETF determines the details of the protocol, and OIF creates an implementation agreement for the protocol that includes hardware. ITU is the only organization in the world that executes formal standardization of Internet technology. Figure 10.10 shows the relationship between standardization, protocol, and standardization organization.

Standardization

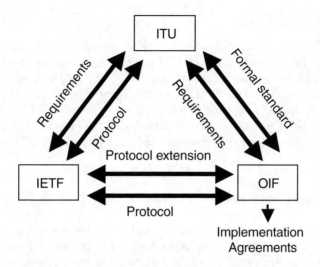

FIGURE 10.9 Relationship between IETF, OIF, and ITU-T.

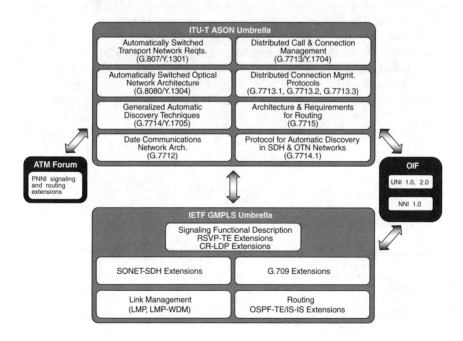

FIGURE 10.10 Relationship between standardization, protocol, and standardization organization.

10.4 ATM FORUM

The ATM Forum [8] was established in 1991 to promote industrial collaboration and to accelerate the introduction of ATM products and services by quickly establishing specifications to secure the interoperability of ATM. With more than 150 companies as members of the ATM Forum, this organization has expanded its scope of activity to include broadband technologies not limited to ATM in an effort to promote the prevalence of broadband technologies that include ATM while continuing to provide multiservice solutions based on comprehensive ATM technology.

10.4.1 COMMITTEES

The ATM Forum comprises three committees.

> *Technical Committee*: This committee meets four times per year to carry out its mission of creating interoperable ATM specifications. Figure 10.11 shows the organizational structure of the ATM Forum, focusing on the details of the working groups in the Technical Committee. Specifications approved by the board of directors are released on the ATM Forum's Web site (http://www.atmforum.com/standards/approved.html). The Technical Committee collaborates with other international standardization organizations, other forums, and entities in the communications industry to provide internationally accepted interoperable ATM specifications. The committee held its first joint meeting with the Business Content Delivery (BCD) Forum (http://www.bcdforum.org) and the MPLS Forum (http://www.mplsforum.org) at its January 2003 meeting in Santa Clara, California.
>
> *Market Awareness Committee*: This committee is concerned with advertising through mass media and organizing international conferences. Its mission

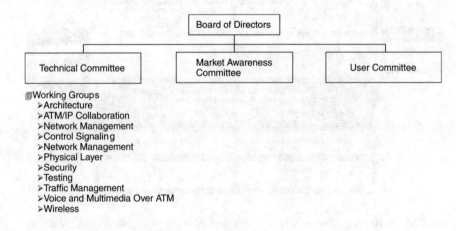

FIGURE 10.11 ATM forum organizational chart.

is to promote the advantages of ATM technology and to pioneer new markets. The committee's major activities include the Broadband Exchange Program and the ATM roadmap.

- Broadband Exchange Program: Besides improving the QoS (quality of service) functions of ATM, the program also sponsors a one-day seminar on the day before the Technical Committee meets, with attendees discussing the future of ATM as an infrastructure of broadband communication systems. In this seminar, one theme is picked from the six key areas (converged network services, next-generation networks, optical networking, 3G/4G networks, homeland security and public-safety networks, and content-delivery networks) for discussion. (See http://www.atmforum.com/meetings/broadband.html.)
- ATM roadmap: The Market Awareness Committee creates a "roadmap for the future" to highlight the subjects and direction of future development of ATM technology so that ATM will share a complementary role with other related technologies and promote development of the broadband society. Figure 10.12 shows an outline of the ATM Forum's roadmap. (See http://www.atmforum.com/news/roadmap_media_summary.html.)

User Committee: This committee, composed of ATM end users, was established in 1993. As the name implies, its mission is to understand the user's demands and expectations for ATM technologies and services. The committee is independent of ATM system vendors, so it serves as a filter to prioritize and direct the creation of ATM specifications.

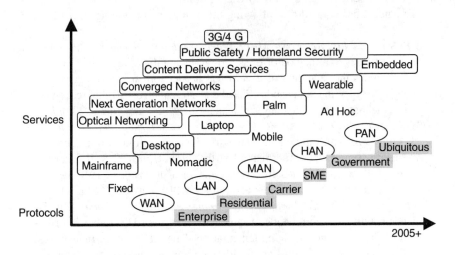

FIGURE 10.12 Outline of ATM Forum's road map.

10.4.2 Future Activity

The fundamental specifications of ATM technologies were fixed by the Anchorage agreement in 1996 Anchorage meeting. Current specifications are under the purview of such working groups as the ATM&IP Collaboration Working Group, the Control Signaling Working Group, and the Voice and Multimedia over ATM Working Group. (For the state of each work group's activity on standardization, see http://www.atmforum.com/ standards/specwatch.html.)

The ATM Forum is focused on improving QoS functions utilizing the characteristic features of ATM, but it also has an eye on future needs. The characteristic features of ATM include:

- Ability to accommodate voice traffic (efficiency is twice that of voice over IP [VoIP])
- Ability to transmit real-time video with high quality
- Compatibility with mobile environment
- Superior security
- High reliability and a recovery function that can transmit mission-critical traffic
- Consolidation of multiple traffic with different qualities of service

10.5 MPLS FORUM

The MPLS Forum [9] was established in 2000 with the aim of developing application technologies, improving interoperability for MPLS protocols from multiple vendors, improving and extending MPLS protocols, creating implementation agreements (IA), and developing education programs. As of April 2005, there were 45 member companies in the MPLS Forum.

To date, it has created the following three implementation agreements (IA).

MPLS Forum 1.0: Voice over MPLS Implementation Agreement: This agreement relates to a method of transferring voice traffic directly with MPLS. It includes the definition of a VoMPLS header format supporting various payload types, including audio, dialed digits (Digital Tone Multi Frequency), channel-associated signaling, and a silence-insertion descriptor (2001.7).

MPLS Forum 2.0: MPLS-PVC User-to-Network Interface Implementation Agreement: This agreement relates to a UNI (user-to-network interface) specification of PVC (permanent virtual circuit) in MPLS. It describes the UNI protocols used for connecting to a provisioned PVC service between a user's network device and an MPLS network. (2002.9)

MPLS Forum 3.0: LDP Conformance Implementation Agreement: This IA deals with the recommended test coverage for the LDP (Label Distribution Protocol) protocol implementation. The LDP Conformance Test Coverage document aims at providing comprehensive test coverage scenarios based on various MPLS LDP RFCs/Internet drafts. (2002.12)

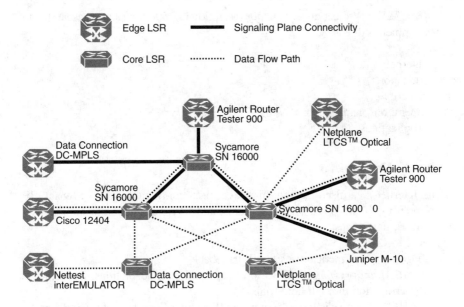

FIGURE 10.13 GMPLS demo network technology.

In October 2000, the MPLS Forum carried out a massive demonstration of interoperability in a NGN (next generation network) consisting of LSRs from several different vendors using the topology shown in Figure 10.13. In addition, in February 2003, the MPLS Forum held its MPLS World Congress in Paris, where it performed another interoperability demonstration. The hot topics of this event were virtual private networks (VPNs) and fast rerouting. Both of these are reliable, secure, and novel network technologies. Discussions on such subjects as OAM, VPN, and traffic engineering have been focused on their highly reliable technologies.

10.6 WIDE PROJECT

The WIDE Project [10], launched in 1988, was originally conceived as an effort to establish a widely integrated distributed environment (WIDE): a new, broad-scale computer environment based on operating systems and communications technology. The goal was to connect computers throughout the world in the hope of creating a distributed system designed to serve mankind on a massive scale. After 2002, the WIDE Project had arrived at the stage where IP v.6, the newest Internet protocol, is now making automobiles, household appliances, and a myriad of other products accessible via the Internet. The initial vision of a large-scale, wide-area distributed computing environment is becoming a reality in the form of a ubiquitous network that is accessible anywhere, at any time.

The WIDE Project covers mainly the following five topics as a research and development project:

Internet
Transport
Security
Operations/management
Applications

10.6.1 Internet Area

The Internet area discusses the third-layer protocol, including IP traceback, IP v.6 mobility, IP v.6/IPsec stack, programs, and IP v.6 evaluation technology.
Working Groups:

IP traceback: Practical uses for IP traceback
IP v.6 Mobility (v6mob): IP v.6 protocols and their adaptations, including verification through field trials
KAME Project: Development of a free IP v.6 and IPsec stack for (Berkeley Software Distribution) variants
LAbel SwiTching (LAST): Experimental facility named "Internet Building" using label switching
TAHI Project: Research and development of IP v.6 evaluation technology

10.6.2 Transport Area

The transport area discusses the fourth and upper-layer protocol and the second layer. The topics include DNS protocol as well as capability and operational issues on the Internet with (Digital Video Broadcasting Return Channel via Satellite)
Working Groups:

DNS: Various topics on DNS, including protocol, implementation, operation, measurement, analysis, etc.
DVB-RCS (RCS): Capability and operational issues on the Internet with DVB-RCS, which provides two-way satellite communication with a small earth station
Integrated distributed environment with overlay network (IDEON): Research and development of overlay network with infrastructure for P2 application
TACA: Technologies for home networks

10.6.3 Security Area

This area discusses security on the Internet. The topics include general IPsec and security of IP v.6.

Working Groups:

IPsec: General issues regarding IPsec
moCA: Members-oriented certification authority
Cross-site scripting (xss): Various solutions for cross-site scripting vulnerability
Security of IP v.6: Security issues of new segments and deployment areas on IP v.6

10.6.4 OPERATIONS/MANAGEMENT AREA

This area discusses Internet and server management/operation technology. The topics include Telephone Number Mapping (ENUM) operation trials and discussion of the ENUM server/client.
Working Groups:

ENUM: Operational trials for ENUM and discussion of ENUM server/client
IRC: Development of IRC and operation of WIDE IRC servers
MAWI: Measurement and analysis of traffic

10.6.5 APPLICATIONS AREA

This area discusses deployment, testing, and adaptation of Internet applications. The topics include development of a light-speed and lightweight mail reader, development of (I Am Alive) system, and development of an Internet environment embracing in-car connectivity.
Working Groups:

Cue: Development of light-speed and lightweight mail reader
IAA development (IAA-DEV): Discussion and development of IAA system to enhance the robustness and capability of the current version of IAA System
InternetCAR (iCAR): Development of an Internet environment embracing in-car connectivity

10.7 PHOTONIC INTERNET LABORATORY

Since 2002, the Photonic Internet Laboratory (PIL) [11] has been delineating the world-class photonic-GMPLS (Generalized Multiprotocol Label Switching), which utilizes broadband, cost-effective photonic technology to implement IP-centric managed networks. PIL is a new consortium for researching the GMPLS protocol and achieving a de facto standard in this area. It is supported by the Ministry of Public Management, Home Affairs, Posts and Telecommunications (MPHPT) in Japan. Its members are creating leading-edge GMPLS code modules and testing them at the shared lab site. The experimental results, new ideas, and protocols are being forwarded to standardization bodies such as IETF and OIF. This section describes the results of the world's first MPLS/GMPLS multiregion (multilayer), multiroute, multivendor interoperability test.

10.7.1 PIL Organization

PIL was founded in September 2002 to promote research on and the development of the next-generation photonic network and to encourage global standardization activities. PIL currently consists of eight companies and a university: Nippon Telegraph and Telephone Corp. (NTT), NEC Corp., Fujitsu Laboratories, Ltd., Furukawa Electric Co., Ltd., Mitsubishi Electric Corp., Oki Electric Industry Co., Ltd., Hitachi, Ltd. PIL activities are supported by research and development aimed at establishing international technical standards as part of the Strategic Information and Communications R&D Promotion Scheme of the MPHPT, which is funding selected IT activities.

There are two working groups (WGs). The technical test WG assesses leading-edge protocol code modules developed by member companies, and the standardization strategy WG is responsible for technical discussions on standardization proposals. It has discussed and submitted 42 standardization proposals to IETF and OIF as of April 2005. All of the proposed contributions are posted on PIL's Web site (www.pilab.org).

The framework of the standardization strategy WG is shown in Figure 10.14. As shown, each member company researches and develops new ideas, protocols, and running code modules. The topics tackled by PIL are organized by the steering committee. Each proposal is discussed and tested by PIL.

PIL member strategies and topics are shown in Figure 10.15. Each member has a different direction or strategies for standardization. PIL promotes the core part of the protocol and extensions. In addition, the next step in standards

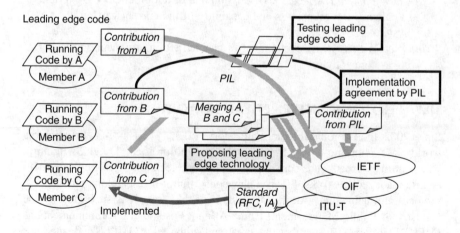

FIGURE 10.14 PIL's standardization approach.

Standardization

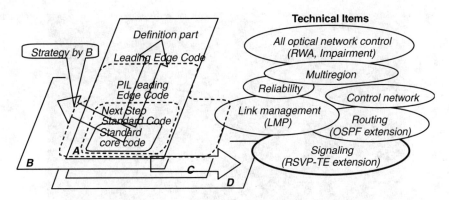

FIGURE 10.15 Leading-edge code for next-generation photonic network.

submission and in leading-edge code developments are key activities of PIL. Technical items are also listed in Figure 10.15. PIL has already tested the signaling and routing parts of GMPLS. We intend to cover the multilayer (region) traffic-control framework, reliability (protection and restoration), control-network issues, and MPLS/GMPLS interworking.

10.7.2 MPLS-GMPLS, Multilayer, Multiroute Interworking Tests

PIL members, NTT, NEC Corp., Fujitsu Laboratories, Ltd., Furukawa Electric Co., Ltd., and Mitsubishi Electric Corp. successfully concluded the world's first MPLS-GMPLS interoperability tests using multilayer, multiroute GMPLS signaling/routing protocols at the MPLS2003 meeting held in Washington, DC, in October 2003. A photograph of this test is shown in Figure 10.16. MPLS2003 (www.mpls2003.com) Sponsored by ISOCORE (www.ISOCORE.com) had more than 600 attendees from major service providers, carriers, and vendors. In addition, key IETF members joined in the discussions held at the PIL booth.

Beginning with the MPLS2003 demonstration, Hitachi, Ltd., joined the team in interworking the MPLS network with the GMPLS networks. These results were demonstrated in January 2004 at the Gigabit Network Symposium 2004 in Tokyo, Japan. The trial network setup is shown in Figure 10.17 and Figure 10.18. It represents a large-scale backbone network connecting two metro networks. Metro networks 1 and 2 use MPLS technology and are constructed around MPLS routers produced by Fujitsu Ltd. and Furukawa (GeoStream R920 and FITELnet-G21, respectively). The backbone network use the GMPLS technology as implemented in four kinds of equipment:

HIKARI routers by NTT (IP and wavelength)
GMPLS routers (FITELnet-G80) by Furukawa (IP and TDM)

FIGURE 10.16 PIL booth at MPLS 2003.

TDM cross-connects (SpectralWave U-Node) by NEC (TDM and wavelength)

Optical cross-connects by Mitsubishi Electric, Hitachi, Fujitsu Laboratories, and NTT (wavelength and optical fiber)

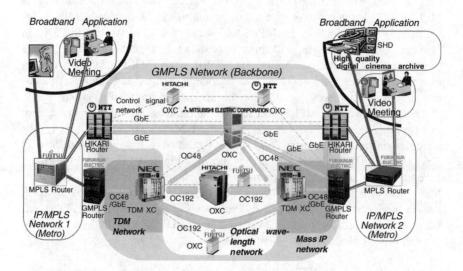

FIGURE 10.17 Trial network setup using GMPLS and MPLS.

FIGURE 10.18 Booth of JGN 2004.

The expressions in parentheses at the end of each item indicate the function and the layer handled by each company's equipment. GMPLS control software that can handle at least two layers was developed and implemented by each company. The various pieces of GMPLS equipment are connected using control links (to exchange control messages holding routing information) and data links with rates exceeding 1 gigabit (Gigabit Ether, OC-48, and OC-192). Because a wide variety of network equipment (e.g., IP routers and cross-connects) will coexist in a GMPLS network, it is essential to verify interoperability between multivendor network equipment.

We have successfully realized the distribution and collection of routing information and the setup of paths in the multilayer and multivendor environment. It is possible to use the various communication paths provided (packet, TDM, wavelength, and fiber) according to the communication quality demanded.

In addition, this GMPLS backbone network supports two functions: seamless connection to Metro networks and high reliability. The former is realized by MPLS and GMPLS interworking technology. This technology was verified in a multivendor environment in October 2003, a world first. This experiment successfully transmitted high-resolution video, such as digital cinema, across MPLS and GMPLS networks. It confirms the feasibility of establishing an economical and highly reliable IP network that offers high-resolution video transmission at reasonable cost.

REFERENCES

1. ITU Association of Japan; available on-line at http://www.ituaj/jp/.
2. International Telecommunication Union; available on-line at http://itu.int/ITU-T/.
3. Internet Engineering Task Force; available on-line at http://www.ietf.org.
4. Esaki, H., What is IETF; available on-line at http://rfc-jp.nic.ad.jp/what_ is_ietf/ietf_abstract.html.

5. Optical Internetworking Forum; available on-line at http://www.oiforum.com.
6. Okamoto, S., Image of Optical Cross-Connect System Viewing from the Trend of Standardization in OIF, IECIE technical report PS2002-3, IECIE.
7. Afferton, T., OIF Overview, Federal Communications Commission, Washington, DC, 2002; available on-linea http://www.fcc.gov/oet/tac/Tom_Afferton_9.18.02_OIF_forFCC91802_Final.ppt.
8. ATM Forum; available on-line at http://www.atmforum.com.
9. MPLS Forum; available on-line at http://www.mplsforum.org.
10. WIDE Project; available on-line at http://www.wide.ad.jp.
11. PIL; available on-line at http://www.pilab.org.

APPENDIX 10.1 ITU TOPICS

Question	Title
ITU-T Study Group 13 Structure (Study Period 2001–2004)	
WP 1/13: Project Management and Coordination	
12/13	Global Coordination of Network Aspects
15/13	General Network Terminology, Including IP Aspects
WP 2/13: Architectures and Internetworking Principles	
1/13	Principles, Requirements, Frameworks, and Architectures for an Overall Heterogeneous Network Environment
5/13	Network Interworking Including IP Multiservice Networks
10/13	Core Network Architecture and Interworking Principles
13/13	Interoperability of Satellite and Terrestrial Networks
14/13	Access Architecture Principles and Features at the Lower Layers for IP-Based and Other Systems
16/13	Telecommunication Architecture for an Evolving Environment
WP 3/13: Multiprotocol Networks and Mechanisms	
2/13	ATM Layer and Its Adaptation
3/13	OAM and Network Management in IP-Based and Other Networks
11/13	Mechanisms To Allow IP-Based Services Using MPLS To Operate in Public Networks
WP 4/13: Network Performance and Resource Management	
4/13	Broadband- and IP-Related Resource Management
6/13	Performance of IP-Based Networks and the Emerging Global Information Infrastructure
7/13	B-ISDN/ATM Cell Transfer and Availability Performance
8/13	Transmission Error and Availability Performance
9/13	Call-Processing Performance
ITU-T Study Group 15 Structure (Study Period 2001–2004)	
WP 1/15: Network Access	
2/15	Optical Systems for Access Networks
3/15	Support for Recommendations Specifying Systems Based on ISDN Physical Layers
4/15	Transceivers for Customer Access and In-Premises Phone-Line Networking Systems on Metallic Pairs

Standardization

WP 2/15: Network Signal Processing
5/15   Compression and Classification in Signal-Processing Network Equipment
6/15   Speech Enhancement in Signal-Processing Network Equipment
7/15   Voice-Gateway Equipment
8/15   Interaction Aspects of Signal-Processing Network Equipment

WP 3/15: OTN Structure
9/15   Transport Equipment and Network Protection/Restoration
10/15   ATM and Internet Protocol (IP) Equipment
11/15   Signal Structures, Interfaces, and Interworking for Transport Networks
12/15   Technology-Specific Transport-Network Architectures
13/15   Network Synchronization and Time-Distribution Performance
14/15   Network Management for Transport Systems and Equipment

WP 4/15: OTN Technology
15/15   Characteristics and Test Methods of Optical Fibers and Cables
16/15   Characteristics of Optical Systems for Terrestrial Transport Networks
17/15   Characteristics of Optical Components and Subsystems
18/15   Characteristics of Optical Fiber Submarine Cable Systems

WP 5/15: Projects and Promotion
1/15   Access Network Transport
19/15   General Characteristics of Optical Transport Networks

APPENDIX 10.2   IETF WORKING GROUPS

A2.1   Applications Area

www.ietf.org/html.charters/**acap**-charter.html	Application Configuration Access Protocol
www.ietf.org/html.charters/**apex**-charter.html	Application Exchange
www.ietf.org/html.charters/**calsch**-charter.html	Calendaring and Scheduling
www.ietf.org/html.charters/**cdi**-charter.html	Content Distribution Internetworking
www.ietf.org/html.charters/**crisp**-charter.html	Cross Registry Information Service Protocol
www.ietf.org/html.charters/**ediint**-charter.html	Electronic Data Interchange-Internet Integration
www.ietf.org/html.charters/**fax**-charter.html	Internet Fax
www.ietf.org/html.charters/**ftpext**-charter.html	Extensions to FTP
www.ietf.org/html.charters/**geopriv**-charter.html	Geographic Location/Privacy
www.ietf.org/html.charters/**imapext**-charter.html	Internet Message Access Protocol Extension
www.ietf.org/html.charters/**impp**-charter.html	Instant Messaging and Presence Protocol
www.ietf.org/html.charters/**ipp**-charter.html	Internet Printing Protocol
www.ietf.org/html.charters/**ldapbis**-charter.html	LDAP (v3) Revision

www.ietf.org/html.charters/**ldapext**-charter.html	LDAP Extension
www.ietf.org/html.charters/**ldup**-charter.html	LDAP Duplication/Replication/Update Protocols
www.ietf.org/html.charters/**msgtrk**-charter.html	Message Tracking Protocol
www.ietf.org/html.charters/**nntpext**-charter.html	NNTP Extensions
www.ietf.opes/html.charters/**opes**-charter.html	Open Pluggable Edge Services
www.ietf.org/html.charters/**prim**-charter.html	Presence and Instant Messaging Protocol
www.ietf.org/html.charters/**provreg**-charter.html	Provisioning Registry Protocol
www.ietf.org/html.charters/**rescap**-charter.html	Resource Capabilities Discovery
www.ietf.org/html.charters/**simple**-charter.html	SIP for Instant Messaging and Presence Leveraging Extensions
www.ietf.org/html.charters/**tn3270e**-charter.html	Telnet TN3270 Enhancements
www.ietf.org/html.charters/**trade**-charter.html	Internet Open Trading Protocol
www.ietf.org/html.charters/**usefor**-charter.html	Usenet Article Standard Update
www.ietf.org/html.charters/**vpim**-charter.html	Voice Profile for Internet Mail
www.ietf.org/html.charters/**webdav**-charter.html	WWW Distributed Authoring and Versioning
www.ietf.org/html.charters/**xmpp**-charter.html	Extensible Messaging and Presence Protocol

A2.2 General Area

www.ietf.org/html.charters/**ipr**-charter.html	Intellectual Property Rights
www.ietf.org/html.charters/**nomcom**-charter.html	Operation of the IESG/IAB Nominating and Recall Committees

A2.3 Internet Area

www.ietf.org/html.charters/**atommib**-charter.html	AToM MIB
www.ietf.org/html.charters/**dhc**-charter.html	Dynamic Host Configuration
www.ietf.org/html.charters/**dnsext**-charter.html	DNS Extensions
www.ietf.org/html.charters/**eap**-charter.html	Extensible Authentication Protocol
www.ietf.org/html.charters/**idn**-charter.html	Internationalized Domain Name
www.ietf.org/html.charters/**ifmib**-charter.html	Interfaces MIB
www.ietf.org/html.charters/**ipcdn**-charter.html	IP over Cable Data Network
www.ietf.org/html.charters/**ipoib**-charter.html	IP over InfiniBand
www.ietf.org/html.charters/**iporpr**-charter.html	IP over Resilient Packet Rings

Standardization

www.ietf.org/html.charters/**ipv6**-charter.html	IP Version 6 Working Group
www.ietf.org/html.charters/**itrace**-charter.html	ICMP Traceback
www.ietf.org/html.charters/**l2tpext**-charter.html	Layer Two Tunneling Protocol Extensions
www.ietf.org/html.charters/**magma**-charter.html	Multicast & Anycast Group Membership
www.ietf.org/html.charters/**mobileip**-charter.html	IP Routing for Wireless/Mobile Hosts
www.ietf.org/html.charters/**nemo**-charter.html	Network Mobility
www.ietf.org/html.charters/**pana**-charter.html	Protocol for Carrying Authentication for Network Access
www.ietf.org/html.charters/**pppext**-charter.html	Point-to-Point Protocol Extensions
www.ietf.org/html.charters/**send**-charter.html	Securing Neighbor Discovery
www.ietf.org/html.charters/**zeroconf**-charter.html	Zero Configuration Networking

A2.4 Operations and Management Area

www.ietf.org/html.charters/**aaa**-charter.html	Authentication, Authorization, and Accounting
www.ietf.org/html.charters/**adslmib**-charter.html	ADSLMIB
www.ietf.org/html.charters/**bmwg**-charter.html	Benchmarking Methodology
www.ietf.org/html.charters/**bridge**-charter.html	Bridge MIB
www.ietf.org/html.charters/**disman**-charter.html	Distributed Management
www.ietf.org/html.charters/**dnsop**-charter.html	Domain Name Server Operations
www.ietf.org/html.charters/**entmib**-charter.html	Entity MIB
www.ietf.org/html.charters/**eos**-charter.html	Evolution of SNMP
www.ietf.org/html.charters/**hubmib**-charter.html	Ethernet Interfaces and Hub MIB
www.ietf.org/html.charters/**ipfix**-charter.html	IP Flow Information Export
www.ietf.org/html.charters/**mboned**-charter.html	MBONE Deployment
www.ietf.org/html.charters/**multi6**-charter.html	Site Multihoming in IP v.6

www.ietf.org/html.charters/**nasreq**-charter.html	Network Access Server Requirements
www.ietf.org/html.charters/**policy**-charter.html	Policy Framework
www.ietf.org/html.charters/**psamp**-charter.html	Packet Sampling
www.ietf.org/html.charters/**ptomaine**-charter.html	Prefix Taxonomy Ongoing Measurement & Inter Network Experiment
www.ietf.org/html.charters/**rap**-charter.html	Resource Allocation Protocol
www.ietf.org/html.charters/**rmonmib**-charter.html	Remote Network Monitoring
www.ietf.org/html.charters/**sming**-charter.html	Next Generation Structure of Management Information
www.ietf.org/html.charters/**snmpconf**-charter.html	Configuration Management with SNMP
www.ietf.org/html.charters/**snmpv3**-charter.html	SNMP Version 3
www.ietf.org/html.charters/**v6ops**-charter.html	IP v.6 Operations

A2.5 Routing Area

www.ietf.org/html.charters/**bgmp**-charter.html	Border Gateway Multicast Protocol
www.ietf.org/html.charters/**forces**-charter.html	Forwarding and Control Element Separation
www.ietf.org/html.charters/**idmr**-charter.html	Inter-Domain Multicast Routing
www.ietf.org/html.charters/**idr**-charter.html	Inter-Domain Routing
www.ietf.org/html.charters/**isis**-charter.html	Intermediate System to Intermediate System for IP Internets
www.ietf.org/html.charters/**manet**-charter.html	Mobile Ad-hoc Networks
www.ietf.org/html.charters/**msdp**-charter.html	Multicast Source Discovery Protocol
www.ietf.org/html.charters/**ospf**-charter.html	Open Shortest Path First IGP
www.ietf.org/html.charters/**pim**-charter.html	Protocol Independent Multicast
www.ietf.org/html.charters/**rip**-charter.html	Routing Information Protocol
www.ietf.org/html.charters/**rpsec**-charter.html	Routing Protocol Security Requirements
www.ietf.org/html.charters/**ssm**-charter.html	Source-Specific Multicast
www.ietf.org/html.charters/**udlr**-charter.html	UniDirectional Link Routing
www.ietf.org/html.charters/**vrrp**-charter.html	Virtual Router Redundancy Protocol

Standardization

A2.6 SECURITY AREA

www.ietf.org/html.charters/**idwg**-charter.html	Intrusion Detection Exchange Format
www.ietf.org/html.charters/**inch**-charter.html	Extended Incident Handling
www.ietf.org/html.charters/**ipsec**-charter.html	IP Security Protocol
www.ietf.org/html.charters/**ipsp**-charter.html	IP Security Policy
www.ietf.org/html.charters/**ipsra**-charter.html	IP Security Remote Access
www.ietf.org/html.charters/**kink**-charter.html	Kerberized Internet Negotiation of Keys
www.ietf.org/html.charters/**krb**-wg-charter.html	Kerberos WG
www.ietf.org/html.charters/**msec**-charter.html	Multicast Security
www.ietf.org/html.charters/**openpgp**-charter.html	An Open Specification for Pretty Good Privacy
www.ietf.org/html.charters/**pkix**-charter.html	Public-Key Infrastructure (X.509)
www.ietf.org/html.charters/**sacred**-charter.html	Securely Available Credentials
www.ietf.org/html.charters/**sasl**-charter.html	Simple Authentication and Security Layer
www.ietf.org/html.charters/**secsh**-charter.html	Secure Shell
www.ietf.org/html.charters/**smime**-charter.html	S/MIME Mail Security
www.ietf.org/html.charters/**stime**-charter.html	Secure Network Time Protocol
www.ietf.org/html.charters/**syslog**-charter.html	Security Issues in Network Event Logging
www.ietf.org/html.charters/**tls**-charter.html	Transport Layer Security
www.ietf.org/html.charters/**xmldsig**-charter.html	XML Digital Signatures

A2.7 SUB-IP AREA

www.ietf.org/html.charters/**ccamp**-charter.html	Common Control and Measurement Plane
www.ietf.org/html.charters/**gsmp**-charter.html	General Switch Management Protocol
www.ietf.org/html.charters/**ipo**-charter.html	IP over Optical
www.ietf.org/html.charters/**mpls**-charter.html	Multiprotocol Label Switching
www.ietf.org/html.charters/**ppvpn**-charter.html	Provider Provisioned Virtual Private Networks
www.ietf.org/html.charters/**tewg**-charter.html	Internet Traffic Engineering

A2.8 Transport Area

www.ietf.org/html.charters/**avt**-charter.html	Audio/Video Transport
www.ietf.org/html.charters/**dccp**-charter.html	Datagram Congestion Control Protocol
www.ietf.org/html.charters/**diffserv**-charter.html	Differentiated Services
www.ietf.org/html.charters/**enum**-charter.html	Telephone Number Mapping
www.ietf.org/html.charters/**ieprep**-charter.html	Internet Emergency Preparedness
www.ietf.org/html.charters/**ippm**-charter.html	IP Performance Metrics
www.ietf.org/html.charters/**ips**-charter.html	IP Storage
www.ietf.org/html.charters/**iptel**-charter.html	IP Telephony
www.ietf.org/html.charters/**malloc**-charter.html	Multicast-Address Allocation
www.ietf.org/html.charters/**megaco**-charter.html	Media Gateway Control
www.ietf.org/html.charters/**midcom**-charter.html	Middlebox Communication
www.ietf.org/html.charters/**mmusic**-charter.html	Multiparty Multimedia Session Control
www.ietf.org/html.charters/**nfsv4**-charter.html	Network File System Version 4
www.ietf.org/html.charters/**nsis**-charter.html	Next Steps in Signaling
www.ietf.org/html.charters/**pilc**-charter.html	Performance Implications of Link Characteristics
www.ietf.org/html.charters/**pwe3**-charter.html	Pseudo Wire Emulation Edge to Edge
www.ietf.org/html.charters/**rddp**-charter.html	Remote Direct Data Placement
www.ietf.org/html.charters/**rmt**-charter.html	Reliable Multicast Transport
www.ietf.org/html.charters/**rohc**-charter.html	Robust Header Compression
www.ietf.org/html.charters/**rserpool**-charter.html	Reliable Server Pooling
www.ietf.org/html.charters/**seamoby**-charter.html	Context Transfer, Handoff Candidate Discovery, and Dormant Mode Host Alerting
www.ietf.org/html.charters/**sigtran**-charter.html	Signaling Transport
www.ietf.org/html.charters/**sip**-charter.html	Session Initiation Protocol
www.ietf.org/html.charters/**sipping**-charter.html	Session Initiation Proposal Investigation

Standardization

www.ietf.org/html.charters/**speechsc**-charter.html

www.ietf.org/html.charters/**spirits**-charter.html

www.ietf.org/html.charters/**tsvwg**-charter.html

Speech Services Control

Service in the PSTN/IN Requesting InTernet Service

Transport Area Working Group

Index

A

AAL 11,33,38
AAL type 1 39
AAL type 2 39,41
AAL type 3/4 43
AAL type 5 44
AAL1 34
AAL2 34
AAL3/4 34
AAL5 34
ABR 59,89
ABT 60
ABT-DT 60
ABT-IT 60
Acknowledgment packet 25
Aging 91
AIS 63
Alarm-transmission 63
ALL scheme 258
AND scheme 257
Architectural signaling 212
ARP 20
ARPANet 305
ATM communication system 36
ATM Forum 316
ATM transmission system 33
ATM-SVC 11
Autonomous system (AS) 74

B

Backbone 3
Backup methods 276
Backup-SRG 280
BECN 56
Bellman-Ford algorithm 76
Best-effort type 202
BGP 75
BGP speaker 98
BGP-4 95,148
BGP-4 router 147
Bidirectional path signaling 209
Binary search method 186
Binary tree 109
B-ISUP 11,13

Broadcasting address 71
Bundling 213
Burstlike nature 2
BXCQ 234

C

CAM 183,188
CBR 57
CIDR 72,109
Circular type 85
Class A 18
Class B 18
Class C 18
Class D 18
Client-server model 221
Cloud-route (CR) 244
CLP 37
Common-buffer-type-switch 165
Communication by TCP/IP 22
Complete priority scheduling 177,178
Component link 214
Confederation 107
Connection admission control (CAC) 47,48,56
Connection-accepting control 7
Connectionless communication 5,15
Connection-oriented 33
Connection-oriented communication 5,11,112
Connectivity certification 214
connectivity-verification 217
Conservative 124
Constraint-Based Routing Label Distribution
    Protocol (CR-LDP) 294
Constraint-based shortest psth first (CSPF)
    135,232
Continuity check 65
Control function 159
Control plane (C plane) 11,33,220
Control-channel management 214
Core router 149
Counting to Infinity 78
CPCS 43
Credit window algorithm 51
CR-LDP 135
CRRD 176
Cut through 7,8,244

D

DARPA 305
Data-path function 159
De facto standard 308
Deficit counter 182
Destination address 128
Destination IP address 20
Dijkstra calculation 249
Dijkstra method 84,86
Disjoint path 264
Distance-vector type 75
Distributed control mechanism 228
DLCI 116
don't fragment 70
Downstream path 209
Downstream type 122
DR (Designated Router) 89
DRR (Dedict round-robin) scheduling 165,177
DUAL 79
Dual-cost metrics 79
DXC 192

E

EBGP 96,147
Edge router 4,151
EGP 74
Egress router 8
E-NNI 314
EPD 54
ERO 135
Ethernet cable 4
Experimental RFC 308
Explicit route 130
Explict-router-type LSP 125
Extra traffic type 202

F

Failure management 214,219
Fairness 159,164
FEC 113
Fiber layer 191
Fiber-switching capability (FSC) 291
FIFO input buffer type switch 165,167
FIFO queuing 177
Finish time 180
Fletcher's algorithm 91
Flow 113
Forwarding 159
Forwarding adjacency 232

Forwarding function 161
Forwarding table 71
Fragment offset 70
Frame-relay 35
FSC (fiber-switch capable) 194,232
FTN 117

G

Generalized processor sharing 177,178
GFC 36
GMPLS 193,211
G-PID 207
GPS 178

H

Header checksum 69
Heuristic algorithm 229
Hierarchical clous-router network (HCRN) 223,244
Hierarchical label-switched path (LSP) 243
Hierarchical signaling 212
Hierarchization 212,220
High-end class 159
High-end router 164
HIKARI router 242,290
Historical RFC 308
HOL 168
Hold down 79
Hop count 124
Hop-by-hop-type LSP 125
Host address 19
Host block 71

I

IBGP 96,149
Identify the link 200
IETF 221,305
IGP 74,141
IGRP 75
ILM 117
Independent type 123
Informational RFC 308
Ingress router 8
Initiator 210
I-NNI 314
Interleave 121
Internet 16
Internet Area 306

Index

Internet Engineering Task Force (IETF) 241
Internet protocol 69
Internet-Draft 307
Interworking test 323
IP 69
IP address 16,200
IP datagram communication 112,128
IP forwarding 69
IP header 69
IP network address 19
IP routing 74
IP routing table 159
IP transmission table 109
IPv4 18
IPv6 18,319
IS-IS 75,127
iSLIP 171
ISOC 306
ISOCORE 323
Iteration 171
ITU (International Telecommunication Union) 303
ITU-T 303

L

L2SC (layer-2-switch capable) 232
Label 6,111
Label merge 121
Label pop 117
Label push 117
Label request 206
Label setting 210
Label stack 118
Label swap 6
LAN 159
Leaky bucket method 50
LER (Label-edge router) 113,130
Liberal 124
Limited-wavelength-convertible 253
Linear type 85
Link Bundling 201,214
Link ID 201
Link local identifier 201
Link management protocol 198,213
Link protection 282,287
Link remote identifier 201
Link state 88
Link TLV 201
Link-property correlation 214,216
Link-state 225,229
Link-state advertisement (LSA) 227

Link-state type 75,83
LMP 213
LMP neighbors 215
Lollipop type 85
Longest match 186
Longest-prefix matching 109
Longest-prefix matching search 73
Loop-back function 65
Loose bit 137
Loose specification 131
Low-end-class 160
Lowest-cost route 86
LS Sequence-number field 91
LSA 91,92
LS-age field 91
LSA-update packet 91
LSC (lambda-switch capable) 194,232
LS-checksum field 91
LSP (Label-switched path) 113,130
LSP encoding type 207
LSR (Label-switch router) 113,130,193

M

MAC address 17
Make-before-break 241
MAN 159
Maximum-weighting size-matching algorithm 171
Message Type 137
MID 43
Middle-class 159
Middle-class router 161
MP-BGP 153
M-Plnae 33
MPLS 4,8,33,111
MPLS Forum 318
MP_S 192
MTU 70
Multilayer 232,241
Multilayer label-switched path (LSP) 230
Multilayer LSP network control 231
Multilayer traffic engineering 220
Multimedia 1,3

N

Neighboring router 69
Network block 71
Network-LSA 89
Next hop 112,128
NHLFE 116

NNI 34
Node Protection 283,287
Non-wavelength-convertible (NWC) 253

O

OAM 47,61
OAM cell 37
Object 135
Object length 137
OIF 221,308
On-demand type 123
Opaque LSA 201
Operations & Management Area 306
Optical cross-connect (OXC) 225
Optical LSP 233
Optical supervisory channel (OSC) 291
Optical UNI 221
Optical-channel (Och) 293
Optical-layer management 293
Ordered type 123
Originating IP address 70
Open Shortest Path First (OSPF)
    74,75,87,127,199,242,255
OTN 293
OUNI 308
Output scheduling 164
Output-buffer-type switch 165
Overlay model 151,219
OXC 192

P

P2P 1
Packet discarding 7
Packet LSP 233
Packet scheduling 177
Packetized general processing sharing 177,179
PAD 42
Path attribute 100
Path message 135,204
Path-vector 124
Path-vector type 75,82
Patricis tree 109,185
Peer model 151,220
Performance-monitoring function 66
PGPS 179
PHP 119
PIL (Photonic Internet Laboratory) 321
PIM 171
Point-to-multipoint 201
Point-to-point 201

Point-to-point Protocol 290
Policy control 135
Policy routing 100
Pop 8
Port 213
PPD 54
Preemption 132
Prefix 72,182
Priority 133
Priority control 7,47,51,56
Processing the ERO object 140
Protection 285
Protocol identifer 70
PSC (packet-switch capable) 193,232
Push 8
PVC (Permanent virtual connection) 46

Q

Q.2931 11
QoS (quality of service) 5,47,227

R

RAI 63
RD 177
Refresh message 205
Region 195
Required rate 164
Resending based on judgment of time-out
    26
Restriction of wavelength 212
RESV message 204
RIP 75
Risk-shared link group 204
Round trip 26
Round trip time 25
Round-robin 173
Route 134
Route and wavelength assignment (RWA)
    223
Route control 7
Route lookup 182
Route reflector 106
Route selection 103
Router LSA 88,201
Router TLV 201
Routing 127
Routing Area 306
Routing control 47,57
Routing loop 69
Routing protocol 159

Index

Routing table 159,182
RRO 140
RSVP protocol version number 137
RSVP-TE 135,204,255

S

Scalability 244
Scheduling 169
Scheduling function 159
SDH/SONET 191
Sequential number 25,26
Service-quality 165
Service-quality class 165
Session object 144
Shaping control 47
Shared restoration 285
Shared-risk link group (SRLG) 223
SHIM header 8,115
shim label 214
Shortest-path 135
Signaling 323
Site 151
Sliding window algorithm 51
Source IP address 20
Source routing 130
Split horizon /poison reverse 79
Spontaneous type 123
SRAM 166
SRG-disjoint 285
SRTS 39
SSCOP 13
SSCS 14
Standard Track RFC 307
Starvation 169
Strict specification 131
sub TLV 201
Sub-IP Area 306
Summary-LSA 90
Swap 117
Switched virtual connection (SVC) 46
Switching capability 195,232
Switching function 159
Switching type 206
System configuration 159
System management 159

T

TCAM (Ternay CAM) 188
TCP packet 17
TCP protocol 23

TCP/IP 305
TDM (time-division-multiplex capable) 193,232
TDM layer 191
TE link 199,213
TE metric 201
Telecom Standardization Sector 303
Three-stage cross network 175
TLV 201
Token backet 180
Topology-driven timing 115
TOS 70
Traffic demand 227
Traffic engineering 7,127,191,223
Traffic engineering technology 47
Traffic parameter 133
Traffic trunk 132
Traffic-control 6
Traffic-driven 114
Traffic-engineerin (TE) 244
Traffic-shaping control 53
Transport layer 22,23
Trie structure 184
Triggered update 79,82
TTL 69,115
Tunnel-ID 142

U

U plane (User plane) 11,33
UBR 59
UBR+ 59
UNI 34
Unnumbered link 200
UPC (traffic) control 7,49,57
Upstream path 210
Upstream type 122
Usage parameter control (UPC) 47,49

V

VBR 57
VCI/VPI 33
Version 70
Virtual circuit 112,128
Virtual-network topology (VNT) 223,234
Virtual-wavelength-path (VWP) 291
VOQ 165,168
VPI/VCI 116
VPN 127,150
VPNs 319
VRF 153

VSR 309

W

WAN 159
Wavelength converters 254
Wavelength-convertible (WC) 253,254
Wavelength-path layer 191
WDM (Wavelength-division multiplexing) 4,191
WDRR 181
Weighted-SRLG (WSRLG) 223,266
WFQ 179

WIDE Project 319
Wind control 27
Wire speed 164
WRR 165,181
WRR (Weighted round-robin) scheduling 165,177
WWW 1

X

x.25 36
3R functions 254